Understanding the National Electrical Code

Third Edition

Mike Holt

Delmar Publishers

an International Thomson Publishing company I(T)P®

Albany • Bonn • Boston • Cincinnati • Detroit • London • Madrid
Melbourne • Mexico City • New York • Pacific Grove • Paris • San Francisco
Singapore • Tokyo • Toronto • Washington

NOTICE TO THE READER

Cover Design: Courtesy of Brucie Rosch

Delmar Staff
Publisher: Alar Elken
Acquisitions Editor: Mark Huth
Developmental Editor: Jeanne Mesick
Project Editor: Megeen Mulholland
Production Coordinator: Toni Hansen
Art/Design Coordinator: Cheri Plasse
Editorial Assistant: Dawn Daughtery
Production Manager: Mary Ellen Black

For more information, contact Delmar, 3 Columbia Circle, PO Box 15015, Albany, NY 12212-0515; or find us on the World Wide Web at http://www.delmar.com

International Division List

Japan:
Thomson Learning
Palaceside Building 5F
1-1-1 Hitotsubashi, Chiyoda-ku
Tokyo 100 0003 Japan
Tel: 813 5218 6544
Fax: 813 5218 6551

Australia/New Zealand:
Nelson/Thomson Learning
102 Dodds Street
South Melbourne, Victoria 3205
Australia
Tel: 61 39 685 4111
Fax: 61 39 685 419

UK/Europe/Middle East:
Thomson Learning
Berkshire House
168-173 High Holborn
London
WC1V 7AA United Kingdom
Tel: 44 171 497 1422
Fax: 44 171 497 1426

Latin America:
Thomson Learning
Seneca, 53
Colonia Polanco
11560 Mexico D.F. Mexico
Tel: 525-281-2906
Fax: 525-281-2656

Canada:
Nelson/Thomson Learning
1120 Birchmount Road
Scarborough, Ontario
Canada M1K 5G4
Tel: 416-752-9100
Fax: 416-752-8102

Asia:
Thomson Learning
60 Albert Street, #15-01
Albert Complex
Singapore 189969
Tel: 65 336 6411
Fax: 65 336 7411

Library of Congress Cataloging-in-Publication Data
Holt, Charles Michael.
    Understanding the national electrical code / Mike Holt.—3rd ed.
        p.   cm.
    Includes bibliographical references and index.
    ISBN 0-7668-0350-3 (alk. paper)
    1. National Fire Protection Association. National Electrical Code
    (1999) 2. Electric engineering—Insurance requirements—United
    States. 1. Title.
    TK260.H65   1998   98-49614
    621.319'24'0218—dc20   CIP

I dedicate this textbook to my family, and the Lord.

Delmar Publishers Is Your Electrical Book Source!

Whether you're a beginning student or a master electrician, Delmar Publishers has the right book fo you. Our complete selection of proven best-sellers and all-new titles is designed to bring you the most up-to-date, technically-accurate information available.

NATIONAL ELECTRICAL CODE

National Electrical Code® 1999/NFPA
Revised every three years, the *National Electric Code®* is the basis of all U.S. electrical codes.
Order # 0-8776-5432-8
Loose-leaf version in binder
Order # 0-8776-5433-6

National Electrical Code® Handbook 1999/NFPA
This essential resource pulls together all the extra facts, figures, and explanations you need to interpret the 1999 *NEC®*. It includes the entire text of the Code, plus expert commentary, real-world examples, diagrams, and illustrations that clarify requirements.
Order # 0-8776-5437-9

Illustrated Changes in the 1999 National Electrical Code®/O'Riley
This book provides an abundantly-illustrated and easy-to-understand analysis of the changes made to the 1999 *NEC®*.
Order # 0-7668-0763-0

Understanding the National Electrical Code®, 3E/Holt
This book gives users at every level the ability to understand what the *NEC®* requires, and simplifies this sometimes intimidating and confusing code.
Order # 0-7668-0350-3

Illustrated Guide to the National Electrical Code®/Miller
Highly-detailed illustrations offer insight into Code requirements, and are further enhanced through clearly-written, concise blocks of text that can be read very quickly and understood with ease. Organized by classes of occupancy.
Order # 0-7668-0529-8

Interpreting the National Electrical Code®, 5E/Surbrook
This updated resource provides a process for understanding and applying the *National Electrical Code®* to electrical contracting, plan development, and review.
Order # 0-7668-0187-X

Electrical Grounding, 5E/O'Riley
Electrical Grounding is a highly illustrated, systematic approach for understanding grounding principles and their application to the 1999 *NEC®*.
Order # 0-7668-0486-0

ELECTRICAL WIRING

Electrical Raceways and Other Wiring Methods, 3E/Loyd
The most authoritive resource on metallic and nonmetallic raceways, provides users with a concise, easy-to-understand guide to the specific design criteria and wiring methods and materials required by the 1999 *NEC®*.
Order # 0-7668-0266-3

Electrical Wiring Residential, 13E/Mullin
Now in full color! Users can learn all aspects of residential wiring and how to apply them to the wiring of a typical house from this, the most widely-used residential wiring book in the country.
Softcover Order # 0-8273-8607-9
Hardcover Order # 0-8273-8610-9

House Wiring with the NEC®/Mullin
The focus of this new book is the applications of the *NEC®* to house wiring.
Order # 0-8273-8350-9

Electrical Wiring Commercial, 10E/Mullin and Smith
Users can learn commercial wiring in accordance with the *NEC®* from this comprehensive guide to applying the newly revised 1999 *NEC®*.
Order # 0-7668-0179-9

Electrical Wiring Industrial, 10E/Smith and Herman
This practical resource has users work their way through an entire industrial building—wiring the branch-circuits, feeders, service entrances, and many of the electrical appliances and sub-systems found in commercial buildings.
Order # 0-7668-0193-4

Cables and Wiring, 2E/AVO
This concise, easy-to-use book is your single-source guide to electrical cables—it's a "must-have" reference for journeyman electricians, contractors, inspectors, and designers.
Order # 0-7668-0270-1

ELECTRICAL MACHINES AND CONTROLS

Industrial Motor Control, 4E/Herman and Alerich
This newly-revised and expanded book, now in full color, provides easy-to-follow instructions and essential information for controlling industrial motors. Also available are a new lab manual and an interactive CD-ROM.
Order # 0-8273-8640-0

Electric Motor Control, 6E/Alerich and Herman
Fully updated in this new sixth edition, this book has been a long-standing leader in the area of electric motor controls.
Order # 0-8273-8456-4

Introduction to Programmable Logic Controllers/Dunning
This book offers an introduction to Programmable Logic Controllers.
Order # 0-8273-7866-1

Technician's Guide to Programmable Controllers, 3E/Cox
Uses a plain, easy-to-understand approach and covers the basics of programmable controllers.
Order # 0-8273-6238-2

Programmable Controller Circuits/Bertrand
This book is a project manual designed to provide practical laboratory experience for one studying industrial controls.
Order # 0-8273-7066-0

Electronic Variable Speed Drives/Brumbach
Aimed squarely at maintenance and troubleshooting, *Electronic Variable Speed Drives* is the only book devoted exclusively to this topic.
Order # 0-8273-6937-9

Electrical Controls for Machines, 5E/Rexford
State-of-the-art process and machine control devices, circuits, and systems for all types of industries are explained in detail in this comprehensive resource.
Order # 0-8273-7644-8

Electrical Transformers and Rotating Machines/Herman
This new book is an excellent resource for electrical students and professionals in the electrical trade.
Order # 0-7668-0579-4

Delmar's Standard Guide to Transformers/Herman
Delmar's Standard Guide to Transformers was developed from the best-seller *Standard Textbook of Electricity* with expanded transformer coverage not found in any other book.
Order # 0-8273-7209-4

DATA AND VOICE COMMUNICATION CABLING AND FIBER OPTICS

Complete Guide to Fiber Optic Cable System Installation/Pearson
This book offers comprehensive, unbiased, state-of-the-art information and procedures for installing fiber optic cable systems.
Order # 0-8273-7318-X

Fiber Optics Technician's Manual/Hayes
Here's an indispensible tool for all technicians and electricians who need to learn about optimal fiber optic design and installation as well as the latest troubleshooting tips and techniques.
Order # 0-8273-7426-7

A Guide for Telecommunications Cable Splicing/Highhouse
A "how-to" guide for splicing all types of telecommunications cables.
Order # 0-8273-8066-6

Premises Cabling/Sterling
This reference is ideal for electricians, electrical contractors, and inspectors needing specific information on the principles of structured wiring systems.
Order # 0-8273-7244-2

ELECTRICAL THEORY

Delmar's Standard Textbook of Electricity, 2E/Herman
This exciting full-color book is the most comprehensive book on DC/AC circuits and machines for those learning the electrical trades.
Order # 0-8273-8550-1

Industrial Electricity, 6E/Nadon, Gelmine, and Brumbach
This revised, illustrated book offers broad coverage of the basics of electrical theory and industrial applications. It is perfect for those who wish to be industrial maintenance technicians.
Order # 0-7668-0101-2

EXAM PREPARATION

Journeyman Electrician's Exam Preparation, 2E/Holt
This comprehensive exam prep guide includes all of the topics on journeyman electrician competency exams.
Order # 0-7668-0375-9

Master Electrician's Exam Preparation, 2E/Holt
This comprehensive exam prep guide includes all of the topics on master electrician's competency exams.
Order # 0-7668-0376-7

REFERENCE

ELECTRICAL REFERENCE SERIES
This series of technical reference books is written by experts and designed to provide the electrician, electrical contractor, industrial maintenance technician, and other electrical workers with a source of reference information about virtually all of the electrical topics that they encounter.

Electrician's Technical Reference—Motor Controls/Carpenter
Electrician's Technical Reference—Motor Controls is a source of comprehensive information on understanding the controls that start, stop, and regulate the speed of motors.
Order # 0-8273-8514-5

Electrician's Technical Reference—Motors/Carpenter
Electrician's Technical Reference—Motors builds an understanding of the operation, theory, and applications of motors.
Order # 0-8273-8513-7

Electrician's Technical Reference—Theory and Calculations/Herman
Electrician's Technical Reference—Theory and Calculations provides detailed examples of problem-solving for different kinds of DC and AC circuits.
Order # 0-8273-7885-8

Electrician's Technical Reference—Transformers/Herman
Electrician's Technical Reference—Transformers focuses on the theoretical and practical aspects of single-phase and 3-phase transformers and transformer connections.
Order # 0-8273-8496-3

Electrician's Technical Reference—Hazardous Locations/Loyd
Electrician's Technical Reference—Hazardous Locations cover electrical wiring methods and basic electrical design considerations for hazardous locations.
Order # 0-8273-8380-0

Electrician's Technical Reference—Wiring Methods/Loyd
Electrician's Technical Reference—Wiring Methods covers electrical wiring methods and basic electrical design considerations for all locations, and shows how to provide efficient, safe, and economical applications of various types of available wiring methods.
Order # 0-8273-8379-7

Electrician's Technical Reference—Industrial Electronics/Herman
Electrician's Technical Reference—Industrial Electronics covers components most used in heavy industry, such as silicon control rectifiers, triacs, and more. It also includes examples of common rectifiers and phase-shifting circuits.
Order # 0-7668-0347-3

RELATED TITLES

Common Sense Conduit Bending and Cable Tray Techniques/Simpson
Now geared especially for students, this manual remains the only complete treatment of the topic in the electrical field.
Order # 0-8273-7110-1

Practical Problems in Mathematics for Electricians, 5E/Herman
This book details the mathematics principles needed by electricians.
Order # 0-8273-6708-2

Electrical Estimating/Holt
This book provides a comprehensive look at how to estimate electrical wiring for residential and commercial buildings with extensive discussion of manual versus computer-assisted estimating.
Order # 0-8273-8100-X

Electrical Studies for Trades/Herman
Based on *Delmar's Standard Textbook of Electricity,* this new book provides non-electrical trades students with the basic information they need to understand electrical systems.
Order # 0-8273-7845-9

Contents

Preface

INTRODUCTION

Understanding the National Electrical Code was written to provide insight into and understanding of many of the technical rules of the *NEC*. Informal and relaxed in style the book contains clear graphics and examples that apply to the 1990s.

The scope of this book is based on the fundamental rules of the *NEC* (Article 90 through Chapter 4), that apply to systems rated 600 volt, nominal, or less. Specifically, those installations that operate at a voltage of: 120/240, 208Y/120, 480Y/277, 208, 240, and 480, and 600 volts [220-2 voltages].

This book cannot clear up confusing, conflicting, or controversial *Code* rules, but it does put each rule in sharper focus to help you understand its intended purpose. We all would like a specific rule for every installation, but this is not practical. The number of possible applications and combinations of electrical installations can be endless.

Although new products, materials, and installation techniques are continually being developed, the *NEC,* which is revised only every 3 years, is not able to accommodate all of these rapid changes. Fortunately, the *Code* allows the electrical inspector the authority to permit alternate methods, and even to waive the *NEC* rules when necessary. See, for instance, Section 90-4.

This book also contains tips on proper electrical installations, advice, or cautions to possible conflicts or confusing *Code* rules, and warnings of dangers related to improper electrical installations.

HOW TO USE THIS BOOK

This book is intended as a study guide to be used with the *NEC*, not a replacement for the *NEC*. Keep a copy of the 1999 *National Electrical Code* handy. As you read through this book, review the graphics and examples in relation to your *Code* book and discuss it with others. This book contains many cross-references to other related *Code* rules to help you develop a better understanding of the *Code* and includes my comments about each rule. These cross-refences are identified by a Code Section number in brackets such as "Section 90-4" which would look like "[90-4]."

> **Note.** This book follows the *NEC* format but it does not cover all of the *NEC* rules. In addition, you may notice that not all Articles, Sections, Subparts, Exceptions, or FPN's, have been included in this book. Do not be concerned if you see that Exception No. 2 and Exception No. 7 are explained, but others are not.

Each unit of this book contains a list of objectives, explanations of the *NEC*, detailed graphics (a picture is worth a thousand words), a summary, and practice questions. As you come to **bold-faced** words, turn to the Unit 2 Article 100 Definitions in the front of the book to gain a better understanding of the term. When you see an italicized word, turn to the Glossary in the back of this book to review those industry terms that are not contained in Article 100. At the beginning of each Unit, I listed definitions and glossary terms that must be understood to properly apply the *Code* requirements for that unit.

I suggest that before you get too deep into the book, scan the entire book to identify those Sections that discuss the kinds of rules you might need or want to understand better. For instance, if you're interested in the installation of receptacles and lighting outlets, you should focus on that unit [Article 410], and not necessarily on the rules that apply to motors [Article 430] or transformers [Article 450].

As you progress through the *NEC* and this book, it is quite possible that you will find some rules or comments that I made that you don't understand. Also you may not agree with me. Sometimes, you'll need to understand a rule in Chapter 3 before you can fully understand a rule in Chapter 2. Please, don't get frustrated. Highlight in yellow the section in this book you're having problems with and discuss it with your supervisor, inspector, coworkers, and others. Maybe they'll have some additional feedback. After you have completed the book, go back and review the book once again and see if you now understand those highlighted areas.

ATTITUDE AND GRAY AREAS

A certain attitude is needed to understand the *NEC*. There are many different types of electrical installations, but the *Code* could not begin to consider all of them. However, to be successful in the electrical industry, you must understand the *NEC*. I hope this book will encourage you to learn more about the *Code*. I have to say that the more I learn, the more I realize how much there still is to learn.

As this book progresses, you will develop some insight into these rules. Yes, there are gray areas, but generally the meaning is quite clear.

DIFFERENT INTERPRETATIONS

Electricians, contractors, some inspectors, and others love arguing *Code* interpretations and discussing *Code* requirements. As a matter of fact, discussing the *NEC* and its application is a great way to increase your knowledge of the *Code* and its intended use. The best way to discuss *Code* requirements with others is by referring to a specific Section in the *Code*, rather than by talking in vague generalities.

I have taken great care in researching the *Code* rules in this book, but I'm not perfect. If you feel that I have made an error, please let me know by contacting me directly at *Mike@mikeholt.com*, or *1-888-NEC CODE*.

ABOUT THE AUTHOR

Mike Holt worked his way up through the electrical trade from an apprentice electrician to a master electrician and electrical inspector. Mike did not complete high school due to adverse circumstances and dropped out after completing 11th grade. Realizing that success depends on one's education, he immediately attained his GED, and ten years later he attended the University of Miami's Graduate School for a Master's in Business Administration (MBA).

Mike is nationally recognized as one of America's most knowledgeable electrical trainers and has touched the lives of thousands of electricians, inspectors, contractors, and engineers.

Mike Holt resides in Central Florida, is the father of seven children, and has many outside interests and activities. He is a National Barefoot Waterskiing Champion, has set five barefoot waterski records, and is currently ranked No. 2.

Mike enjoys white water rafting, racquetball, playing his guitar, and spending time with his family. His commitment to God has helped him develop a lifestyle that balances family, career, and self.

ACKNOWLEDGMENTS

I would like to say thank you to all the people in my life who believed in me, and even those who did not. There are many people who played a role in the development and production of this book. I will start with Mike Culbreath (Master Electrician), who helped me transform my words and visions into lifelike graphics. I could not have produced such a fine book without his help.

Next, Levi McConnell, of Charmed Life Publishing, for the electronic production and typesetting. A very special thanks to Brooke Stauffer from NECA, for copyediting the text, and to Elzy R. Williams, P.E., Adjunct Professor at John Tyler Community College, and Craig H. Matthews, P.E., from Austin Brockenbrough and Associates, L.L.P, for their incredible job of proofing this textbook for technical correctness.

To my beautiful wife, Linda, and my seven children, Belynda, Melissa, Autumn, Steven, Michael, Meghan, and Brittney, thank you for loving me.

Also thanks to all those who helped me in the electrical industry, including *Electrical Construction and Maintenance* magazines for my first "big break." To Joe McPartland, "my mentor," who was there to help and encourage me. Joe, I'll never forget to help others as you helped me.

In addition, I would like to thank Joe Salimando the publisher of *Electrical Contractor* and Dan Walters of the *National Electrical Contractors Association* (NECA) for my second "big break," and for putting up with all of my crazy ideas.

I would like to also thank James Stallcup, Dick Lloyd, Mark Ode, D.J. Clements, Joe Ross, John Calloggero, Tony Selvestri, and Marvin Weiss for being special people in my life.

The final personal thank you goes to Sarina, my long-time friend and office manager. Thank you for covering the office for me while I spend so much time writing books, doing seminars, and producing videos and software. Your love and concern for me has helped me through many difficult times.

Delmar Publishers as well as the author would also like to thank those individuals who reviewed the manuscript and offered invaluable suggestions and feedback. Their assistance is greatly appreciated.

The following individuals provided detailed critiques of the manuscript and offered valuable suggestions for improvement:

Charles Trout
Electrical Code Consultant
Boca Raton, FL

Robert Blakely
Mississippi Gulf Coast Community College
Gulfport, MS

Darren Dickenson
Dunwoody Institute
Minneapolis, MN

Larry Killebrew
Mid-America Vocational Technical School
Wayne, OK

John Penley
Albert Lea Technical College
Albert Lea, MN

Charles Plimpton
New Hampshire Technical College
Laconia, NH

John P. Cox
Brevard Community College
Cocoa, FL

William Elarton
Los Angeles Trade and Technical School
Los Angeles, CA

William Bickner
Electrical Consultant
Stillwater, OK

David Gelhauf
Tri-County Vocational Technical School
Glouster, OH

Larry C. Phillips
Minneapolis Technical College
Minneapolis, MN

Rodney Stanley
Morehead State University
Morehead, KY

Craig H. Matthews, P.E.
Senior Electrical Engineer
Austin Brockenbrough and Associates, L.L.P.

Tim De Donder
Kaw Area Vocational Technical School

Brooke Stauffer
National Electrical Contractors of America

Elzy R. Williams, P.E.
Professor at John Tyler Community College

Rovene Quigley
Electrician

The *National Electrical Code*

The *NEC* is intended to be used by experienced persons having an understanding of electrical terms, theory, and trade practices. Such persons include electrical contractors, electrical inspectors, electrical engineers, and qualified electricians. The *Code* was not written to serve as an instructive or teaching manual [90-1(c)] for untrained persons.

Learning to use the *NEC* is like learning to play the game of chess. You first must learn the terms used to identify the game pieces, the concepts of how each piece moves, and the layout of how the pieces are placed on the board. Once you have this basic understanding of the game, you're ready to start playing the game. But all you can do is make crude moves because you really don't understand what you're doing. To play the game well, you'll need to study the rules, understand the subtle and complicated strategies, and then practice, practice, practice.

The same with the *Code*. Learning the terms, concepts, and layout of the *NEC* gives you just enough knowledge to be dangerous. Perhaps most difficult are the subtle meaning within the *Code* rules.

There are thousands of different applications of electrical installations, and there is not a specific *Code* rule for every application. To properly apply the *NEC*, you must understand the safety related issue of the rule and then apply common sense.

NEC TERMS AND CONCEPTS

The *NEC* uses many technical terms and expressions. It's crucial that you understand the meanings of basic words like ground, grounded, grounding, and neutral. If you don't understand basic terms used in the *Code*, you won't understand the rule itself.

It is not only the technical words that require close attention in the *NEC*, even the simplest words can make a big difference. The word "or" can imply alternate choices for equipment, wiring methods, and other requirements. Sometimes "or" can mean any item in a group. The word "and" can be an additional requirement, or any item in a group.

> **Note.** Electricians, engineers, and other trade-related professionals have created their own terms and phrases (slang or jargon). One of the problems with the use of slang terms is that the words mean different things to different people.

Understanding the safety related concepts behind *NEC* rules means understanding how and why things work the way they do (electrical theory). How does a bird sit on an energized power line without getting fried? Why, when we install a lot of wires close together, do we reduce the amount of current that each conductor can carry? Why can't a single current-carrying conductor be installed within a metal raceway? Why does the *NEC* permit a 40 ampere circuit breaker to protect motor circuit conductors that are only rated 20 ampere? Why are bonding jumpers sometimes required for metal raceways containing 480Y/277 volt circuits, but not for 120/240 volt circuits?

If you understand why or how things work, you have a better chance of understanding the *NEC* rules.

THE NEC STYLE

Contrary to popular belief, the *NEC* is a fairly well organized document, although parts of it are somewhat vague. Understanding the *NEC* structure and writing style is extremely important to understand and use the *Code* book effectively.

The *National Electrical Code* is organized into 12 types of components.

1. Chapters (major categories)

2. Articles (individual subjects)

3. Parts (divisions of an Article)

4. Sections, Lists, and Tables (*Code* rules)

5. Exceptions (*Code* rules)

6. Fine Print Notes (explanatory material, not mandatory *Code* language)

7. Definitions (*Code* rules)   ~~100 - Article~~

8. Superscript Letter ˣ

9. Marginal Notations, *Code* changes (I) and deletions (•)

10. Table of Contents

11. Index

12. Appendices

1. *Chapters.* There are nine chapters and each chapter contains Articles. The nine chapters fall into four groupings:

- General Rules: Chapters 1 through 4

- Specific Rules (Hazardous locations, signs, control wiring): Chapters 5 through 7

- Communication Systems (Telephone, Radio/Television, and Cable TV Systems): Chapter 8

- Tables: Chapter 9

2. *Articles.* The *NEC* contains approximately 125 Articles. An Article covers a specific subject as in the following examples:

Article 110 – General Requirements
Article 250 – Grounding
Article 300 – Wiring Methods
Article 430 – Motors
Article 500 – Hazardous (classified) Locations
Article 680 – Swimming Pools
Article 725 – Control Wiring
Article 800 – Communication Wiring

3. *Parts.* When an Article is sufficiently large, the Article is subdivided into Parts. For example, Article 250 contains nine parts, including:

- *Part A.* General
- *Part B.* Circuit and System Grounding
- *Part C.* Grounding Electrode System

CAUTION: The "Parts" of a Code Article are not included in the Section numbers. Because of this, we have a tendency to forget what "Part" the Code rule is relating to. For example, Table 110-34 gives the dimensions of working space clearances in front of electrical equipment. If we are not careful, we might think that this table applies to all electrical installations. But Section 110-34 is located in Part C Over-600 volt Systems of Article 110! The working clearance rule for under-600 volt systems is located in Part A of Article 110, in Table 110-26.

4. *Sections, Lists, and Tables.*

Sections. Each actual *Code* rule is called a Section and is identified with numbers, such as Section 225-26. A *Code* Section may be broken down into subsections by letters in parentheses, and numbers in parentheses may further break down each subsection. For example, the rule that requires all receptacles in a bathroom to be GFCI protected is contained in Section 210-8(a)(1).

Note. Many in the electrical industry incorrectly use the term "Article" when referring to a *Code* Section.

Lists. The 1999 *NEC* has changed the layout of some Sections that contain lists of items. If a list is part of a numeric subsection, such as Section 210-52(a)(2), then the items are listed as a., b., c., etc. However, if a list is part of a Section, then the items are identified as (1), (2), (3), (4), etc.

Tables. Many *Code* requirements are contained within Tables which are a systematic list of *Code* rules in an orderly arrangement. For example, Table 300-15 lists the burial depths of cables and raceways.

5. *Exceptions.* Exceptions are in italicized and provide an alternative to a specific rule. There are two types of exceptions; mandatory and permissive. When a rule has several exceptions, those exceptions with mandatory requirements are listed before those written in permissive language.

(a) *Mandatory Exception.* A mandatory exception uses the words "shall" or "shall not." The word "shall" in an exception means that if you are using the exception, you are required to do it in a particular way. The term "shall not" means that you cannot do something.

(b) *Permissive Exception.* A permissive exception uses such words as "shall be permitted," which means that it is accepted to do it in this way.

6. *Fine Print Note, (FPN), [90-5].* A Fine Print Note contain explanatory material, intended to clarify a rule or give assistance, but it is not *Code* requirement. FPNs often use the term "may," but never "shall."

7. *Definitions.* Definitions are listed in Article 100 and throughout the *NEC*. In general, the definitions listed in Article 100 apply to more than one *Code* Article, such as "branch circuit," which is used in many Articles.

Definitions at the beginning of a specific Article applies only to that Article. For example the definition of a "Swimming Pool" is contained in Section 680-4 because this term applies only to the requirements of Article 680 – Swimming Pools.

Definitions located in a Part of an Article apply only to that Part of the Article. For example the definition of "motor control circuit" applies only to Article 430, Part F.

Definitions located in a *Code* Section apply only to that *Code* Section. For example the definition of "Festoon Lighting" located in Section 225-6(b) applies only to the requirements contained in Section 225-6.

8. *Superscript Letter* ^{x}. The superscript letter x indicates that the material was extracted from other technical standards published by the NFPA. Appendix A, at the back of the *Code* book, identifies the NFPA documents and the Section(s) from which the material was extracted.

9. *Changes and Deletions.* Changes and deletions to the *NEC* are identified in the margins of the 1999 *NEC* in the following manner: A vertical line (|) marks changes and a bullet (•) identifies deletion of a a *Code* rule. Many rules in the 1999 *NEC* were relocated. The place from which the *Code* rule was removed has a bullet (•) in the margin, and the place where the rule was inserted has a a vertical line (|) in the margin.

10. *Table Of Contents.* The Table of Contents located in the front of the *Code* book displays the layout of the Chapters, Articles, and Parts as well as their location in the *Code* book.

11. *Index.* We all know the purpose of an index, but it's not that easy to use. You really need to know the correct term. Often it's much easier to use the Table of Contents.

12. *Appendices.* There are four appendices in the 1999 *NEC*:

• Appendix A – Extract Information

- Appendix B – Ampacity Engineering Supervision

- Appendix C – Conduit and Tubing Fill Tables

- Appendix D – Electrical Calculation Examples

HOW TO FIND THINGS IN THE CODE

How fast you find things in the *NEC* depends on your experience. Experienced *Code* users often use the Table of Contents instead of the index.

For example, what *Code* rule indicates the maximum number of disconnects permitted for a service?

Answer. You need to know that Article 230 is for Services and that it contains a Part F. Disconnection Means. If you know this, using the Table of Contents, you'll see that the answer is contained at page 66.

People frequently use the index which lists subjects in alphabetical order. It's usually the best place to start for specific information. Unlike most books, the *NEC* index does not list page numbers; it lists Sections, Tables, Articles, parts, and appendices by their Section number.

> **Note.** Many people say the *Code* takes them in circles, and sometimes it does. However, this complaint is often heard from inexperienced persons who don't understand electrical theory, electrical terms and electrical practices.

CUSTOMIZING YOUR CODE BOOK

One way for you to get comfortable with your *Code* book is to customize it to meet your needs. This you can do by highlighting, underlining *Code* rules, and using convenient tabs.

Highlighting and Underlining. As you read through this book, highlight in the *NEC* book those *Code* rules that are important to you such as yellow for general interest, and orange for rules you want to find quickly. As you use the Index and the Table of

Contents, highlight terms in those areas as well. Underline or circle key words and phrases in the *NEC* with a red pen (not a lead pencil) and use a 6 inch ruler to keep lines straight and neat.

Because of the new format of the 1999 *NEC* (8½ × 11), I highly recommend that you highlight in green the Parts of at least the following Articles.

Article 230 – Services

Article 250 – Grounding

Article 410 – Fixtures

Article 430 – Motors

Author's Comment. Trust me; you'll be glad you did.

Tabbing the NEC. Tabbing the *NEC* permits you to quickly access *Code* Article, Section, or Tables. However, too many tabs will defeat the purpose.

Experience has shown that the best way to tab the *Code* book is to start by placing the last tab first and the first tab last (start at the back of the book and work your way toward the front). Install the first tab, then place each following tab so that they do not overlap the information of the previous tab.

The following is a list of Articles and Sections I most commonly refer to. Place a tab only on the Sections or Articles that are important to you.

Tab#	Description	Page
1	Index	621
2	Examples: Appendix D	609
3	Raceway Fill Tables: Appendix C	585
4	Conductor Area: Table 5 of Chapter 9	564
5	Raceway Area: Table 4 of Chapter 9	562
6	Satellites and Antennas: Article 820	545
7	Communication: Article 800	533
8	Fiber Optic Cables: Article 770	527
9	Fire Alarms: Article 760	519
10	Control Circuits: Article 725	510
11	Emergency Circuits: Article 700	501
12	Pools, Spas, and Fountains: Article 680	476
13	Electric Signs: Article 600	433
14	Marinas: Article 555	431
15	Mobile/Manufactured: Article 550	398
16	Carnivals, Circuses, and Fairs: Article 525	387
17	Health Care Facilities: Article 517	359

Unit 1

Article 90
NEC Introduction

OBJECTIVES

After studying this unit, the student should be able to understand:
- the types of occupancies and areas that must comply with the *NEC*.
- the types of occupancies and areas that are exempt from the *NEC*.
- who enforces the Code and what their responsibilities are.
- the purpose of listing and labeling electrical equipment.

NATIONAL ELECTRICAL CODE
NFPA 70

Article 90 is the Introduction to the *National Electrical Code*. As with most introductions, this Article is often skipped. To understand the *NEC* and its application better, it is very important that you thoroughly read and review this introductory Article.

Definitions and Glossary Terms

To better understand the *NEC* rules contained in this unit, review the following:

Definitions, Article 100 - Unit 2
Buildings
Conductors
Equipment
Labeled
Listed
Premises wiring

Glossary Term
Authority having jurisdiction

ARTICLE 90 **INTRODUCTION**

90-1 **Purpose**

(a) **Practical Safeguarding.** The purpose of the *NEC* is the protection of persons and property by minimizing the risks caused by the use of *electricity*, Fig. 1-1.

(b) **Adequacy.** The *Code* is intended for the application of safety. When the rules of the *NEC* are

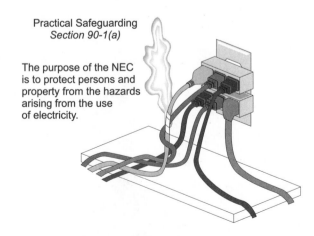

Practical Safeguarding
Section 90-1(a)

The purpose of the NEC is to protect persons and property from the hazards arising from the use of electricity.

Fig. 1-1 Practical Safeguarding

complied with, an installation is expected to be essentially free from hazards. However, when installations comply with the *NEC*, this does not mean that the electrical system will be efficient, convenient, adequate for good service, or that it will work properly.

> Fine Print Note (FPN): Hazards can occur because of overloading of circuits and improper installation of **equipment**. These problems often arise because the original installation did not provide for future expansion, which is not required by the *Code*.

CAUTION: The *NEC* does not contain any rule that requires consideration for future expansion of electrical use. The *NEC* is concerned solely with safety; but the electrical designer must be concerned with safety, efficiency, convenience, good service, and future expansion. Often, electrical systems are designed and installed that exceed *NEC* requirements. However, the inspector does not have the authority to require installations to exceed the *NEC* requirements, unless additional requirements have been adopted by local ordinance.

(c) Intention. The *Code* is not a how-to book; it is not intended as a design specification or an instruction manual for untrained persons.

90-2 Scope

The *National Electrical Code* is not intended to apply to all electrical installations. Subsection (a) explains what is covered and subsection (b) explains what is *not* covered by the *NEC*.

(a) Covered. The *Code* covers most electrical installations, including:

(1) Most **buildings**, mobile homes, recreational vehicles, floating buildings, yards, carnivals, parking and other lots, and private industrial substations. Also covered are **conductors** and **equipment** that

connect to the supply of electricity, conductors and equipment outside on the premises, and the installation of fiber optic cable.

(5) Installation within buildings used by the electric utility, such as office buildings, warehouses, garages, machine shops, and recreational buildings that are not an integral part of a generating plant, substation, or control center must be installed in accordance with the *NEC*, Fig. 1-2.

(b) Not Covered. The *Code* does not cover:

(1) Installations in cars, trucks, boats, ships, planes, electric trains, or underground mines.

(2) Self-propelled mobile surface mining machinery and their attendant electrical trailing cables.

(3) Railway power, signaling and communications wiring.

(4) Communications equipment under the exclusive control of a communication utility, such as telephone and cable TV companies, are covered by their own wiring and equipment rules and are not required to comply with the *NEC*. However, the

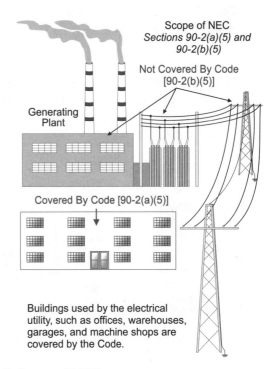

Fig. 1-2 Scope of NEC

interior and exterior wiring of phone, communications, and *CATV* not under the exclusive control of communication utilities must comply with *NEC* Chapter 8.

(5) Installations, including associated lighting, under the exclusive control of electric utilities for the purpose of power generation, distribution, control, transformation, and transmission are not required to comply with the *NEC*. This includes installations located in buildings used exclusively by utilities for such purposes, outdoors on property owned or leased by the utility, or on public highways, streets, roads, etc., or outdoors on private property by established rights such as easements.

Note. Any wiring installation such as lighting fixtures on private property without established rights (such as easements) and not intended for the purpose of communications, metering, generation, control, transformation, transmission, and distribution of electric energy, must be installed according to the *NEC* requirements, even if installed by the electric utility, Fig. 1-3.

90-3 Code Arrangement

The *Code* is divided into an Introduction and nine chapters. These chapters are divided as follows:

General Rules. The Introduction and Chapters 1 through 4 apply in general to all installations, which represents the scope of this book.

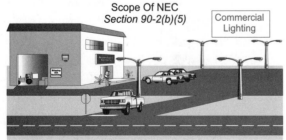

Scope Of NEC
Section 90-2(b)(5)

Commercial Lighting

Commercial outdoor lighting, such as parking lots, falls under the jurisdiction of the National Electrical Code.

Fig. 1-3 Scope of NEC

Special Rules. Chapters 5 through 7 apply to special occupancies, equipment, or conditions, and may modify the general rules of Chapters 1 through 4. Examples include: aircraft hangers, health care facilities, x-ray equipment, etc.

Communications Systems. Chapter 8 contains the requirements for communication circuits such as telephones, satellite dishes, TV antennas, and CATV systems. The requirements within Chapter 8 are independent of Chapter 1 through 7 requirements, unless a specific reference in Chapter 8 is made to a requirement in those Chapters. Chapter 8 covers communications.

Tables. Chapter 9 consists of tables that are used for raceway sizing and conductor fill and voltage drop.

90-4 Enforcement

The *NEC*, while purely advisory, is intended to be a document that can be adopted by governmental bodies and other inspection departments. It is up to these bodies, states, counties, cities, etc., to adopt the *NEC* as a legal requirement for electrical installations.

The enforcement of complying with the *NEC* falls under the *authority having jurisdiction*. For the purposes of this book, the authority having jurisdiction (AHJ) will be considered the electrical inspector. Generally, the electrical inspector is employed by some government agency and is responsible to an advisory council or board for his or her decisions or rulings.

An inspector's authority and responsibilities include:

Interpretation of the NEC Rules. The inspector is responsible to interpret the *NEC* rules. This means that the inspector must have a specific rule upon which to base his/her interpretations. If an inspector rejects your installation, you have the right to know the specific *NEC* rule that you violated, but naturally, we must be realistic in the recognition that we must often submit to a higher authority, Fig. 1-4.

Interpretation Of NEC Rules By Inspector
Section 90-4

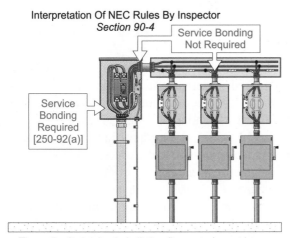

The authority having jurisdiction (inspector) must base
his/her interpretations on specific rules of the NEC.

Fig. 1-4 Interpretation of NEC Rules by Inspector

Note. The art of getting along with the electrical inspector, is knowing what the *Code* says and when to choose your battles.

Approval of Equipment and Materials. The electrical inspector is the person who decides the approval of equipment. However, if equipment is listed by a qualified electrical testing laboratory (listing agency), its internal wiring does not need to be reinspected at the time of installation [90-7].

> **WARNING:**
> Only the inspector can approve equipment [90-4]. He or she can reject the use of any equipment and can approve nonlisted equipment. The primary basis of equipment approval by the inspector is listing and labeling by qualified testing laboratories [90-7 and 110-2].

Waiver of Rules. Waiver of specific requirements of the *Code* or permitting alternate methods. When an installation does not comply with normal *NEC* rules, the inspector may waive specific requirements of the *Code* or permit alternate methods. This is permitted only where it is ensured that equivalent electrical safety can be achieved.

Waiver of New Code Requirements. Waiver of new *Code* requirements. If the 1999 *Code* requires materials, products, or construction that are not yet available, the inspector may allow materials, products, and construction methods that were acceptable in the 1996 *Code*.

Note. It takes time for manufacturers to redesign, manufacture, and distribute new products to meet new *Code* requirements.

Equipment Installation. It is the inspector's responsibility to ensure that the electrical equipment is installed to the equipment listing or labeling instructions. The inspector is also responsible for detecting any field modification of equipment. Listed equipment may not be modified in the field without the approval of the listing agency or the electrical inspector [90-7, 110-3(b)].

90-5 Mandatory Rules and Explanatory Material

(a) Mandatory Rules. Rules that identify actions that are specifically required or prohibited are characterized by the use of the terms "shall" or "shall not."

Section 110-3(b) states that listed or labeled equipment "shall" be installed, used, or both, in accordance with any instructions included in the listing or labeling.

(b) Permissive Rules. Rules which identify actions that are allowed but not required, such as options or alternative methods are characterized by the use of the terms "shall be permitted" or "shall not be required." A "permissive rule" is often an exception to the general requirement.

Section 250-102(d) states that the equipment **bonding jumper** can be installed inside or outside of a raceway or enclosure.

(c) Explanatory Material. Explanatory material, such as references to other standards, references to related Sections of the *Code*, or information related to a *Code* rule is included in the form of a Fine Print Note (FPN). Fine Print Notes are informational only and are not to be enforced. Most

FPN's contain a reference to another related *Code* Section.

90-6 Formal Interpretations

National Fire Protection Association (*NFPA*) formal interpretation procedures are listed in the Regulations Governing Committee Projects, available from the *NFPA* Secretary of the Standards Council. This is a very time-consuming process and is rarely done. Not only that, but formal interpretations are not legally binding.

90-7 Examination of Equipment for Product Safety

Product safety evaluation is done by nationally recognized independent testing laboratories that publish lists of equipment that meet nationally recognized test standards. Products and materials that are **listed**, **labeled**, or identified by a responsible and respected organization, are the basis of approval by the electrical inspector. National testing laboratories decrease the need for inspectors to reinspect or evaluate the electrical equipment at the time of installation.

SUMMARY

The purpose of the *NEC* is for the protection of persons and property. ☐ The *NEC* is a safety standard. ☐ The *NEC* is not intended as a design specification or an instruction manual for untrained persons. ☐ The *NEC* covers most electrical installations, but not all. ☐ The *NEC* does not cover cars, trucks, boats, ships, planes, underground mines, trains, utility controlled communication equipment, and utility power distribution locations. ☐ The *NEC* covers commercial parking lot lighting installed by the electric utility on private property. ☐ The *NEC* is divided into an Introduction and nine Chapters. ☐ Product evaluation is done by nationally recognized independent testing laboratories, not the electrical inspector. ☐ The authority having jurisdiction (AHJ) approves the use of products and enforces the requirements of the *NEC*, but they cannot make-up their own rules. ☐ Mandatory *Code* rules use the word shall. ☐ Permissive Rules identify actions that are allowed but not required, such as options or alternative methods. ☐ Explanatory material is contained in the Fine Print Notes (FPNs). ☐ Product safety evaluation is done by nationally recognized independent testing laboratories.

UNIT 1 REVIEW QUESTIONS

1. List the types of occupancies and areas that electrical installations must comply with the *NEC*.

2. List the types of occupancies and areas that are exempt from the *NEC* installation requirements.

3. Who enforces the *NEC* and what are their responsibilities?

4. What is the purpose of listing and labeling of electrical equipment?

5. In your own words explain the meaning of the following Article 100 definitions:

    Building
    Conductors
    Equipment
    Listed
    Labeled
    Premises Wiring

NEC CHAPTER 1
GENERAL REQUIREMENTS

Scope of Chapter 1 Articles

Unit 2   *ARTICLE 100*   DEFINITIONS

Unit 3   *ARTICLE 110*   REQUIREMENTS FOR ELECTRICAL
                             INSTALLATIONS

Chapter 1 of the *NEC* contains the general *Code* rules. After reviewing Units 1-3 of this book, you will develop a better understanding of the terms used in the *NEC* and the *NEC* writing style.

ARTICLE 100

Article 100 – Definitions contains the definitions of those terms used throughout the *Code*. In general, only those terms used in two or more Articles are defined in Article 100. Definitions of standard terms, such as volt, voltage drop, ampere, impedance, or resistance are not listed in the *Code* because they do not affect its proper application. See the glossary in the back of this book for general and trade terms. Part A of this Article contains definitions intended to apply throughout the *Code*. Part B contains definitions applicable only to systems operating at over 600 volt.

Note. Part B. Over 600 Volt Nominal, is not covered in this book.

ARTICLE 110

Article 110 contains the general requirements for electrical installations.
Part A. General
Part B. 600 volt nominal or less

Notes:

Unit 2

Article 100
Definitions

OBJECTIVES

After studying this unit, the student should be able to understand:
- the definitions that are most important for the proper application of the Code.
- the definitions that are most important to understand many of the NEC rules.

PART A. **GENERAL**

Article 100 of the NEC contains definitions intended to apply whenever the terms are used throughout the NEC. A definition that is listed in a specific Article applies to that Article only, and a definition given at the beginning of a Part only applies to that Part.

The official dictionary of the NFPA is the *IEEE Standard Dictionary of Electrical and Electronic Terms.*

Accessible, Equipment: Admitting close approach and you can get to it. It's not guarded by locked doors, elevation, or other effective means.

Accessible, Readily or Readily Accessible: Capable of being reached quickly without having to climb over or remove obstacles. This means that you should not have to use portable ladders, chairs, etc., to reach the equipment. The *NEC* does not give a specific height, but Section 380-8(a) does permit switches to be located up to 6 feet 7 inches (2 meters) above the working platform, Fig. 2-1.

Accessible, Wiring Methods: Not permanently closed in by the building structure or finish and capable of being removed or exposed without damaging the building structure or finish. Wiring methods installed above a dropped ceiling are considered accessible because the ceiling panels are removable without damaging the building structure. See the definitions for "concealed" and "exposed."

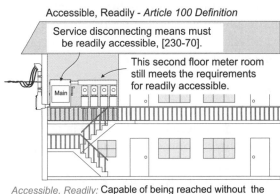

Accessible, Readily - *Article 100 Definition*

Service disconnecting means must be readily accessible, [230-70].

This second floor meter room still meets the requirements for readily accessible.

Main

Accessible, Readily: **Capable of being reached without the use of portable ladders or climbing over or moving obstacles.**

Fig. 2-1 Accessible, Readily

All wiring within outlet boxes must be accessible [370-29]; fluorescent fixtures mounted over outlet boxes must permit access to the box [410-14(b)].

Ampacity: The *amperes* that a conductor can carry continuously without exceeding its temperature rating as adjusted for *ambient temperatures* above 86°F, and or conductor bundling. Since electric *current* flowing through a conductor produces heat (I^2R), the ampacity of a conductor must be adjusted to accommodate ambient temperature and conductor bundling. See Sections 310-10, 310-15 and Table 310-16.

Appliance: Electrical equipment (generally other than industrial) is normally built in standardized sizes. Some examples are ranges, ovens, cooktops, refrigerators, drinking water coolers, beverage dispensers, and many listed cord-and plug-connected units. Motors, air-conditioning equipment, and electric space heating units are not appliances. The *NEC* has specific Articles for their requirements, such as Article 424 – Electric Space-Heating Equipment, Article 430 – Motors, and Article 440 – Air-conditioning and Refrigeration Equipment.

Approved: Many people think that if equipment is listed then it is approved for use. This is not the case. Approval means acceptable to the *authority having jurisdiction (AHJ)*, which is usually the electrical inspector. If an installation requires equipment or materials that are not listed by a nationally recognized testing laboratory (NRTL), that doesn't mean that the equipment or material cannot be installed or used. The decision on the approval of equipment is solely left to the judgment of the authority having jurisdiction. After you review Sections 90-4, 90-7 and 110-2, this concept will become clearer.

> **Note.** Some local and state codes have a specific "written" requirement that equipment and material must be listed in order to be approved for use.

Attachment Plug (Cap) (Plug Cap): The male device at the end of a flexible cord that plugs into a receptacle. The use of cords is limited by the *NEC;* see Sections 210-50(a), 305-4, 400-7, 410-14, 410-30, 422-32, and 645-5.

Bathroom: An area including a basin with one or more of the following: a toilet, a tub, or a shower.

Bonding (Bonded): To tie together and to make as electrically one. The grounding of all metal enclosures of electrical wiring and equipment is accomplished by bonding together all metal, non-current-carrying parts of the electrical system.

Bonding Jumper: A reliable conductor to ensure electrical conductivity between metal parts required to be *electrically connected.*

Bonding Jumper, Equipment: A conductor connection between two or more portions of the equipment grounding conductor.

Bonding Jumper, Main: A conductor that connects the equipment grounding conductor to the grounded (neutral) conductor at the service equipment, by a screw, strap, wire, or conductor according to the requirement of Section 250-102. The purpose of the main bonding jumper is to provide a *low impedance* path for *fault-current* back to the *transformer* [250-28].

Branch Circuit [Article 210]: The conductors between the final overcurrent device and the outlet(s).

Branch Circuit, Appliance: The term used to describe the 20 ampere circuit that supplies energy to one or more outlets to which appliances are to be connected. This applies specifically to the two small appliance circuits required in Section 210-11(c)(1) for the kitchen and dining area receptacles in a dwelling unit.

Branch Circuit, General Purpose: These circuits are rated 125 volt, 15 or 20 ampere [210-23(a)] and supply a number of outlets for lighting and appliances. These circuits are permitted to supply convenience receptacles and lighting outlets.

> **Note.** There is no limit to the number of lights or receptacles on a general-purpose branch circuit in dwelling units, see Section 220-3(10).

Branch Circuit, Individual: This circuit supplies only one power-consuming piece of equipment and the circuit can be of any size.

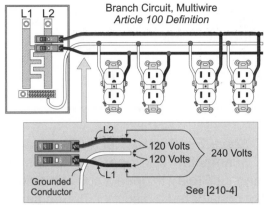

Branch Circuit, Multiwire
Article 100 Definition

L2
120 Volts
120 Volts   240 Volts
L1
Grounded
Conductor   See [210-4]

Multiwire Branch Circuit: A circuit with two or more ungrounded (hot) conductors having a voltage potential between them, and a grounded (neutral) conductor that has equal potential between each ungrounded conductor.

Fig. 2-2 Branch Circuit, Multiwire

Cabinet - *Article 100 Definition*

Cover with Door   Cabinet   Panelboard

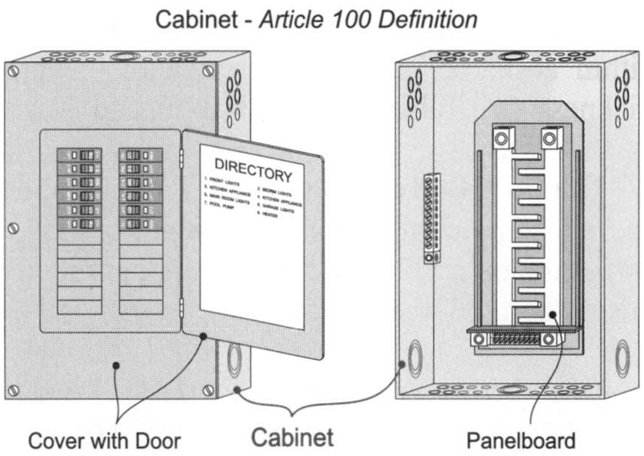

Cabinet: A surface or flush mounted enclosure that a cover with a door is attached to. They are often used to enclose panelboards and overcurrent protection devices. They are often called "Panels" by electricians.

Fig. 2-3 Cabinet

Branch Circuit, Multiwire: A branch circuit with two or more ungrounded (hot) conductors with a neutral or grounded conductor. The hot (ungrounded) conductors must have a potential voltage difference between them and must have an equal potential (voltage) to the grounded (neutral) conductor. There are special *Code* rules for multiwire branch circuits; see Sections 210-4 and 300-13(b) in this book, Fig 2-2.

Advantages of multiwire branch circuits. A 120/240 single-phase system (three wire) multiwire branch circuit can serve the same function as two separate 2-wire circuits (four wires). There are 25 percent fewer conductors, the *voltage drop* is reduced by as much as 50 percent, and there is a possible reduction in conduit size.

Disadvantages of multiwire branch circuits. Multiwire branch circuits require that the grounded (neutral) conductor be continuous and it cannot be opened when the circuits are in use. If the grounded (neutral) conductor is opened when loads are on, a "floating neutral" will develop. Depending on the *resistance* and configuration of the loads on the circuit, the operating voltage for each appliance can be significantly different. Instead of each load operating at 120 volt, one load could be operating at 200 volt and the other at 40 volt. Often the equipment will be damaged; see Section 300-13(b).

Building: A structure that stands alone or is cut off from other structures by fire walls or approved fire doors. Often the building code contains the specific requirements for identifying a building.

Cabinet [Article 373]: An enclosure designed either for surface or flush mounting with a trim permitting a swinging door to be hung. Section 373-8 permits conductor splices within a cabinet or cutout box. See the Glossary for the definition of a cutout box, Fig. 2-3.

Circuit Breaker [Article 240, Part G]: The most common types of circuit breakers are inverse time (IT), adjustable (electronically controlled), and instantaneous trip/motor circuit protector. Most of the examples we will cover in this book apply to the use of inverse time circuit breakers (also called "molded case" circuit breakers).

Concealed: Inaccessible because of the structure or finish of the building. However, wiring behind panels designed to allow access is considered "exposed."

Conductor, Bare: Having no electrical insulation. Bare conductors are generally used for the grounding conductor [250-119] and bonding conductor [250-102(a) and 680-22].

Conductor, Covered: A conductor encased within material not recognized as electrical insulation. An example of covered is the soft green material on the ground wire in appliance cords and some of the larger cables, such as 6/3 NM cable.

nductor, Insulated: A conductor encased ~~code~~-recognized material with the proper ~~...~~s. See Tables 310-13, 310-16, and 402-3.

Conduit Body (LB's, LL's, LR's, T's, etc.): These fittings are used for convenience or long runs for pulling ease [370-28]. Splices are permitted in a conduit body [370-16(c)].

Connector, Pressure (Solderless): A device that establishes a connection by means of mechanical pressure, such as the screw terminals for wire on switches, receptacles, circuit breakers, neutral and grounding bars in panels, etc. All conductors must be terminated in an approved manner according to Section 110-14.

Continuous Load: A load where the current is expected to continue for three hours or more, such as store lighting, parking lot lighting, signs, etc. Continuous loads are not limited to lighting; see Sections 210-19(a), 384-16(d), 422-10(a) and 422-13.

Controller: See Section 430-81.

Demand Factor: Ratio of the amount of connected load, which will be operating at the same time, to the total amount of connected load on the circuit. See Article 220, Parts B and C.

Device: A unit of an electrical system that is intended to carry, but not utilize (consume), electric energy. Receptacles, most types of switches, circuit breakers, fuses, and similar equipment are considered devices because they do not consume (utilize) electricity.

Disconnecting Means: A device or group of devices that disconnects the main power supply conductors. Switches, attachment plugs and receptacles, automatic and nonautomatic circuit breakers, knife switches, and safety switches are types of equipment that might qualify as a disconnecting means, Fig. 2-4.

Disconnects are required for:

Appliances	422, Part D
Electric space heating	424-19
Electric duct heaters	424-65
Motor control conductors	430-74
Motor controllers	430-102(a)
Motors	430-102(b)
Air-conditioning	440-14
Refrigeration equipment	440-14
Swimming pool equipment	680-12

> **WARNING:**
> *Shunt Trip Breaker Not A Disconnect.* A pushbutton that operates an electromagnetic relay to open a circuit breaker is not considered a disconnect. See service and motor disconnect requirements in Sections 230-70 through 230-83 and 430-101.

Dwelling: There are four different types of dwellings identified by the *NEC*. You can never take any rule or definition lightly; *Code* rules apply to specific types of occupancies, and there are significant differences between each of the following:

Dwelling Unit. A unit that has provisions for cooking, sleeping and bathroom facilities. There are many rules that apply specifically to dwelling units such as GFCI protection of receptacles [210-8], location of receptacle outlets [210-52], and location of lighting outlets [210-70(a)].

Multifamily Dwelling. A building with three or more dwelling units. Receptacles are not required for outdoor locations in a multifamily

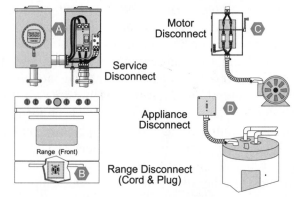

Disconnecting Mean - *Article 100 Definition*

Disconnecting Means: **A device or other means that disconnects circuit conductors from their source of supply.**

Fig. 2-4 Disconnecting Means

dwelling, but, if receptacles are installed outdoors, they must be GFCI protected [210-8(a)(3)].

One-Family Dwelling. A building consisting of only one dwelling, such as a house. Depending on the local building code, town houses that have a separation between the units by a fire rated wall are often considered separate buildings. In this case, each unit of a group of town houses is a one-family dwelling.

Two-Family Dwelling. A building consisting of two dwelling units, often called a duplex. Two-family dwelling units require receptacles to be installed outside, and be **accessible** at grade level [210-52(e)]. This rule does not apply to multifamily dwelling units.

Effectively Grounded: Intentionally connected to earth through a ground connection or connection of sufficiently low impedance. The path must have the current-carrying capacity to prevent a build-up of voltage that may result in conditions hazardous to people. Metal frames of buildings, when they are in direct contact with the foundation structural steel or with the earth, are considered effectively grounded.

Enclosure: Panels, auxiliary gutters, boxes, and motor control equipment are just some types of enclosures. The fence or walls surrounding an installation to prevent personnel from accidentally contacting energized parts, or to protect the equipment from physical damage, are also enclosures.

Energized: Electrically connected to a source of power (potential difference).

Equipment: A general term that includes every type of electrical component, such as a device, fitting, appliance, fixture, etc.

Exposed Live Parts: Energized parts not suitably guarded, isolated or insulated; see Section 110-27(a).

Exposed, Wiring Methods: On, or attached to, the surface of a building including the wiring behind dropped ceiling panels.

Feeder: All conductors between a source of power from a battery, a solar voltaic system, generator, transformer, or converter windings and the final branch circuit overcurrent protection device(s). Conductors from a generator or transformer are considered "feeder conductors," not "service conductors."

Garage: The portion of a building in which self-propelled vehicles are intended to be kept for use, storage, rental, repair, exhibition, or demonstration.

> **Note.** Receptacles can be installed at any height in a dwelling unit garage.

Grounded: The intentional connection to earth, or to some conducting body of the internal wiring system. This is generally done at the electrical service [250-24(a)], or at a separately derived system [250-30(a)(1)].

Grounded (Neutral) Conductor: A circuit conductor that is intentionally connected (grounded) to earth ground. This conductor is often called the *neutral wire* and is identified by the use of white or gray coloring on the insulation [200-6].

This conductor carries the unbalanced load between the *phase* conductor currents [220-22] and is grounded at the service location [250-24(a)] and at separately derived systems [250-30(a)(2)]. This is covered in more detail in Articles 200 and 250.

> **Note.** Some systems have a phase (hot) conductor grounded (corner grounded *delta connected* system); this conductor is also considered a grounded conductor. The subject of corner grounded delta systems is beyond the scope of this book.

Grounding Conductor (Equipment): A conductor that connects noncurrent-carrying metal parts of equipment and ultimately terminates at the service equipment enclosure or source of a separately derived system. Grounding conductors must provide a permanent and continuous low impedance path for fault-current to ensure the operation of the overcurrent protection devices.

The grounding conductor path can consist of a separate conductor, raceways, enclosures, boxes, housings, frames of motors, and all noncurrent-carrying metal parts of equipment associated with an electrical system, see Sections 250-118 and 250-119.

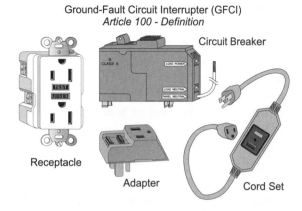

Fig. 2-5 Ground-Fault Circuit Interrupter (GFCI)

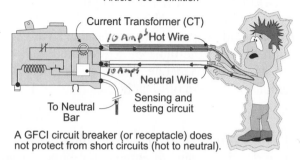

Fig. 2-6 Simplified Ground-Fault Circuit-Interrupter (GFCI)

5 mil Amp - Trips.

Note. The grounding conductor can be bare or insulated conductor; but if insulated, the conductor must be identified by the color green [250-119]. Many electricians call the grounding conductor a *bond wire*.

Grounding Electrode Conductor: The conductor used to connect the grounding electrode (normally the ground rod) to the grounded (neutral) or grounding conductor at the service equipment [250-24(a)], or separately derived system (transformer or generator) [250-30(a)(2)]. This conductor is sized according to Section 250-66.

Ground-Fault Circuit-Interrupter (GFCI): A device for the protection of personnel that functions to deenergize a circuit when the current to ground exceeds some predetermined value required to operate the overcurrent protective device. These devices are available as receptacles and circuit breakers, Fig. 2-5.

GFCI protection is required for replacement of nongrounding-type receptacles with grounding-type receptacles [210-7(d)], dwelling units [210-8], and swimming pools [Article 680].

Note. A GFCI protection device operates on the principle that a hot and neutral current runs through a sensitive *coil* (current transformer). If a *ground-fault* is encountered on the *load side* of the device, current on the hot and neutral will be different, thereby activating the shunt-trip mechanism. A GFCI device responds to ground-fault current of 5 milliampere (5/1,000 *ampere*), Fig. 2-6.

> **WARNING:**
> If a person touches the hot and neutral of a GFCI protected circuit at the same time, the GFCI device will not sense an unbalance of current and will not trip, possibly resulting in injury or death, Fig. 2-6.

Identified (Equipment): Recognizable as suitable for a specific use or condition, such as direct burial, by listing and labeling from a Nationally Recognized Testing Laboratory (NRTL). For example, no more than one conductor under a *terminal*, unless identified otherwise [110-14].

In Sight (Within Sight From): Visible and not more than 50 feet distant. See Sections 422-31(b), 430-102(a) and 440-14.

Interrupting Rating: The highest short circuit current at rated voltage that the device can safely interrupt, such as 10,000, 22,500, or 65,000 amperes RMS, see Section 110-9.

Labeled: Equipment or materials with a label, symbol, or other identifying mark, applied by a Nationally Recognized Testing Laboratory acceptable to the inspector. Labeling and listing equipment provides the basis for inspector approval of the equipment; see Sections 90-4, 90-7, 110-2, and 110-3 for more details. Many are familiar with the testing laboratory labels or marking on electrical equipment, which may be in the form of a sticker,

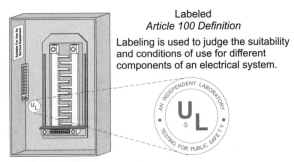

Labeled
Article 100 Definition
Labeling is used to judge the suitability and conditions of use for different components of an electrical system.

Labeled: Equipment or materials that have a label, symbol, or mark of a Nationally Recognized Testing Laboratory (NRTL). Note: Sometimes the label is on the box or instructions that come with the material.

Fig. 2-7 Labeled

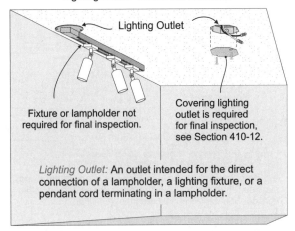

Lighting Outlet - *Article 100 Definition*

Lighting Outlet

Fixture or lampholder not required for final inspection.

Covering lighting outlet is required for final inspection, see Section 410-12.

Lighting Outlet: An outlet intended for the direct connection of a lampholder, a lighting fixture, or a pendant cord terminating in a lampholder.

Fig. 2-8 Lighting Outlet

decal, or printed label, or molded or stamped into the product, Fig. 2-7.

Lighting Outlet: A lighting outlet is an opening in the electrical system for the connection of a lighting fixture, *lampholder*, or pendant cord terminating in a lampholder. Where the *Code* requires a lighting outlet to be installed [210-70], what this means is that a box opening must be provided, not that a lighting fixture be installed, Fig 2-8.

Listed: Equipment or materials that are on lists published by Nationally Recognized Testing Laboratories, such as Underwriters Laboratories (UL), Electrical Testing Laboratories (ETL), Factory Mutual (FM), Canadian Standards Association (CSA), etc., acceptable to the authority having jurisdiction, that are concerned with product evaluation and maintain periodic inspection of the production of listed equipment or material.

The listing must state that appropriate designated standards have been met, or the material/equipment has been tested and found suitable for use in a specified manner. The *Code* does not always require electrical equipment to be listed, but some rules do specifically require listed material/equipment [90-7, 110-2, 110-3, 250-8, 250-70, 250-118].

Live Parts: Exposed electrical parts that present a shock hazard.

Location: Location plays an important part in the type of equipment used:

Damp Location. Partially protected locations under canopies, marquees, roofed open porches, and like locations, and interior locations subject to moderate degrees of moisture, such as some basements, some barns, and some cold-storage warehouses.

Note. Electrical inspectors generally don't consider the bathroom area to be a damp location. But the area within the bathtub and shower space is often considered a wet location up to about 7 feet, and a damp location above 7 feet.

Wet Location. Wet locations include locations underground, in concrete slabs and masonry in direct contact with the earth, locations subject to saturation with water, and locations exposed to weather and unprotected.

Dry Location. An area not normally subjected to dampness or wetness, but which may be subject temporarily to dampness or wetness, such as a building under construction.

Neutral Conductor: Not defined in the *National Electrical Code;* see glossary for the definition.

Nonlinear Load: A load where the wave shape of the steady state current does not follow the wave shape of the applied voltage, Fig 2-9.

The *NEC* contains very few requirements for

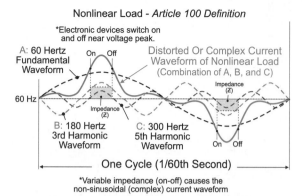

Nonlinear Load - *Article 100 Definition*

Nonlinear Load: A load where the wave shape of the steady state current does not follow the wave shape of the applied voltage. See the (FPN) following the definition.

Fig. 2-9 Nonlinear Load

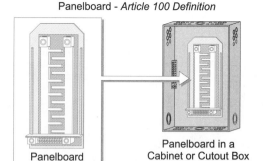

Panelboard - *Article 100 Definition*

Panelboard: A panel or group of panels designed to form a single panel. It is used to control power to light, heat, or power circuits. It is designed to be places in a cabinet or cutout box.

Fig. 2-10 Panelboard

nonlinear loads [210-4(a) FPN, 220-22 FPN 2, 310-15(b)(4)(c), and 450-3 FPN 2].

> **FPN:** The *NEC* does not identify when a load becomes nonlinear, but typical single-phase nonlinear loads are electronic equipment, copy machines, laser printers, electric discharge lighting such as fluorescent and high pressure sodium lights. Three-phase nonlinear loads include Uninterruptible power supply (UPS) and motor variable frequency speed drives (VFD).

Outlet: A point in the wiring system at which electric current is taken to supply utilization equipment (receptacle, lighting, appliance, etc.). This means the point at which the receptacles and lighting fixtures are installed, not the receptacle or the fixture itself.

Overcurrent: Current greater than the rated current of equipment or conductors. It may result from overload, *short-circuit*, or ground-fault.

Overload: Operation of equipment or conductors in excess of ampacity rating. For example, No. 8 THHN wire (**ampacity** of 40 amperes) containing a 60 ampere load.

Panelboard [Article 384]**:** A distribution point for circuits that contains the branch circuit or feeder protection devices. The slang term for this is *guts*, Fig. 2-10.

Premises Wiring: The wiring on the interior and exterior of a building, on the load side of the service point and separately derived systems (transformers and generators). This includes control and signaling circuits, but it does not include wiring internal to appliances, fixtures, motors, controllers, motor control centers, air-conditioning equipment, or other similar equipment.

> **Note.** Premises wiring includes the wiring on the load side of computer power distribution units.

Qualified Person: A person familiar with construction, operation, and hazards associated with electrical wiring; see Section 110-26 for an example of its use.

Raceway: A term meaning an enclosure designed for installation of conductors, cables, or busbars. Raceways include the following:

Code Name	*Article*
Electrical nonmetallic tubing	331
Rigid metal conduit	345
Intermediate metal conduit	346
Rigid nonmetallic conduit	347
Electrical metallic tubing	348
Flexible metal conduit	350
Liquidtight flexible conduit	351
Surface raceways	352
Underfloor raceways	354
Cellular metal floor raceways	356
Cellular concrete floor raceways	358
Wireways	362
Busways	364

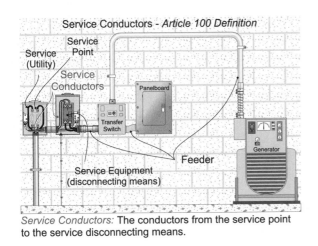

Service Conductors - *Article 100 Definition*

Service Conductors: The conductors from the service point to the service disconnecting means.

Fig. 2-11 Service Conductors

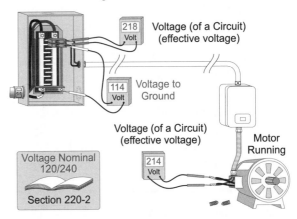

Voltage - *Article 100 Definition*

Voltage (of a Circuit) (effective voltage)

Voltage to Ground

Voltage (of a Circuit) (effective voltage)

Motor Running

Voltage Nominal 120/240

Section 220-2

Note: Nominal voltage is an assigned value of a circuit or system for the purpose of convenience, such as 120/240. The voltage (of a circuit) and voltage to ground are actual voltages.

Fig. 2-12 Voltage

Raintight: Equipment constructed or protected to prevent heavy rain from entering.

Receptacle: A female contact device installed at an outlet for the electrical connection of a single attachment plug. A single receptacle is a single contact device with no other contact device on the same *yoke*, and a multiple receptacle is two or more contact devices each an individual device, mounted on a common yoke.

Receptacle Outlet: An outlet opening in a box for the installation of one or more receptacles. Receptacles are required for dwelling units [210-52], guest rooms [210-60(a)], and heating, air-conditioning, and refrigeration equipment [210-63].

Remote Control Circuit: An electric circuit that controls another circuit through a relay or solid state device.

Separately Derived System: A premises wiring system that derives its power from a battery, solar photovoltaic system, or a generator, transformer, or converter winding. Separately derived systems have no direct electrical connection to supply conductors originating in another system [250-30].

Service: Conductors and equipment for delivering electric energy from the serving utility to the wiring system of the premises served. Conductors and equipment supplied from a battery UPS system, a solar voltaic system, generator, transformer, or

phase converter are not considered a service. See the definition for "Feeder." The general rule is that only one service is permitted for each building [230-2(a) and 230-40].

Service Conductors: Service conductors only originate from the service point and terminate at service equipment (disconnect). Conductors supplied from back-up generators, battery UPS systems, transformers, etc. are considered "feeders," Fig. 2-11.

Service-Drop: The overhead conductors from the last pole or aerial support to and including the splices. These conductors are generally the utility company's responsibility [Article 230, Part B].

Service Equipment: The main means of cutoff (disconnecting means) for the service supply conductors [Article 230, Part F].

Service-Lateral: Underground service conductors from the electric utility transformer to the first point of connection to the service-entrance conductors [Article 230, Part C].

Service Point: The point of connection between the facilities of the serving electric utility and the premises wiring.

Special Permission: Written consent from the authority having jurisdiction (inspector).

Voltage (of Circuit): The highest voltage between any two conductors, Fig. 2-12.

Voltage, Nominal: A standard designation such as 120/240, 208Y/120, and 480Y/277. Section 220-2(a) lists nominal voltages (600 volts) for branch circuit, feeder, and service calculations, Fig. 2-12.

Voltage to Ground: For grounded systems, the highest voltage between any hot conductor and grounded (neutral) conductor, such as 120 volt of a 208Y/120 volt system. For ungrounded circuits, the greatest voltage between the given conductor and any other conductor of the circuit, Fig. 2-12.

Weatherproof: Constructed or protected that exposure to the weather will not interfere with successful operation.

SUMMARY

Article 100 contains definitions that are important for the proper application of the *NEC*. ☐ Definitions can also be found throughout the *NEC* in specific Articles. ☐ Common terms, such as volt, voltage drop, ampere, impedance, and resistance are not defined in the *NEC*. ☐ A definition that is listed in a specific Article applies to that Article only. Example, Section 384-14 defines a lighting and appliance branch circuit panelboard, which applies only to Article 384.

REVIEW QUESTIONS

1. In your own words explain the meaning of the following Article 100 definitions:

    Accessible, Wiring Methods
    Accessible, Readily or Readily Accessible
    Approved
    Branch Circuit [Article 210]
    Building
    Concealed
    Conductor
    Connector, Pressure (Solderless)
    Device
    Dwelling
    Enclosure
    Energized
    Equipment
    Exposed, Wiring Methods
    Feeder
    Ground-Fault Circuit-Interrupter (GFCI)
    Identified (Equipment)
    In Sight (Within Sight From)
    Labeled
    Lighting Outlet
    Listed
    Live Parts
    Outlet
    Raceway
    Receptacle
    Receptacle Outlet
    Service
    Special Permission

Unit 3

Article 110
Requirements for Electrical Installations

OBJECTIVES

After studying this unit, the student should be able to understand:
- the guidelines for approval of equipment.
- the reasons why equipment must have an interrupting rating sufficient for the available fault-current.
- the rules that apply to conductor terminations.
- the rules that apply to conductor splices.
- the rules that apply to working space for electrical equipment.
- the rules that apply to working space entrances.
- the rules that apply to identification of disconnects and circuits.

Definitions and Glossary Terms

To better understand the *NEC* rules contained in this unit, review the following:

Definitions, Article 100 - Unit 2

Accessible	Approved
Branch circuit	Building
Circuit breaker	Conductors
Controller	Device
Disconnect	Dry locations
Enclosure	Energized
Equipment	Exposed
Feeder	Identified
Interrupting rating	Labeled
Listed	Nominal voltage
Raceways	Voltage to ground

Glossary Terms

AL	Back stabbing
Circular mils	Corrosion
CU	Derating
Exothermic welding	Fuse
Gray area	Junction box
Line side	Oxidation
Splice	Torque

PART A.   GENERAL

110-2   Approval Of Equipment

All electrical **equipment** prior to installation must be **approved** by the authority having jurisdiction, which is generally the electrical inspector. The

Approved - *Section 110-2*

All equipment and materials required or permitted by the NEC are acceptable only if approved by the authority having jurisdiction.

Approved: Acceptable to the authority having jurisdiction.

Fig. 3-1 Approved

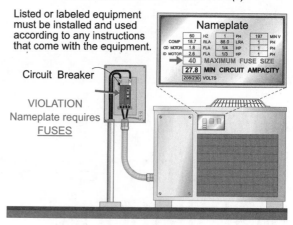

Installation Instructions - *Section 110-3(b)*

Listed or labeled equipment must be installed and used according to any instructions that come with the equipment.

Circuit Breaker

VIOLATION
Nameplate requires
FUSES

Fig. 3-2 Installation Instructions

principal basis for the approval of equipment by the inspector is listing and/or labeling by Nationally Recognized Testing Laboratories [90-7], Fig. 3-1.

But, listing or labeling does not mean the inspector will automatically permit its use.

110-3 Examination, Identification, Installation, And Use Of Equipment

(a) Guidelines for Approval. This subsection provides guidelines for the inspector when judging equipment for approval. Generally, most equipment is considered suitable for use and installation when **listed** and **labeled**. The *NEC* does not list and/or label equipment but depends on Nationally Recognized Testing Laboratories to provide the necessary testing and evaluation of electrical equipment.

(b) Installation and Use According to Instructions. Equipment must be installed, and/or used according to its listed and labeled instructions. Failure to follow these instructions, such as torquing of terminals, sizing of **conductors**, installing no more than one conductor under a *terminal* unless identified for use with more than one conductor [110-14(a)], and so on, is a violation of this *Code* rule.

Equipment is often listed with specific conditions of use, operation, or installation. It is important that equipment be installed according to the instructions limiting the use, operation, or installation. As an example, some air-conditioning equipment nameplates are marked "Maximum Fuse Size"; this means that the equipment must be protected by *fuses* only, not **circuit breakers**, Fig. 3-2.

110-4 Voltages

Equipment must be installed on a system whose **nominal voltage** does not exceed the voltage rating of the equipment, Fig. 3-3.

In addition, electrical equipment must not be connected to a nominal voltage source that is less than the equipment rated [110-3(b)].

Section 220-2(a) provides the nominal voltages that must be used to calculate branch circuit, feeder, and service loads. Some *Code* rules refer to voltage between conductors and other rules refer to **voltage to ground**.

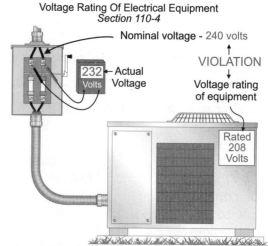

Voltage Rating Of Electrical Equipment
Section 110-4

Nominal voltage - 240 volts

232 Volts ← Actual Voltage

VIOLATION
↓
Voltage rating of equipment

Rated 208 Volts

The voltage rating of electrical equipment cannot be less than the nominal voltage of the circuit to which it is connected.

Fig. 3-3 Voltage Rating of Electrical Equipment

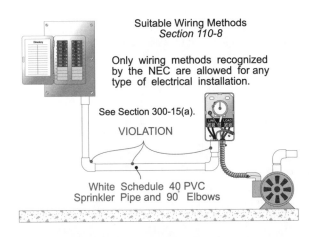

Fig. 3-4 Suitable Wiring Methods

The *NEC* contains voltage limitations for branch circuits [210-6] in dwelling units [410-80(b)]. Most *Code* Articles have specific rules for systems over 600 volt. Article 490 applies only for equipment "Over 600 Volt Systems." Article 300, Part B applies to wiring methods over 600 volts.

110-5 Copper Conductors

When the *NEC* refers to a conductor size, it's in reference to the material of copper (*CU*), unless aluminum is specifically specified in the rule. However, when aluminum is used instead of copper, the conductor size must be increased because aluminum has a higher opposition to the flow of electrons (*resistance*) as compared to copper.

110-6 Conductor Sizes

Conductor sizes are expressed in *American Wire Gauge* (AWG) or *circular mils*. kcmil (thousand circular mils) is used in all *Code* tables for conductor sizes larger than No. 4/0.

110-7 Conductor Insulation

All wiring must be free from *short-circuits* and *ground-faults*. The most common cause of problems in electrical systems is the failure of *insulation* due to excessive heat, moisture, corrosive chemicals, ultraviolet rays from the sun, and/or physical (mechanical) damage.

When a conductor is no longer used (has been abandoned), the end of the conductor must be covered with an insulation equivalent to that of the conductors or with an insulating **device** identified for the purpose, such as a twist-on wire connector (*wirenut*) [110-14(b)].

110-8 Wiring Methods

Articles 318 through 384 list the wiring methods that are considered suitable according to the *NEC*. Article 300 applies to all wiring methods except as modified by Chapters 5, 6, and 7.

Not all wiring methods can be installed in all locations. For example, white schedule 40 *PVC* sprinkler pipes and their associated fittings cannot be used for electrical wiring, Fig. 3-4.

110-9 Interrupt Protection Rating

Overcurrent protection devices, such as circuit breakers and fuses, are intended to *open* the circuit at fault levels. Accordingly, they are required to have an **interrupting rating** sufficient for the maximum possible *fault-current* available on the *line side* terminals of the equipment.

If the overcurrent protection device is not rated for the available fault-current, it could explode while attempting to clear the fault, and/or the downstream equipment could suffer serious damage, causing possible hazards to people. *UL*, *ANSI*, IEEE, NEMA, manufacturers, and other organizations have considerable literature on how to properly size overcurrent protection and methods for calculating *available short-circuit current*.

> **DANGER:**
> If the available *fault-current* exceeds the **controller's** short-circuit current rating, the controller could literally explode, endangering persons and property. To solve this problem, a current-limiting protection device (fast-clearing *fuse*) can be used to reduce the let-thru energy to less than 5,000 ampere.

110-10 Short-Circuit Current Rating

Electrical equipment shall have a *short-circuit current rating* that permits the circuit overcurrent

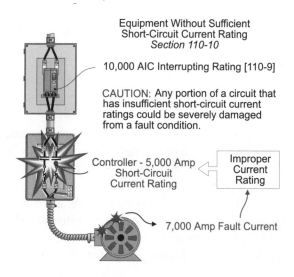

Fig. 3-5   Equipment Without Sufficient Short-Circuit Current Rating

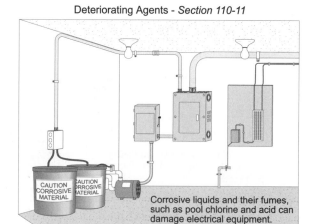

Fig. 3-6 Deteriorating Agents

protection device to clear short-circuit or ground-faults without extensive damage to the electrical components of the circuit, Fig. 3-5.

> **Note.** If you are familiar with short-circuit fault calculations and Microsoft Excel, you can download a spreadsheet template at: *www.mikeholt.com.*

Section 110-10 specifies that electrical equipment shall have sufficient short-circuit current rating to permit the circuit protection device to clear a fault "without extensive damage" to the electrical components of the circuit. There has been debate over the years as to what's considered "without extensive damage." To remove the confusion, a sentence was added to the 1999 NEC to specify that equipment installed and used in accordance with their listings are understood to comply with the "without extensive damage" rule of this Section.

110-11 Deteriorating Agents

All electrical equipment must be suitable for the environment or condition. The electrician must consider the possible presence of corrosive gases, fumes, vapors, liquids, or chemicals. In addition, consideration must be given concerning possible exposure to wet or damp locations, ultraviolet rays from the sun, physical (mechanical) damage, excessive temperature, or any other agent that could have a detrimental effect on electrical equipment and conductors, Fig. 3-6.

> **FPN No. 2:** Some spray cleaning and lubricating compounds contain chemicals that cause severe reactions with plastics and result in deterioration of electrical equipment.

Electrical equipment approved for use in **dry locations** must be protected from the weather during the **building** construction period.

110-12 Mechanical Execution Of Work

Electrical equipment must be installed in a neat and workmanlike manner. What is a neat and workmanlike manner? This is a judgment call for the inspector. Some inspectors have enough authority to insist that **raceways** and boxes should be straight and cables are neat and secured. Even work that will be covered by the structure of the building is subject to this rule. A sloppy job indicates lack of proper training, supervision, or an uncaring attitude. Workmanship should be a matter of pride that electricians have in themselves and the trade.

(a) Unused Openings. Unused openings in all electrical equipment must be closed with an approved fitting that provides protection equivalent to the wall of the equipment [370-18 and 373-4], Fig. 3-7.

Unused Openings - *Section 110-12(a)*

Panel Filler   Knockout Filler   Plug Filler

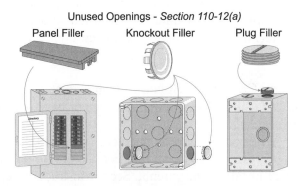

Note: Also see Sections 370-18 and 373-4.

Fig. 3-7 Unused Openings

(c) Protection of Internal Parts. Internal parts of electrical equipment must not be damaged or contaminated by foreign material, such as paint, plaster, cleaners, etc. Precautions must be taken to provide protection from the detrimental effects of paint on the internal parts of **panelboards** and receptacles, Fig. 3-8.

Note. The exterior of electrical equipment can be painted, but you must be careful not to paint over the nameplate on the equipment.

Damaged parts of electrical equipment that may adversely affect the safe operation or the mechanical strength of the equipment cannot be installed. Damaged parts would include such items as cracked insulators, arc shields not in place, overheated fuse clips, and damaged or missing switch or circuit breaker handles.

CAUTION: Caution should be exercised to protect electrical equipment from conditions that could result in damage or contamination of internal parts. Materials (liquid sprays) specified by the manufacturer or listed for the purpose can be used.

110-13 Mounting And Cooling Of Equipment

(a) Mounting. Electrical equipment shall be firmly secured to the surface on which it is mounted [370-23].

(b) Cooling. Electrical equipment that depends on natural air circulation for cooling must be installed according to the manufacturer's instructions. For example, the air flow cooling vents on the equipment should not be blocked. Some floor mounted equipment requires open space around and above for the dissipation of heat. Some transformers have special requirements for cooling [450-9].

110-14 Electrical Conductor Termination

Terminal Conductor Material. Terminals for copper or aluminum must be **identified** for the material of the conductor. Terminals suitable only for aluminum conductors must be marked *AL*. Terminals suitable for both copper and aluminum wire must be marked as *CU/AL* or *CO/ALR* [380-14(c) and 410-56(b)].

Copper and Aluminum Mixed. Dissimilar conductor materials must not make contact in a terminal or splicing device, unless the device is identified for the purpose. Dissimilar conductor materials in contact with each other causes *corrosion* that degrades the conductor, increasing the resistance for the connection or *splice*. This in turn

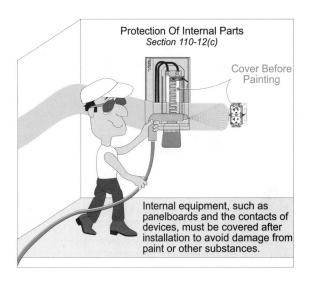

Protection Of Internal Parts
Section 110-12(c)

Cover Before Painting

Internal equipment, such as panelboards and the contacts of devices, must be covered after installation to avoid damage from paint or other substances.

Fig. 3-8 Protection of Internal Parts

may cause dangerous overheating of the termination or splice

Note. Only one manufacturer produces a twist-on wire connector that is **listed** for the joining of copper and aluminum conductors together.

Inhibitors and Compounds. Inhibitors and other compounds that prevent *oxidation* on aluminum conductors or terminals must be suitable for the use and shall not have an adverse effect on the conductors, their insulation, or equipment terminals.

FPN: Tightening Torques. Most electrical equipment requires the conductor terminations to be torqued; these instructions are generally included with the equipment instructions. Failure to *torque* terminals is a violation of Section 110-3(b), which requires that equipment be installed according to its instructions.

(a) Terminations. Connection of conductors to set-screw or compression terminals must be tight and made without damage to the conductors. Push-in type (**back stabbing**) terminations are approved and are commonly used for the wiring of duplex receptacles and snap switches.

One Wire Per Terminal Terminals in general are listed to accept no more than one wire per termination. Terminals that permit more than one wire per termination must be identified, either within the equipment instructions or on the terminal itself, Fig. 3-9.

Note. Be sure you follow the equipment installation instructions [110-3(b)]. They

include information to identify the materials permitted, torque requirement, the number of conductors permitted under a terminal, and how to wrap the conductor around a screw post.

(b) Conductor Splices. Conductors must be spliced using a listed splicing device or by *exothermic welding*.

When using wire connectors (wirenuts), it is not necessary to twist the conductors together before installing the wirenut. See the instructions on the bag or box. However, many electricians prefer to twist the conductors together, particularly for 14 and No. 12 Solid. There is nothing wrong with twisting the conductors together, but this is not a *Code* rule.

> CAUTION: Because of the dangers associated with opening the neutral wire on a **multiwire branch circuits,** it is a good practice to twist the grounded (neutral) conductors together. This is not important for the hot wires; see Section 300-13(b) in this book.

Splice Insulation. In addition, the splice and the free ends of all conductors must be covered with an insulation equal to that of the conductor. The rule applies to both **energized** and de-energized conductors.

If a conductor is no longer used or needed, it can be left in the **enclosure**, but the ends of the conductor must be insulated. This prevents the **exposed** end of the conductor from touching energized parts which can create an electrical hazard.

Splices Underground. Direct burial single conductors (Type UF or USE) can be spliced underground without the use of a *junction box*, but the splicing means used must be listed for direct burial installation; see Sections 300-5(e) and 300-15(b). Multiconductor Type UF or Type USE cable installed underground can be spliced with listed splice kits that encapsulate the conductors and the splice.

(c) Conductor Size. Conductors shall be sized to the lowest temperature rating of any *terminal*, device, or conductor. When *derating* conductor ampacities, use the **ampacity** based on the conduc-

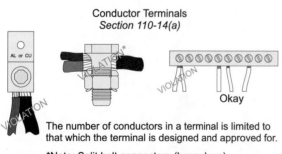

Conductor Terminals
Section 110-14(a)

Okay

The number of conductors in a terminal is limited to that which the terminal is designed and approved for.

*Note: Split bolt connectors (bugs, lugs) are only rated for two conductors.

Fig. 3-9 Conductor Terminals

Using Higher Conductor
Ampacity for Derating
Section 110-14(c)

Raceway contains 8 No. 12 THHN
Current-Carrying Conductors

For derating purposes, such as ambient temperature
or more than 3 current-carrying conductors, *the
higher ampacities of Table 310-16 can be used.*

Example: Determine the ampacity of the 8 No. 12 THHN.

Table 310-16: Ampacity of No. 12 THHN, 90°C = 30 amperes

30 amperes x 0.7 =   21 amperes for each No. 12

Fig. 3-10 Using Higher Conductor Ampacity for
Derating

tor's insulation temperature rating, not the ampacity
at the terminal rating.

Question. What is the ampacity of eight No. 12
THHN conductors in the same raceway; see Table
310-15(b)(2) and Table 310-16, Fig 3-10?

Answer. 21 ampere
Step 1. No. 12 THHN ampacity 30 ampere

Step 2. 30 ampere × 0.70 = 21 ampere

(1) Terminals Rated 60°C. Equipment terminals rated 100 ampere or less (circuit breakers,
fuses, and so forth), and **pressure connector** terminals for No. 14 through No. 1 conductors, shall use:

(a) Conductors sized to the 60°C temperature
rating as listed in Table 310-16; see Table 3-1.

(b) Conductors with temperature ratings
greater than 60°C can be used, but the size of the
conductors shall be based on the 60°C temperature
terminal rating.

Question. What size THHN conductor is
required for a 50 ampere circuit?

Answer.   No. 6

Conductors must be selected according to
the temperature rating of the equipment (60°C);
this requires a No. 6 conductor as selected from
Table 310-16.

Table 3-1
Conductor Sizing Based on Terminal Rating

Terminal Ampacity	Terminal 60°C	Terminal 75°C
15	14	14
20	12	12
30	10	10
40	8	8
50	6	8
100	1	3

(c) Equipment terminations and pressure connectors, listed and identified for use with terminals
rated 75°C, can use conductors sized according to
75°C temperature rating as listed in Table 310-16.

Question. What size THHN conductor is
required for a 50 ampere circuit if the listed equipment has 75°C terminals, Fig. 3-11?

Answer.   No. 8
Conductors must be selected according to the
temperature rating of the equipment rated 75°C;
this requires a No. 8 conductor as selected from
Table 310-16.

(2) Terminals Rated 75°C. Terminals for
equipment rated over 100 ampere, and pressure connector terminals for conductors larger than No. 1,
shall use conductors sized according to:

(a) 75°C temperature rating as listed in Table
310-16; see Table 3-1.

Question. What size THHN conductor is
required for an air-conditioner whose nameplate
specifies a minimum conductor ampacity of 103

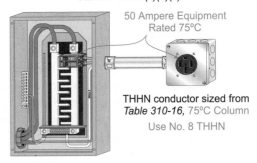

Conductor Sizing Based on Terminal Ratings
Section 110-14(c)(1)(c)

50 Ampere Equipment
Rated 75°C

THHN conductor sized from
Table 310-16, 75°C Column
Use No. 8 THHN

Fig. 3-11 Conductor Sizing Based on Terminal
Ratings

ampere and a maximum overcurrent protection of 150 ampere?

Answer.   No. 2

Conductors must be selected according to the temperature rating of the equipment rated 75°C; this requires a No. 2 conductor as selected from Table 310-16.

(b) Conductors with temperature ratings greater than 75°C can be used, but the size of the conductor shall be based on the 75°C temperature terminal rating; see Table 3-1.

Question. What size THHN conductor is required for a 225 ampere feeder?

Answer.   No. 4/0

Conductors must be selected according to the temperature rating of the equipment (75°C). This requires a No. 4/0 conductor as selected from Table 310-16.

(c) Equipment terminations and pressure connectors listed and identified for use with terminals rated 90°C or greater can use conductors sized according to 90°C temperature rating, unless the equipment is marked otherwise. In practice this practically never happens.

110-21   Manufacturer's Markings

Electrical equipment must be marked with the manufacturer's identification. Additional markings shall be voltage, current, wattage, or other ratings as might be required by other *Code* Sections. These markings are required to withstand the environment in which the equipment is installed.

110-22   Identification

Disconnecting Means. Where the purpose is not evident, **disconnecting means** for appliances, electric space heating, electric duct heating, motors, air-conditioning, services, etc., must be legibly marked to indicate their purpose, Fig. 3-12.

Circuits. Each **branch circuit** and **feeder** must be legibly marked to indicate its purpose [384-13], Fig. 3-12.

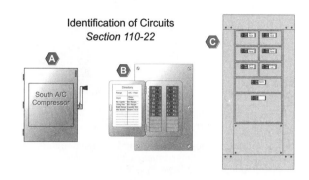

Fig. 3-12 Identification of Circuits

Series Combination Rated Equipment. When an installation is designed to comply with Sections 110-9 and 110-10 by utilizing a series rating system, proper field marking shall be provided to indicate that the equipment is part of a series rated system and that only identified components can be used (particularly for replacement). This requirement is not new; it's just that the text was changed to require the field marking to read:

Caution -- Series Combination System Rated
xxx,xxx Ampere
Identified Replacement Components Required

Note. "xxx,xxx" represents the actual numerical value, such as 22,000.

The available fault-current level must be indicated so that proper component replacement can be made. See Section 240-86 for more information on series-rated overcurrent protection devices. See Section 240-86(a) for more information on series rating markings for **circuit breakers**.

Part B.   600 Volt, Nominal or Less

110-26   Access And Working Space

For the purpose of safe equipment operation and maintenance, all electrical equipment such as panelboards, motor control centers, or disconnects must have sufficient access and working space. Enclosures housing electrical apparatus that are controlled by lock and key are considered accessible to **qualified persons**, Fig. 3-13.

Note. See Sections 230-91 and 240-24 on the rules governing the accessibility of

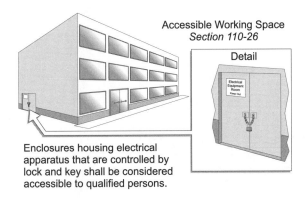

Fig. 3-13 Accessible Working Space

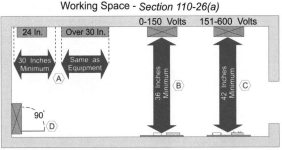

A: The minimum width of working space is 30 inches. If the equipment is over 30 inches, it is the same as the equipment.
B: The minimum depth of working space for any condition of 0 to 150 volts to ground is 3 feet.
C: The minimum working space for 151 to 600 volts to ground is 42 inches.
D: Working space shall permit at least a 90° opening of equipment doors or hinged panels.

Fig. 3-14 Working Space

overcurrent protection devices to building occupants.

(a) Working Space. Working space is required for equipment that may need examination, adjustment, servicing, or maintenance while **energized**.

The phrase "while energized" is the root of many debates; a typical argument is that electric power to almost all equipment can be turned off, somewhere. So if it is not energized, then the working clearance requirement of this Section is not required. A situation where working clearances might not apply would be equipment such as fixtures, receptacles, switches, raceways, boxes, and junction boxes.

This is a *gray area* in the *NEC;* I suggest that you always check with your inspector if you have any doubts.

(1) Working Space Depth. Working space depth shall be measured from the enclosure opening, Fig. 3-14.

Depth, 0-150 Volt to Ground. The working space depth shall not be less than 3 feet measured from the **enclosure** front, Fig. 3-14.

Depth, Over 150 Volt to Ground. The working space depth shall not be less than 3 feet measured from the **enclosure** front, except for the following two conditions:

Condition 2. If the equipment contains **live parts** on one side and concrete or brick on the opposite side, the working space depth shall be no less than 3½ feet, Fig. 3-14.

Condition 3. If the equipment contains live parts on one side and live parts on the opposite side, the working depth between the enclosures fronts shall not be less than 4 feet.

Exception No. 1: Rear and Sides. Working space is not required on the rear or the sides, if there are no renewable or adjustable parts such as fuses or switches, and where terminations are **accessible** from locations other than the back or sides of the equipment. Where rear access is required to work on de-energized equipment on the back of enclosed equipment, a minimum working space of 30 inches horizontally shall be provided. This requirement of 30 inches makes it easier to service electrical connections and equipment from the back when the system is de-energized. When the equipment must be serviced "while energized," compliance with the working space of Table 110-26(a) must be accomplished.

(2) Working Space Width. The working space shall be a minimum of 30 inches wide, but in no case less than the width of the equipment. This working space distance is measured from either, left to right, right to left, or simply from the centerline of the equipment, Fig. 3-15.

90° Opening of Equipment. In all cases the working space must be of sufficient width, depth, and height to permit at least a 90° opening of all equipment doors.

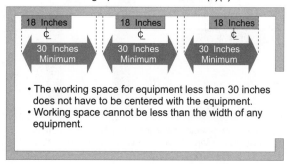

Fig. 3-15 Working Space

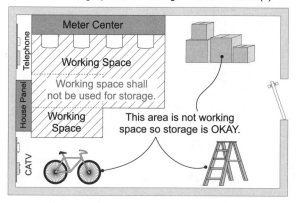

Fig. 3-17 Using Higher Conductor Ampacity for Deriving

(3) Working Space Headroom. The working space must have a minimum headroom from the floor to a height of not less than 6 feet 6 inches for service equipment, switchboards, and motor control equipment; see 110-26(e).

Equipment such as raceways, cables, wireways, **cabinets**, panels, etc. can be located above or below other electrical equipment, where the associated equipment does not extend more than six inches from the front of the electrical equipment. The 6 inch dimension should not compromise the ability of the installer to work with his or her hands on the panel or other electrical equipment located above or below other electrical equipment.

This rule permits a 12 inch wireway to be located under/over a 6 inch deep panel, or a 4 inch

time clock could to be mounted above/below a panel, Fig. 3-16.

(b) No Storage. The working space must be clear at all times and there shall be no storage in the working space area. People who service energized parts must not be subjected to any additional dangers by working about, around, over or under bicycles, boxes, crates, appliances, and other impediments, Fig. 3-17.

Note. Low-voltage and limited-energy equipment must be installed so as not to encroach on the working space requirements of the power equipment.

(c) Entrance. All electrical equipment must have one entrance of sufficient area to give access to the working space. Check with your inspector for what he or she considers "sufficient area." This is another *gray area* in the *Code*. Equipment rated 1,200 ampere or more and over 6 feet wide must have at least one entrance at each end; each entrance must not be less than 24 inches wide and 6½ feet high, Fig. 3-18.

Exception No. 1: Unobstructive Exit. Only one entrance is required where the location permits a continuous and unobstructed way of exit travel.

Exception No. 2: Double Workspace. Only one entrance is required where the working space required by Section 110-26(a) is doubled. However, the equipment must be located so that the edge of

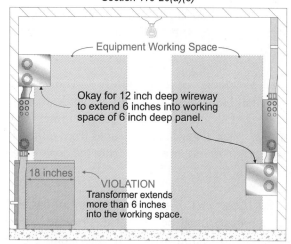

Fig. 3-16   Equipment Extending into Working Space

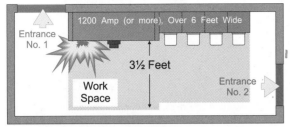

Location Of Entrances To Working Space
Section 110-26(c)

1200 Amp (or more), Over 6 Feet Wide

Entrance No. 1

3½ Feet

Work Space

Entrance No. 2

General Rule: For equipment rated 1200 amperes or more and over 6 feet wide, an entrance is required at each end.

Fig. 3-18 Location of Entrances to Working Space

the entrance is no closer to the equipment than the working space distance as listed in Table 110-26(a), Fig. 3-19.

(d) Illumination. Service equipment, switchboards, panelboards, or motor control centers located indoors must have the working space illuminated. Illumination may be provided by a lighting source located next to the working space, and the illumination must not be controlled by automatic means only.

Note. A switch with a motion sensor can be used, but the use of only a motion detecting device for this purpose is not permitted.

(e) Headroom. The minimum headroom for working spaces about service equipment, panelboards, switchboards or motor control centers shall not be less than 6 feet 6 inches. When the equipment exceeds 6 feet 6 inches, the minimum headroom shall not be less than the height of the equipment.

Exception. The minimum headroom requirement does not apply in existing dwelling units where the service equipment or panelboards do not exceed 200 ampere.

(f) Dedicated Equipment and Foreign Systems Space. No piping, ducts, or equipment foreign to the electrical installation shall be located in the *dedicated equipment space*. The dedicated equipment space is identified as the space the width and depth (footprint) of the equipment from the floor to a height of 6 feet above the equipment, or to the structural ceiling, whichever is lower, Fig. 3-20.

Foreign Systems Space Indoors. No piping, ducts, or equipment foreign to the electrical installation shall be located in the foreign system space unless the electrical equipment is provided with protection from liquids from accidental spillage or leakage from piping systems. The "foreign system space" is identified as the space the width and depth (footprint) of the equipment, from above the equipment to the structural ceiling, Fig. 3-21.

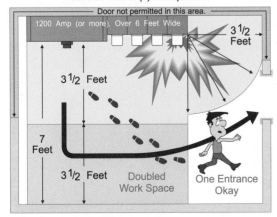

Location Of Entrance To Working Space
Section 110-26(c) Exception 2

Door not permitted in this area.

1200 Amp (or more), Over 6 Feet Wide

3½ Feet

3½ Feet

7 Feet

3½ Feet Doubled Work Space

One Entrance Okay

Where a single entrance is permitted, the entrance must not be closer to the equipment than the required clearance of Table 110-26(a).

Fig. 3-19 Location of Entrance to Working Space

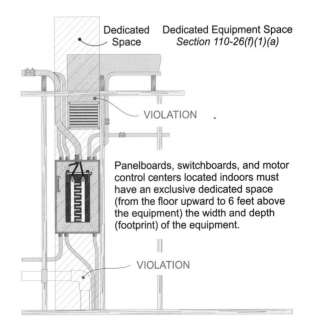

Dedicated Space

Dedicated Equipment Space
Section 110-26(f)(1)(a)

VIOLATION

Panelboards, switchboards, and motor control centers located indoors must have an exclusive dedicated space (from the floor upward to 6 feet above the equipment) the width and depth (footprint) of the equipment.

VIOLATION

Fig. 3-20 Dedicated Equipment Space

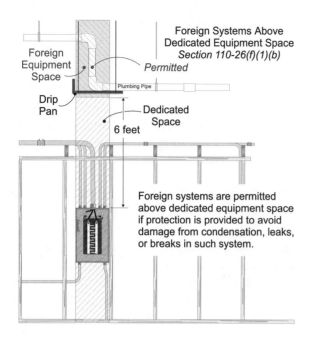

Fig. 3-21   Using Higher Conductor Ampacity for Derating

Fig. 3-22   Prevention of Physical Damage

110-27 **Guarding**

(a) **Prevent Physical Damage.** Electrical equipment must not be installed where it can be exposed to physical damage. Enclosures or guards are required to protect electrical equipment that could be exposed to physical damage, Fig. 3-22.

Note. Exposure to physical damage is subject to interpretation by the inspector; see

Sections 240-24(c) and 300-4 for additional requirements.

(b) **Warning Signs.** Entrances to locations with **exposed live parts** must have warning signs forbidding unqualified persons from entering. No specific warning sign is required in the *NEC* for systems under 600 volt, but for over 600 volt systems Section 110-34 requires that the warning sign read: "Danger-High Voltage-Keep Out."

SUMMARY

Article 110 explains general requirements for electrical installations. ☐ The authority having jurisdiction approves all electrical installations, equipment, and conductors. ☐ Voltage, as used in the *NEC*, means the operating voltage of the circuit; see Article 100 Definition. ☐ Unless otherwise specified in the *NEC*, copper wire is to be used. ☐ Conductor sizes are expressed in American Wire Gauge (AWG) or circular mils. ☐ Electrical equipment must be installed in a neat and workmanlike manner. ☐ Unused openings in any electrical equipment must be closed. ☐ Terminals for copper or aluminum must be identified for the material of the conductor. ☐ Splices and joints must be made by an approved means and are permitted underground with listed splicing devices. ☐ Conductors must be sized based on the terminal rating of the equipment. ☐ Manufacturers are required to provide descriptive markings on electrical equipment. ☐ Disconnects and circuits for all equipment must be legibly marked to indicate their purpose. ☐ Working clearances must be provided for equipment that may need examination or service while energized. ☐ Required working spaces shall not be used for storage. ☐ Dedicated equipment space must be provided up to six feet above switchboards, panelboards, and motor control centers. ☐ Electrical equipment must be protected from physical damage.

REVIEW QUESTIONS

1. Explain the *NEC* guidelines to be used for the approval of electrical equipment.

2. Explain the reasons why equipment must have an interrupting rating sufficient for the available fault-current.

3. Explain the rules that apply to conductor terminations.

4. Explain the rules that apply to conductor splices.

5. Explain the rules that apply to working space for electrical equipment.

6. Explain the rules that apply to working space entrances.

7. Explain the rules that apply to "dedicated equipment space" for switchboards, panelboards and motor control centers.

8. Explain the rules that apply to "foreign equipment space" for switchboards, panelboards and motor control centers.

9. Explain the rules that apply to identification of disconnects and circuits.

10. In your own words explain the meaning of the following Article 100 definitions:

   Accessible
   Approved
   Controller
   Device

Disconnect
Equipment
Identified
Interrupting rating
Nominal voltage
Raceways

11. In your own words explain the meaning of the following Glossary terms:

AL
Back stabbing
CU
Gray area
Line side
Oxidation
Splice
Torque

CHAPTER 2
WIRING AND PROTECTION

Scope of Chapter 2 Articles

Chapter 2 of the *NEC* is a general rules chapter as applied to wiring and protection of conductors. The rules in this chapter apply except as modified in Chapters 5, 6, and 7. Along with Chapter 3, it can be considered the heart of the *Code*. Many of the everyday applications of the *NEC* can be found in this chapter. It covers most of the requirements the electrical industry must deal with on a regular basis. If you consider an electrical system in its simplest context, there are:

1. Services   Article 230
2. Feeders   Article 215
3. Branch Circuits   Article 210

All three systems require neutrals [Article 200], conductor sizing [Articles 220 and 310], conductor protection [Article 240], and grounding [Article 250].

ARTICLE 200 – USE AND IDENTIFICATION OF GROUNDED CONDUCTOR

This Article contains some of the requirements and identification rules for the Grounded Conductor (neutral wire).

ARTICLE 210 – BRANCH CIRCUITS

Since branch circuits make up the largest portion of most electrical systems, this is an important Article for understanding how many of the other *Code* rules apply.

Part A. General Provisions
Part B. Branch Circuit Ratings
Part C. Required Outlets

ARTICLE 215 – FEEDERS

This Article covers the rules for installation, minimum size, and ampacity of feeders.

ARTICLE 220 – FEEDER AND SERVICE CALCULATIONS

Electrical calculations are very intimidating, but this Article is important to understand and it is divided into three parts.

Part A. General
Part B. Feeders and Service Calculations
Part C. Optional Calculations

ARTICLE 225 – OUTSIDE WIRING

This Article contains the requirements for proper installation of outside branch circuits and feeders
Part A. General
Part B. More Than One Building or Structure

ARTICLE 230 – SERVICES

Article 230 covers the requirements for service conductors, as well as equipment for the control and protection of services and their installation requirements. It's so large that it's divided into seven parts.

Part A. General
Part B. Overhead Service-Drop
Part C. Underground Service-Lateral
Part D. Service-Entrance Conductors
Part E. Service Equipment
Part F. Disconnecting Means
Part G. Overcurrent Protection

ARTICLE 240 – OVERCURRENT PROTECTION

Overcurrent Protection of a properly grounded electrical system is the first line of defense against the potential of electrical damage to life and property. Overcurrent protection is the automatic shut-off for overloads, short-circuits, and ground-faults. Proper sizing and application of overcurrent protection devices is critical for every electrical system.

Part A. General
Part B. Location
Part C. Enclosures
Part D. Disconnecting and Guarding
Part E. Plug Fuses, Fuseholders, and Adapters
Part F. Cartridge Fuses and Fuseholders
Part G. Circuit Breakers

ARTICLE 250 – GROUNDING AND BONDING

Grounding is one of the most important and least understood Articles in the *NEC*. Safety is the key element and purpose of the *NEC* [90-1(a)]. Proper grounding and bonding is essential for maximum protection of life and property.

Part A. General
Part B. Circuit and System Grounding
Part C. Grounding Electrode System and Grounding Electrode Conductor
Part D. Enclosure and Raceway Grounding
Part E. Bonding
Part F. Equipment Grounding
Part G. Methods of Equipment Grounding

Notes:

Unit 4

Use and Identification of Grounded Conductor

OBJECTIVES

After studying this unit, the student should be able to understand:
- the difference between a grounded and neutral conductor.
- the rules associated with the identification of the grounded (neutral) conductor.
- the use of the white wire.

Definitions and Glossary Terms

To better understand the *NEC* rules contained in this unit, review the following:

Definitions Article 100 - Unit 2
Grounded
Grounded conductor
Separately derived system
Service conductors
Service equipment

Glossary Terms
Different systems
Four-way switch
Harmonic current
Neutral current
Polarity
Three-way switch

GROUNDED CONDUCTOR OR NEUTRAL CONDUCTOR?

A white or natural gray coloring is used exclusively for identifying the grounded conductor (often called the neutral). What is a grounded conductor? What is a *neutral conductor*? Let's take a look at some different circuits and determine the correct term.

Grounded Conductor

At the **service equipment**, transformer or other **separately derived system** location, one conductor must be **grounded** to the earth. The conductor that is connected to the earth is generally called the *neutral wire*. However, since this conductor is connected to the earth, the *NEC* calls it the **grounded conductor**, Fig. 4-1.

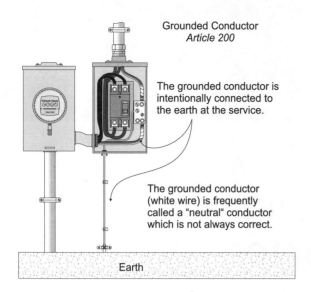

Fig. 4-1 Grounded Conductor

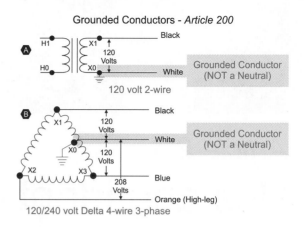

Fig. 4-3 Using Higher Conductor Ampacity for Derating

Neutral Conductor

The term neutral conductor means the conductor that has equal potential difference between it and all of the ungrounded conductors. This would include a 120/240 volt single-phase or a 120/208 three-phase system, Fig. 4-2.

The grounded conductor of a 2-wire 120 volt circuit or a 4-wire, 120/240 volt system is not a neutral conductor because the potential difference between each of the ungrounded conductors to the grounded (neutral) conductor is not the same, Fig. 4-3.

When the electrical trade industry uses the term "neutral" we all know what wire we're talking about; it's the white or gray wire. The proper term for this wire is the "grounded conductor."

When you call the grounded conductor a neutral wire, and it's not a neutral (delta high-leg system), you are using the term "neutral" incorrectly, but hey most people do this (including myself). As a matter of fact, since 1987 the *NEC* has been removing the term neutral in the *Code* and replacing it with the term grounded conductor.

Purpose of Grounded Conductor

The purpose of the grounded (neutral) conductor is to permit line-to-neutral loads, such as 120 and 277 volt circuits, and it serves as a current-carrying conductor to carry return neutral current, Fig. 4-4.

Today's office buildings have large quantities of single-phase **nonlinear loads** such as personal computers and laser printers. These loads produce reflective triplen *harmonic currents* that add on the neutral conductor (instead of canceling). These triplen harmonic neutral currents can cause the neutral conductor to carry as much as 200 percent of its maximum expected load.

In addition, a safety ground wire is not installed with the **service conductors** to electric service equipment (meter locations), therefore the grounded (neutral) conductor serves to provide the low impedance *ground-fault* return path to clear ground-faults.

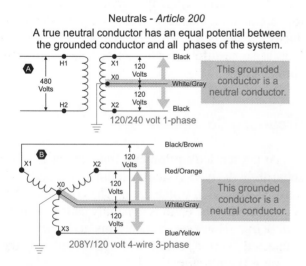

Fig. 4-2 Neutrals

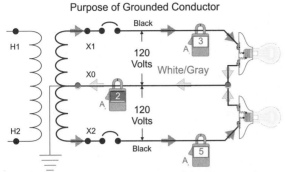

Purpose of Grounded Conductor

The grounded (neutral) conductor permits line-to-neutral loads, such as 120 and 277 volt circuits and serves as a current-carrying conductor to carry return (neutral) current.

Fig. 4-4 Purpose of Grounded Conductor

Grounded Conductor Identification - *Section 200-6(a)*

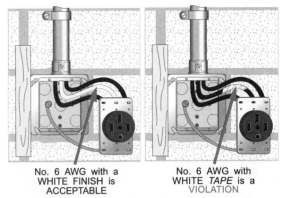

No. 6 AWG with a WHITE FINISH is ACCEPTABLE

No. 6 AWG with WHITE *TAPE* is a VIOLATION

Grounded conductors No. 6 and smaller must have a white or gray outer finish. Re-identification is not permitted.

Fig. 4-5 Grounded Conductor Identification

CAUTION: There are special installations in which a corner-grounded delta system is installed with one phase (hot) conductor grounded. The grounded phase conductor is also a grounded (neutral) conductor, but this topic is beyond the scope of this book.

200-1 Scope

Article 200 contains the requirements for identification of the grounded (neutral) conductor and its terminals.

200-3 Connection to Grounded System

Premises wiring must have the grounded (neutral) conductor electrically connected to the supply system grounded conductor. What this means is that the grounded conductor of the premises wiring (branch circuits, feeders and the service) must be electrically continuous to the utility grounded conductor, or to the grounded conductor of a separately derived system (generally a transformer).

200-6 Identification of the Grounded Conductor

(a) Grounded Conductors. No. 6 or Smaller. Grounded (neutral) conductors No. 6 and smaller

shall have a continuous white or natural gray outer finish along their entire length, Fig. 4-5.

Note. The use of white reidentification tape, paint, or other methods of markings is not permitted, Fig. 4-5.

(b) Grounded Conductors. Larger than No. 6. Grounded (neutral) conductors larger than No. 6 can be identified by distinctive white markings (tape, paint, or other effective means) at the wire terminations. Reidentification can be with white phase tape, but not with gray phase tape, Fig. 4-6.

(d) Different System Grounded Conductors Installed in the Same Raceway or Enclosure. Where conductors of *different systems* are installed

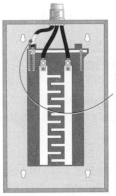

Grounded Conductor Identification *Section 200-6(b)*

Grounded conductors larger than No. 6 AWG can be re-identified by white markings.

Gray tape not permitted.

Fig. 4-6 Grounded Conductor Identification

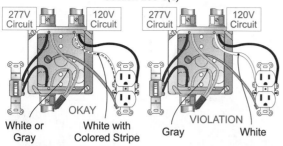

Different System Grounded Conductors
Section 200-6(d)

White or Gray   White with Colored Stripe   Gray   White

When grounded conductors of different systems are within the same box, raceway, cable, gutter, or other enclosure, one grounded conductor must be either white or gray and the other grounded conductor must be white with a colored stripe (not green).

Fig. 4-7 Different System Grounded Conductors

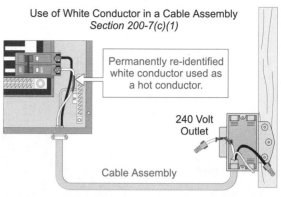

Use of White Conductor in a Cable Assembly
Section 200-7(c)(1)

Permanently re-identified white conductor used as a hot conductor.

240 Volt Outlet

Cable Assembly

A white conductor within a cable assembly can be used as an ungrounded (hot) conductor where permanently re-identified.

Fig. 4-8 Use of White Conductor in a Cable Assembly

in the same **raceway**, cable, box, auxiliary gutter, or other type of enclosure, one system grounded (neutral) conductor, if required, shall have an outer covering of white or gray according to the requirements of Section 200-6(a) or (b).

Each other system grounded (neutral) conductor, if required, shall have an outer covering of white with a readily distinguishable different color stripe (not green) running along the insulation, Fig 4-7.

Note. Proposals to permit the color white and gray for grounded (neutral) conductor identification when circuits for different voltage systems are contained within the same raceway, cable, or enclosure were rejected. The Code Panel felt that the colors white and gray are not always readily distinguishable from each other and should not be used to identify the different system grounded (neutral) conductors when installed in the same raceway, cable, or enclosure.

200-7   Use of White or Natural Gray Conductor Finish for the Hot Wire

(c) White or natural gray conductor finish, tape, or paint can be used only for the grounded (neutral) conductor.

Note. The 1999 NEC does not permit the white or gray conductor to be used for power conductors, even if permanently re-identified, Fig. 4-8.

(1) **Cable Assembly.** White or gray conductor within a cable assembly can be used for the ungrounded conductor, but the white or gray conductor must be permanently re-identified to indicate its use as an ungrounded conductor at each location where the conductor is visible and **accessible**, Fig. 4-9.

(2) **Switches.** White or gray conductor within a cable assembly used for single-pole, three-way or four-way switch loops must be permanently re-identified to indicate its use as an ungrounded conductor at each location where the conductor is visible and accessible, Fig. 4-10.

Note. The 1996 NEC permitted white or natural gray insulated wire as an ungrounded (hot) conductor for *single-pole, three-way and four-way switches* without reidentification.

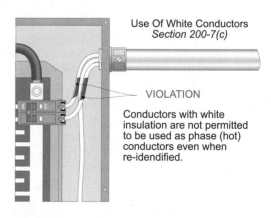

Use Of White Conductors
Section 200-7(c)

VIOLATION

Conductors with white insulation are not permitted to be used as phase (hot) conductors even when re-idendified.

Fig. 4-9 Use of White Conductors

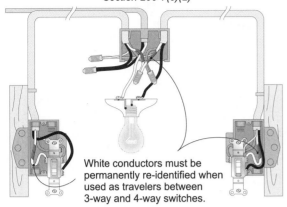

Use of White Conductors Within Cable Assemblies
Section 200-7(c)(2)

White conductors must be permanently re-identified when used as travelers between 3-way and 4-way switches.

Fig. 4-10 Use of White Conductors within Cable Assemblies

Screw Shell Terminal Identification
Section 200-10(c)

The grounded conductor must be connected to the screw shell.

VIOLATION
Reverse Polarity

Note: Correct polarity of a screw shell keeps the screw shell threads from being energized. This reduces the chance of a person touching a "hot" while replacing a lamp.

Fig. 4-11 Screw Shell Terminal Identification

200-9 Terminal Identification

The terminals for the termination of grounded (neutral) conductors must be white in color (really silver because of the metal). The terminals for the termination of the ungrounded conductors must be a color that is readily distinguishable from white (often a brass or copper color). The terminal for the grounding conductor must be a green hexagon-headed or hexagon shaped terminal screw [250-119].

200-10 Identification of Terminals

(c) Screw Shell. The grounded (neutral) conductor must be connected to the *lampholder* screw shell [410-47], Fig. 4-11.

200-11 Polarity

A grounded (neutral) conductor cannot be connected to terminals, or leads, that will cause reversed *polarity* [410-23], Fig. 4-11.

SUMMARY

Article 200 contains the requirements for identification of grounded (neutral) conductors. ☐ Grounded means an intentional connection to earth or to some conducting body of the internal wiring system. See Article 100 definitions. ☐ If the premises wiring system has a grounded (neutral) conductor, it must be electrically connected to a supply system that has a grounded (neutral) conductor. ☐ Grounded conductors No. 6 and smaller must have a continuous white or natural gray insulation. ☐ Grounded conductors larger than No. 6 can be reidentified with the color white markings. ☐ When two different voltage wiring systems are installed in the same raceway or enclosure, one system grounded (neutral) conductor must be white or natural gray, and the other grounded (neutral) conductor must be white with a color stripe (not green). ☐ White insulation, tape, or paint can be used only for the grounded (neutral) conductor. ☐ White wire can be used for the hot wires (ungrounded conductor), switch legs and travelers, if the white wire is permanently reidentified at all accessible and visible locations.

REVIEW QUESTIONS

1. Explain the difference between a grounded and a neutral conductor.

2. Explain the rules associated with identification of the grounded (neutral) conductor.

3. Explain the requirements to use a white wire for the hot conductor.

4. In your own words explain the meaning of the following Article 100 definitions:
   Grounded
   Grounded conductor
   Separately derived system
   Service conductors
   Service equipment

5. In your own words explain the meaning of the following Glossary terms:
   Different systems
   Harmonic current
   Neutral current
   Polarity

Unit 5

Article 210
Branch Circuits

OBJECTIVES

After studying this unit, the student should be able to understand:
- the advantages of multiwire branch circuits.
- the disadvantages and dangers of multiwire branch circuits.
- the rules that apply to the replacement of nongrounding-type receptacles.
- how a GFCI protection device works.
- GFCI protection rules that apply to commercial occupancies.
- the rules that apply to circuits with a single receptacle.
- the rules that apply to receptacle circuits.
- the rules associated with kitchen countertops for dwelling units.
- the rules that apply to receptacles outside dwelling units.
- the rules that apply to heating, air-conditioning, and refrigeration equipment lighting and receptacles for commercial occupancies.
- the rules that apply to dwelling unit lighting and switching.

Definitions and Glossary Terms

To better understand the *NEC* rules contained in this unit, review the following:

Definitions, Article 100 - Unit 2

Accessible	Ampacity
Appliance	Attachment plug
Bathroom	Continuous load
Dwelling unit	Feeder
Garage	Lighting outlet
Outlet	Overcurrent
Panelboard	Readily accessible
Receptacle	Receptacle outlet

Glossary Terms

Arc	Arcing-fault
Crawl space	Fault-current
Harmonic current	High-leg
Neutral conductor	Voltage drop

PART A. GENERAL PROVISIONS

210-1 Scope

Article 210 contains the requirements for **branch circuits**, such as conductors sizing, identification, receptacle GFCI protection, and receptacle and lighting outlet requirements, Fig. 5-1.

47

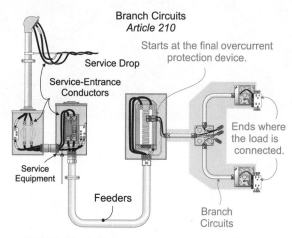

Branch Circuits
Article 210

Starts at the final overcurrent protection device.

Service Drop

Service-Entrance Conductors

Ends where the load is connected.

Service Equipment

Feeders

Branch Circuits

Branch Circuit: (Article 100 Definition) The conductors between the final overcurrent device and the outlet(s).

Fig. 5-1 Branch Circuits

210-2 Other Articles

Other important Articles having specific rules for individual branch circuits include:
Air Conditioning – 440
Appliances – 422
Information Technology Equipment – 645
Fire Alarms – 760
Motors – 430
Signs – 600
Space Heating – 424
Transformers – 450

210-3 Branch Circuit Rating

The branch circuit rating is the maximum rating or setting of the circuit overcurrent protection device. For example, the branch circuit rating of No. 10 THHN (rated 40 ampere as listed on Table 310-16) on a 20 ampere circuit breaker is 20 ampere, Fig. 5-2.

As a general rule circuits that have more than one outlet can only be connected to circuits rated 15, 20, 30, 40, or 50 ampere.

210-4 *Multiwire Branch Circuits*

(a) General. Multiwire branch circuits can be considered as one circuit, or it can be considered as two or three separate branch circuits, depending on its use. For example, the *NEC* requires two small

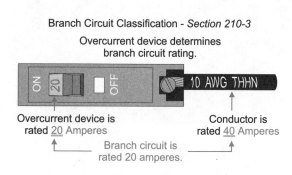

Branch Circuit Classification - *Section 210-3*
Overcurrent device determines branch circuit rating.

ON 20 OFF   10 AWG THHN

Overcurrent device is rated 20 Amperes
Conductor is rated 40 Amperes
Branch circuit is rated 20 amperes.

Fig. 5-2 Branch Circuit Classification

appliance circuits for the **dwelling unit** kitchen counter receptacles [210-52(b)(2) and 210-11(c)(1)]; the two circuits can be supplied by one 120/240 volt multiwire branch circuit.

Originate from Same Panel. All conductors of a multiwire branch circuit shall originate from the same **panelboard**. This is to prevent inductive heating and to reduce *conductor impedance* for *fault-currents*. See Sections 300-3(b) and 300-20(a) for more details on inductive heating.

FPN: *Three-phase*, 4-wire *wye-connected* power systems (208Y/120 and 480Y/277 volt) that supply **nonlinear loads** that can produce unwanted and potentially hazardous *harmonic currents* that add on to the *neutral conductor*. The neutral current can be as much as twice the phase conductor current. The neutral conductor must be sized to carry the excessive harmonic neutral current to prevent the conductor from overheating and possibly causing a fire or equipment damage, Fig. 5-3.

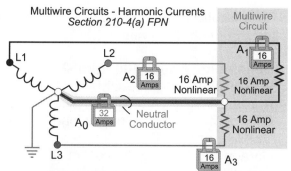

Multiwire Circuits - Harmonic Currents
Section 210-4(a) FPN

Multiwire Circuit

L1   L2   A2 16 Amps   16 Amp Nonlinear   A1 16 Amps   16 Amp Nonlinear

A0 32 Amps   Neutral Conductor   16 Amp Nonlinear

L3   16 Amps A3

Odd triplen harmonic currents add on the neutral conductor of a 3-phase 4-wire "wye" system.

Fig. 5-3 Multiwire Circuits - Harmonic Currents

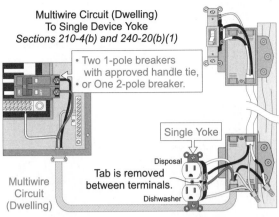

Multiwire Circuit (Dwelling)
To Single Device Yoke
Sections 210-4(b) and 240-20(b)(1)

• Two 1-pole breakers with approved handle tie,
• or One 2-pole breaker.

Single Yoke

Disposal

Tab is removed between terminals.

Dishwasher

Multiwire Circuit (Dwelling)

A multiwire circuit to one yoke (strap) must have means to simultaneously open all ungrounded conductors.

Fig. 5-4 Multiwire Circuit Feeding a Single Strap

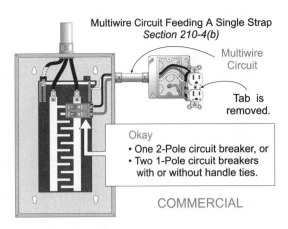

Multiwire Circuit Feeding A Single Strap
Section 210-4(b)

Multiwire Circuit

Tab is removed.

Okay
• One 2-Pole circuit breaker, or
• Two 1-Pole circuit breakers with or without handle ties.

COMMERCIAL

Fig. 5-5 Multiwire Circuits - More than One Voltage System

WARNING:

You can reduce the heating of the neutral wire by installing a separate neutral for each phase or by increasing the size of the neutral conductor. To measure the harmonic current on the neutral wire, you must use a true RMS ammeter, you cannot use the standard average responding clamp-on ammeter.

(b) Dwelling Units. Dwelling unit multiwire branch circuits that terminate on the same mounting strap or *yoke* of a **receptacle** or switch must be provided with a means to simultaneously disconnect all ungrounded circuit conductors, Fig. 5-4.

Note. This does not apply to receptacles in commercial occupancies, Fig. 5-5.

(c) Line-to-Neutral Load. Multiwire branch circuits shall supply only line-to-neutral loads. This rule prevents a 240 volt line-to-line load from being on the same multiwire branch circuit for general use 125 volt receptacles.

Exception No. 1: One Line-to-Line Load. Multiwire branch circuits can supply one line-to-line load, such as a range, dryer, etc., with the use of fused switch, or single-pole circuit breakers with **approved** handle ties [240-20(b)(2)].

Exception No. 2: Line-to-Line and Line-to-Neutral Loads. Multiwire branch circuits can supply

both line-to-neutral and line-to-line loads where all ungrounded conductors of the multiwire branch circuit are *opened* simultaneously by the branch circuit overcurrent protection device, such as a 2- or 3-pole breaker with internal common trip.

Multiwire branch circuits should never be for expensive equipment, such as for stereos, computers, or other expensive equipment; see my comments in Section 300-13(b).

Note. Single-pole circuit breakers with approved handle ties are not permitted for this purpose.

(d) Buildings With More Than One Voltage System. Where more than one nominal voltage system exists in a building, each ungrounded conductor of a multiwire branch circuit, where **accessible**, shall be identified by phase and system. The identification can be by color coding, phase tape, tagging or other approved means. The method used for conductor identification must be posted at each panelboard. Most inspectors require the identification to be posted on the outside of the panelboard, Fig. 5-6.

This rule is primarily for commercial and industrial buildings that have two voltage systems, such as 208Y/120 volt, 3-phase for receptacle circuits and 480Y/277 volt, 3-phase for lighting. For the identification of the *high-leg* conductor; see Sections 215-8 for feeders, 230-56 for service conductors, and 384-3(e) for switchboards or panelboards.

Note. If two system voltages do not exist in a building, then multiwire branch circuit identification is not required. However,

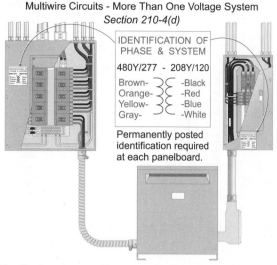

Multiwire Circuits - More Than One Voltage System
Section 210-4(d)

IDENTIFICATION OF
PHASE & SYSTEM

480Y/277 - 208Y/120

Brown-   -Black
Orange-   -Red
Yellow-   -Blue
Gray-   -White

Permanently posted
identification required
at each panelboard.

Identification of ungrounded conductors is required for multiwire
circuits where more than one voltage system exists in a building.

Fig. 5-6 Multiwire Circuits - More than One
Voltage System

identifying the branch circuit conductors is
a common practice and at times a requirement of a local electrical code.

WARNING:

When two voltage systems are installed in
the same raceway (or enclosure), one system's neutral can be white or natural gray.
The other system's neutral must be white
with a readily distinguishable different
color stripe (other than green), or identified
with other effective means; see Sections
200-6(d) and 210-5(a).

210-5   Color Code for Branch Circuits

The *color code* requirement for conductors in
the *NEC* only applies to the white or natural gray
grounded (neutral) conductor [200-6], the green
grounding conductor [250-119 and 310-12(b)],
and the orange high-leg conductor [215-8, 230-56,
and 384-3(e)].

Note. All proposals to add a color code
requirement for the ungrounded conductors
were rejected. The Code Panel's comments
were that "there was insufficient substantia-
tion for such a far reaching proposal and

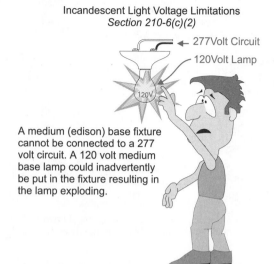

Incandescent Light Voltage Limitations
Section 210-6(c)(2)

← 277Volt Circuit
← 120Volt Lamp

A medium (edison) base fixture
cannot be connected to a 277
volt circuit. A 120 volt medium
base lamp could inadvertently
be put in the fixture resulting in
the lamp exploding.

Fig. 5-7 Incandescent Light Voltage Limitations

that identification of conductors should not
be limited to color coding."

210-6   Branch Circuit Voltage Limitation

(c) 277 and 480 volt Fixtures. Circuits of
277 or 480 volt to lighting fixtures are permitted
only in commercial and industrial occupancies.
When lighting fixtures are connected to 277 or
480 volt circuits, the fixture must be one of the
following:

(1) A listed electric discharge lighting fixture
with any base.

(3) A lighting fixture with a mogul base
screw-shell.

(4) Lampholders other than the screw-shell
type applied within their voltage ratings.

CAUTION: There have been many cases
where persons have wired 120 volt medium-
base type lampholders to a 277 volt circuit. If a
120 volt rated incandescent fixture is installed
on a 277 volt circuit, the lamp will often
explode. When installing incandescent lighting
fixtures to 277 volt circuits, be sure to use a
mogul-base lampholder or electric discharge
lighting fixtures with any base, Fig. 5-7.

210-7 Receptacles

(a) Grounding Type. Only grounding type receptacles can be installed on 15- and 20 ampere branch circuits.

Note. The position of the ground of a receptacle is not specified in the *NEC*. The ground can be up, down, or sideways. Proposals to specify the mounting position of the ground were all rejected, Fig. 5-8.

Single Receptacles. Single **receptacles** shall have an ampere rating not less than the rating of the branch circuit (100 percent). See Section 210-21(b)(1).

Multioutlet Receptacles. Multioutlet receptacles must be installed according to Sections 210-21(b)(2) and (b)(3).

(b) Receptacle Grounding. Receptacles must have the grounding terminals properly grounded. Failure to connect the receptacle grounding terminal to the **equipment grounding conductor** could lead to death or electrical shock. See Sections 250-146 and 250-148.

Exception No. 1: Generators. Receptacles mounted on portable and vehicle-mounted generators installed in accordance with Section 250-34, are not required to be grounded.

Exception No. 2: GFCI Protected Receptacles. Grounding-type receptacles without a grounding conductor can be used to replace nongrounding type receptacles, if the receptacles are GFCI protected [210-7(d)].

(c) Methods of Grounding Receptacles. Receptacles must have the grounding terminal electrically connected to the branch circuit equipment grounding conductor [250-146].

FPN: Isolated Ground Receptacles. An insulated equipment grounding conductor is required for the isolated ground receptacle [250-146(d)], and a separate equipment grounding conductor must ground the metal box [250-148] and the metal cover plate [410-56(d)]. Isolated ground receptacles must be identified by an orange

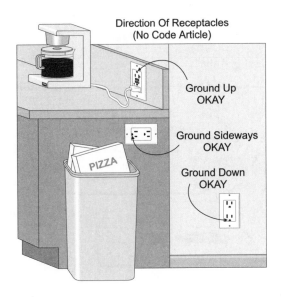

Fig. 5-8 Direction of Receptacles

triangle located on the face of the receptacle [410-56(c)].

(d) Receptacle Replacement. Replacement of receptacles shall comply with (1), (2), and (3) below as applicable.

(1) Where Grounding Means Exist. Where a grounding means exists in the receptacle enclosure, grounding-type receptacles shall replace nongrounding type receptacles and shall be grounded according to Section 210-7(c).

(2) GFCI Protection Required. When receptacles are replaced in locations where **ground-fault circuit-interrupter** protection is required, the replacement receptacles must be GFCI protected. This includes the replacement of receptacles in bathrooms, **garages**, outdoors, *crawl spaces*, unfinished basements, kitchens, wet bar sinks and rooftops, etc.; see Section 210-8 in this textbook for GFCI requirements.

(3) Where No Ground Exists. Where no grounding means exist in the box, such as old NM cable without a ground, nongrounding-type receptacles can be replaced with (a), (b) or (c), see 250-130(c), Fig. 5-9.

(a) A nongrounding type receptacle.

(b) A GFCI-receptacle or a grounding type receptacle fed downstream from a GFCI-receptacle.

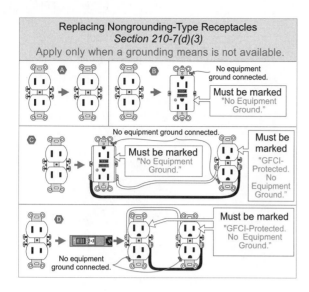

Fig. 5-9 Replacing Nongrounding-Type Receptacles

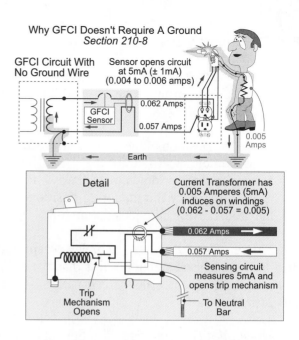

Fig. 5-10 Why GFCI Does Not Require a Ground

These receptacles shall be marked "No Equipment Ground."

(c) A grounding-type receptacle protected with a GFCI circuit breaker. These receptacles shall be marked "GFCI Protected" and "No Equipment Ground."

Note. The GFCI protection will function properly on a 2-wire circuit without an equipment grounding conductor. The equipment grounding conductor serves no purpose in the operation of the GFCI protection device and has no effect on the function of the GFCI test-button.

210-8 GFCI Protected Receptacles

GFCI devices are designed to protect persons from being injured or killed from an electric shock. Dangerous electric shock begins at about 10 milliampere and the GFCI protection device opens the circuit whenever ground-fault current exceeds 5 milliampere, Fig. 5-10.

Since 1971, the *NEC* has been expanding the requirements of GFCI protection to include the following locations:

Subject	Section
Agricultural Building	547-9
Carnivals, Circuses and Fairs	525-18
Commercial Garages	511-10
Elevator Pits	620-85
Health Care Facilities	517-20(a), 517-21
Portable or Mobile Signs	600-10
Roof Top Receptacles	210-8(b)
Swimming Pools	680-6(a)
Temporary Wiring	305-6

(a) **Dwelling Units.** All 125 volt, 15 and 20 ampere receptacles installed in the following locations of dwelling units shall be GFCI protected.

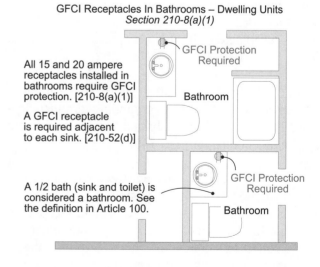

Fig. 5-11   GFCI Receptacles in Bathrooms - Dwelling Units

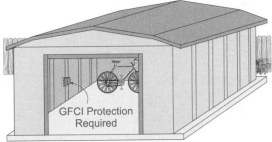

GFCI Protection in Accessory Building – Dwelling
Section 210-8(a)(2)

GFCI Protection
Required

Receptacles located in an accessory building and similar
areas used for storage or work require GFCI protection. *Note:*
The Code does not require receptacles in accessory buildings.

Fig. 5-12 GFCI Protection in Accessory Building -
Dwelling

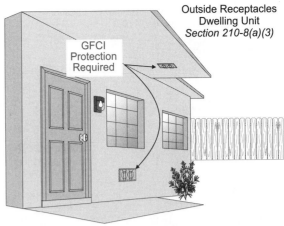

Outside Receptacles
Dwelling Unit
Section 210-8(a)(3)

GFCI
Protection
Required

All 125 volt 15 and 20 ampere receptacles installed outside
of a dwelling unit require GFCI protection. This includes
receptacles over 6 feet 6 inches above grade.

Fig. 5-14 Outside Receptacles - Dwelling Unit

Note. The GFCI protection required by his
Section can be accomplished by the use of a
GFCI circuit breaker or GFCI receptacle.

(1) Bathroom Area Receptacles. All 125
volt, 15 and 20 ampere receptacles installed in the
bathroom must be GFCI protected, Fig. 5-11.

Note. Proposals to permit receptacles for
dedicated equipment to be exempted from
the GFCI protection rules for bathroom
were all rejected because the Code Panel
felt that it was not in the interest of safety to
permit **appliances** (not GFCI protected) in
an area where individuals are, in many
cases, wet.

**(2) Garage and Accessory Building Recep-
tacles.** All 125 volt, 15 and 20 ampere receptacles
installed in garages and in grade-level portions of
unfinished or finished accessory buildings used for
storage or work areas shall have ground-fault circuit-interrupter protection, Fig. 5-12.

Note. At least one receptacle must be
installed in each attached garage, and in
each detached garage with electric power
[210-52(g)]. The *Code* does not require a
receptacle to be installed in accessory buildings, but if a 125 volt, 15 or 20 ampere
receptacle is installed, then it shall be GFCI
protected.

*Exception No. 1: Receptacles Not Readily
Accessible.* GFCI protection is not required
for receptacles that are not readily accessible, such
as a ceiling mounted receptacle for the garage door,
Fig. 5-13.

Exception No. 2: Dedicated Appliances. GFCI
protection is not required for a single receptacle or a
duplex receptacle for two appliances on a dedicated
branch circuit located and identified for a specific
cord- and plug-connected appliance, such as a
refrigerator or freezer.

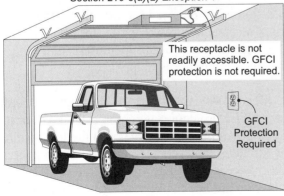

Receptacles Not Readily Accessible In Garages
Section 210-8(a)(2) Exception 1

This receptacle is not
readily accessible. GFCI
protection is not required.

GFCI
Protection
Required

Fig. 5-13 Receptacles Not Readily Accessible
in Garages

(3) Outdoor Receptacles. All 125 volt, 15
and 20 ampere receptacles installed outdoors shall
have ground-fault circuit-interrupter protection for
personnel. This includes receptacles installed under

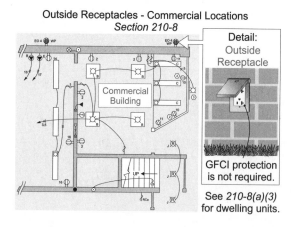

Outside Receptacles - Commercial Locations
Section 210-8

Detail:
Outside
Receptacle

GFCI protection
is not required.

See *210-8(a)(3)*
for dwelling units.

Fig. 5-15 Outside Receptacles - Commercial Locations

GFCI Protection - Kitchen Countertops - *Section 210-8(a)(6)*

Appliance
Receptacle

All countertop receptacles
require GFCI protection.

GFCI Protection
Not Required

Convenience
Receptacle

Island/Peninsular
Countertop

GFCI protection is required for all 125 volt
15 and 20 ampere receptacles that serve
dwelling unit kitchen countertop surfaces.
See *Section 210-52(c)*.

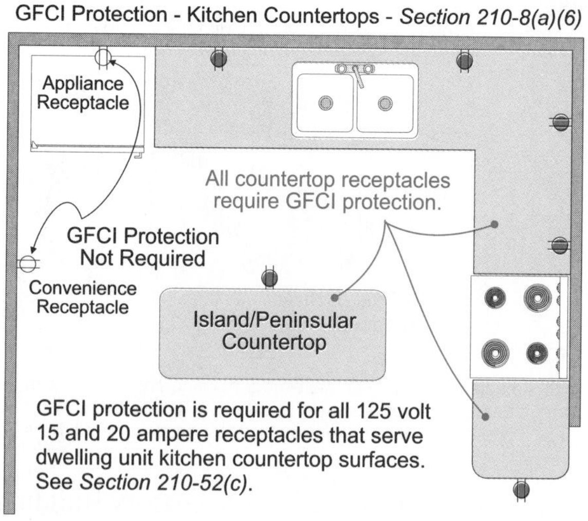

Fig. 5-16 GFCI Protection - Kitchen Countertops

the roof eaves to accommodate Christmas lights, roof top receptacles, and receptacles on balconies, Fig. 5-14.

Receptacles are not required on porches or balconies of **multifamily dwelling** units [210-52(e)], but if they are installed, then they must be GFCI protected.

Note. This rule does not apply to receptacles outside of commercial and industrial occupancies, Fig. 5-15.

All 125 volt, 15 and 20 ampere receptacles, other than those that supply shore power to boats must be protected by ground-fault circuit-interrupters [555-3].

Exception: Snow Melting Equipment. GFCI protection of personnel is not required for fixed electric snow-melting or deicing equipment receptacles that are not readily accessible and are supplied by a dedicated branch circuit is accordance with Section 426-28. GFCI protection is not required because Section 426-28 requires fixed electric snow-melting or deicing equipment must be protected against ground-faults by a GFPE device (equipment ground-fault protector).

Note. Proposals to require GFCI protection for 125 volt, 15 and 20 ampere receptacles installed outdoors of commercial and industrial occupancies were all rejected.

(4) Crawl Space Receptacles. All 125 volt, 15 and 20 ampere receptacles installed in crawl

spaces at or below grade must have ground-fault circuit-interrupter protection for personnel.

Note. U. S. Consumer Product Safety Commission (CPSC) statistics show that people have been shocked or electrocuted while using portable power tools in these areas.

(5) Unfinished Basement Receptacles. Receptacles in unfinished basements used for storage and work spaces, or areas of the basement not intended as habitable rooms. Check with your inspector on the local building code's definition of a habitable room.

Exception No. 1: Receptacles Not Readily Accessible. GFCI protection is not required for receptacles that are not readily accessible.

Exception No. 2: Dedicated Appliance. GFCI protection is not required for a single receptacle or a duplex receptacle for two appliances on a dedicated branch circuit located and identified for a specific cord- and plug-connected appliance.

(6) Kitchen Countertop Surface Receptacles. GFCI protection is required for all 125 volt, 15 and 20 ampere receptacles installed in dwelling unit kitchens that serve countertop surface appliances. Receptacles that are not intended to serve countertop surface appliances such as, convenience receptacles and receptacles for appliances such as, refrigerators and freezers are not required to have GFCI protection, Fig. 5-16.

Wet Bar Receptacles – *Section 210-8(a)(7)*

Receptacle outlets for wet bars must not be located in the face-up position in the work surfaces or countertops of a wet bar sink.

Fig. 5-17 Wet Bar Receptacles

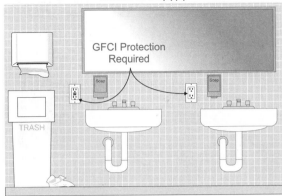

Receptacles Installed in Commercial & Industrial Bathrooms
Section 210-8(b)(1)

All receptacles (if installed) in commercial and industrial bathrooms are required to have GFCI protection for personnel. GFCI protection can be from GFCI receptacles, receptacles fed from a GFCI receptacle, or a GFCI circuit breaker.

Fig. 5-18 Receptacles Installed in Commercial & Industrial Bathrooms

Note. Proposals to require GFCI protection for kitchen countertop receptacles in other than dwelling units were all rejected.

(7) Wet Bar Countertop Receptacles. All 125 volt, 15 and 20 ampere receptacles that serve wet bar countertop surfaces, located within 6 feet of the outside edge of a wet bar sink, must have ground-fault circuit-interrupter protection for personnel. GFCI protection is not required for receptacles that are not intended to serve wet bar countertop surface appliances. Wet bar receptacle outlets must not be installed in a face-up position in the work surfaces or countertops of a wet bar sink, Fig. 5-17.

Note. Requests to add GFCI protection for receptacles within six feet of any sink were all rejected.

(b) GFCI Protection of Receptacles In Other Than Dwelling Units. All 125 volt, 15 and 20 ampere receptacles installed in the following locations shall be GFCI protected.

(1) Bathrooms Area Receptacles. There is no *Code* rule that requires a receptacle to be installed in a commercial or industrial bathroom, but if a receptacle is installed, then it must be GFCI protected, Fig. 5-18.

(2) Rooftop Receptacles. All rooftop receptacles require GFCI protection, Fig. 5-19.

Exception. GFCI protection of personnel is not required for fixed electric snow-melting or deicing

equipment receptacles that are not readily accessible and are supplied by a dedicated branch circuit as covered in Section 426-28.

Note. Section 210-63 requires a receptacle to be installed on commercial and industrial rooftops to serve rooftop equipment.

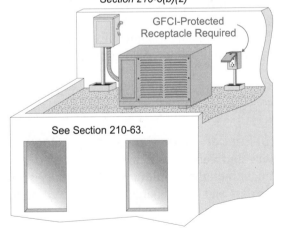

Receptacles Installed On Commercial - Industrial Roofs
Section 210-8(b)(2)

GFCI-Protected Receptacle Required

See Section 210-63.

Fig. 5-19 Receptacles Installed on Commercial - Industrial Roofs

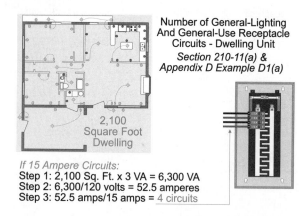

Number of General-Lighting
And General-Use Receptacle
Circuits - Dwelling Unit
*Section 210-11(a) &
Appendix D Example D1(a)*

2,100
Square Foot
Dwelling

If 15 Ampere Circuits:
Step 1: 2,100 Sq. Ft. x 3 VA = 6,300 VA
Step 2: 6,300/120 volts = 52.5 amperes
Step 3: 52.5 amps/15 amps = 4 circuits

Fig. 5-20 Number of General-Lighting and
General-Use Receptacle Circuits - Dwelling Unit

PART B.   BRANCH CIRCUIT RATINGS

210-11   Branch Circuit Requirements

(a) Number of Branch Circuits. The minimum number of branch circuits shall be determined from the total connected load and the size or rating of the circuits used. See Example D1(a) in appendix D at the back of the *National Electrical Code.*

Question – Dwelling Unit. What is the minimum number of 15 ampere receptacle circuits required for general lighting and receptacles for a 2,100 square foot dwelling unit, Fig. 5-20?

Answer.   4 circuits
Step 1. The general lighting and receptacle load is 3 VA per square foot.
2,100 square feet $\times$ 3 VA = 6,300 VA

Step 2. Branch circuit requirement
6,300 VA/120 volt = 52.5 ampere

Step 3. Number of 15 ampere circuits
52.5 ampere/15 ampere = 4 circuits

(b) Feeder   and   Panelboards   Required.
Feeders and panelboards must be installed in all buildings, including buildings that are only shells. The feeders and panelboards must be sized no less than the calculated general lighting load.

Loads Must Be Evenly Proportioned. The general lighting load as determined in Table 220-3(a) shall be evenly proportioned among all of the branch circuits. The *Code* does not limit the number

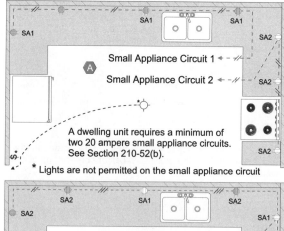

Required Branch Circuits - Small Appliance
Section 210-11(c)(1)

Small Appliance Circuit 1
Small Appliance Circuit 2

A dwelling unit requires a minimum of
two 20 ampere small appliance circuits.
See Section 210-52(b).

* Lights are not permitted on the small appliance circuit

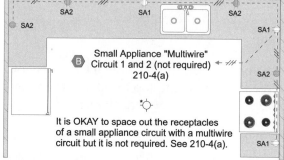

Small Appliance "Multiwire"
Circuit 1 and 2 (not required)
210-4(a)

It is OKAY to space out the receptacles
of a small appliance circuit with a multiwire
circuit but it is not required. See 210-4(a).

Fig. 5-21 Required Branch Circuits -
Small Appliance

of receptacles or lights on any **general purpose branch circuit** (in dwelling units), but it does require the loads to be evenly distributed among the circuits.

(c) Dwelling Unit.

(1) Small   Appliance   Branch   Circuits Required. Two or more 20 ampere **appliance branch circuits** are required for all receptacle outlets in the kitchen, dining room, breakfast room, pantry, or similar dining areas. Lighting outlets or other receptacle outlets cannot be connected to the small appliance branch circuit, Fig. 5-21.

Note. Two small appliance circuits are not required for each separate countertop space.

Other Related Code Sections:

• Receptacles rated 125 volt, 15 or 20 ampere can be used, Section 220-16(a).

• Minimum feeder and service load (1,500 VA), Section 220-16(a).

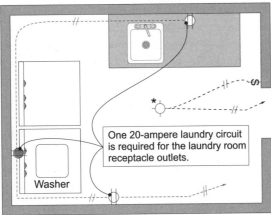

Required Branch Circuits - Laundry Circuit
Sections 210-11(c)(2)

One 20-ampere laundry circuit is required for the laundry room receptacle outlets.

Washer

* Note: Other outlets, such as lights are not permitted on the laundry circuit.

Fig. 5-22 Required Branch Circuits - Laundry Circuit

• Receptacles are required in these areas, Section 210-52(b)(1).

• Receptacles required for kitchen countertop surfaces, Section 210-52(c).

(2) Laundry Branch Circuit Required. One 20 ampere branch circuit is required for the laundry room **receptacle outlet(s)** [210-52(f)]. The laundry room receptacle circuit cannot serve any other **outlet**, such as the laundry room lighting or receptacles in other rooms.

The *NEC* does not require a separate circuit for the washing machine. Only a separate circuit for the laundry room receptacle(s) of which one can be for the washing machine, and the others can be for convenience, Fig. 5-22.

(3) Bathroom Branch Circuit Required. The bathroom receptacle(s) required in Section 210-52(d) must be supplied by a 20 ampere circuit that does not supply any other load. According to UL, 125 volt, 15 ampere receptacles are rated for 20 ampere feed-thru and can be used for this purpose [210-21(b)(2)], Fig. 5-23.

Note. This Section does not require a separate 20 ampere circuit for each bathroom. One 20 ampere circuit can be used to supply multiple bathroom receptacles.

Exception. A dedicated 20 ampere circuit to a single bathroom is permitted to supply the bathroom receptacle outlet(s) and other equipment within the same bathroom, but only if the equipment does not exceed 10 ampere (50 percent of the branch circuit ampere rating) in accordance with Section 210-23(a).

Question No. 1. Can a hydromassage bathtub motor be connected to the bathroom receptacle circuit?

Answer. **Yes**
Yes, if it's not rated more than 10 ampere and the installation instructions do not require a separate circuit for the hydromassage bathtub motor.

Question No. 2. Can a bathroom heater be connected to the bathroom receptacle circuit?

Answer. **Yes**
Yes, if it's not rated more than 10 ampere and the installation instructions do not require a separate circuit for the bathroom heating equipment.

Question No. 3. Can a light outlet, suspended ceiling fan, or bath exhaust fan be on the bathroom receptacle circuit?

Answer. **Yes**

Bathroom Receptacle Circuit(s) - Dwelling
Section 210-11(c)(3)

VIOLATION
Receptacle is not in a bathroom.

20-Ampere Branch Circuit

20A HR

Okay To Other Bathroom Receptacle(s)

15A HR

A minimum of one 20-ampere circuit is required to supply bathroom receptacles. Other outlets are not permitted on the bathroom receptacle circuit.

Fig. 5-23   Bathroom Receptacle Circuit(s) - Dwelling

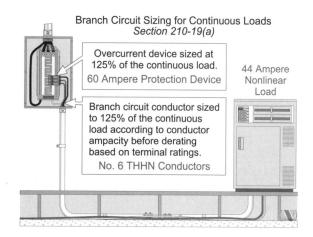

Branch Circuit Sizing for Continuous Loads
Section 210-19(a)

Overcurrent device sized at 125% of the continuous load.
60 Ampere Protection Device

44 Ampere Nonlinear Load

Branch circuit conductor sized to 125% of the continuous load according to conductor ampacity before derating based on terminal ratings.
No. 6 THHN Conductors

Fig. 5-24 Branch Circuit Sizing for Continuous Loads

210-12 Arc-fault Circuit Interrupter (AFCI) Protection

Effective January 1, 2002, all branch circuits that supply 125 volt, 15 and 20 ampere receptacles in dwelling unit bedrooms shall be AFCI protected.

An arc-fault circuit-interrupter is a device intended to provide protection from the effects of *arcing faults* by recognizing characteristics unique to arcing and by functioning to de-energize the circuit when an arc-fault is detected. An AFCI is designed to make a distinction between an unwanted, potentially damaging *arc*, and a condition necessary for continuation of power, such as the arc created when unplugging an appliance under load.

Note. Since the new requirement doesn't go into effect until "after the next *Code* cycle," this issue is a moot point. However, the 1999 *NEC* does not prohibit the installation of AFCI devices before that date.

210-19 Conductor Sizing

(a) Continuous Loads. Branch circuit conductors for **continuous loads** must be sized no less than 125 percent of the continuous loads, plus 100 percent of the noncontinuous loads. Conductors are sized based on the ampacities as listed on Table 310-16 before any **ampacity** adjustment factors according to the terminal temperature rating as listed

in Table 310-16. See Section 110-14(c) for terminal rating rules.

Note. Conductors shall have the ampacity (after any ampacity adjustment) to carry the load and the conductors shall have overcurrent protection in accordance with Section 210-20(a) and 240-3.

Question. What size branch circuit conductor is required for a 4-wire circuit supplying 44 ampere of nonlinear loads continuously? Terminals are rated for 75°C, Fig. 5-24.

Answer.   No. 6
Step 1. Circuit protection device size
44 ampere load × 1.25 = 55 ampere [210-20(a)]
60 ampere protection device [240-6(a)]

Step 2. Size conductor for continuous load
44 ampere × 1.25 = 55 ampere
No. 6 rated 65 ampere at 75°C, Table 310-16

Step 3. Conductor must be rated 44 ampere
75* ampere × 0.80 = 60 ampere

* 90°C ampacity Table 310-16, see 110-14(c)

Step 4. No. 6 conductor must have proper overcurrent protection. A 60 ampere protection device will protect a 60 ampere rated conductor [240-3].

Exception. Where the assembly and the overcurrent protection device is listed for 100 percent continuous load operation, the branch circuit load can be sized at 100 percent of the continuous load. Equipment suitable for 100 percent continuous loading is generally not available under 400 ampere.

FPN No. 1: Conductor ampacity is determined according to Section 310-15, which uses Table 310-16.

FPN No. 2: Branch circuit conductors for motors are sized according to Article 430, not Article 210.

FPN No. 3: Conductor ampacity is affected by *ambient temperature* and conductor bundling. See Sections 310-10 and 310-15 in this book for details and examples.

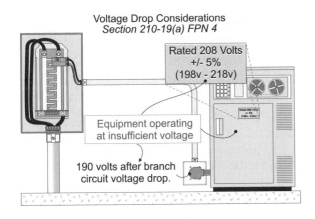

Voltage Drop Considerations
Section 210-19(a) FPN 4

Rated 208 Volts
+/- 5%
(198v - 218v)

Equipment operating
at insufficient voltage

190 volts after branch
circuit voltage drop.

Fig. 5-25 Voltage Drop Considerations

FPN No. 4: *Voltage drop* (see the glossary) should be considered when sizing conductors, but this is not a *Code* requirement. See Section 215-2(d) FPN No. 2 for more information, Fig. 5-25.

(b) Multioutlet Circuit Ampacity. Multioutlet branch circuit conductors that supply receptacles for cord- and plug-connected portable loads shall have an ampacity not less than the rating of the branch circuit; see Sections 210-21(b)(2) and Table 210-24].

(c) Household Range and Cooking Appliance Conductor Sizing. Branch circuit conductors that supply household ranges, wall-mounted ovens, counter-mounted cooking units (cooktops), and other household cooking appliances, must have a conductor ampacity not less than the rating of the branch circuit overcurrent device [220-4(c)].

Ranges rated 8 3/4 kW or more. The minimum branch circuit rating for ranges rated 8 3/4 *kW* or more shall not be less than 40 ampere.

Exception No. 1: Household Cooking Taps. Tap conductors are permitted from a 50 ampere branch circuit for ranges, wall-mounted ovens, and cooktops. The tap conductors must be as short as possible and have an ampacity not less than 20 ampere and shall be sized sufficient to carry the load as calculated according to Section 220-19, Note 4.

Question. What size tap conductor is required for a 6 kW cooktop?

Answer. No. 10

$I = P/E$, $I = 6000$ watts/240 volt, $I = 25$ ampere
No. 10 conductor, Table 310-16

Exception No. 2: Reduced Size Neutral. The neutral conductor can be smaller than the ungrounded conductors for cooking appliances rated at least 8 3/4 kW. The ampacity of the neutral conductor shall be no smaller than a No. 10 (copper), or 70 percent of the branch circuit rating as determined by Section 220-19, Note 4.

Question. What is the minimum size neutral conductor for an 11 kW range?

Answer. No. 10
Step 1. Table 220-19
Column A = 8 kW
$I = P/E$, $I = 8,000$ watts/240 volt = 33 ampere

Step 2. Neutral sized at 70 percent
33 ampere × 0.70 = 23 ampere
No. 10 conductor, Table 310-16

(d) Minimum Conductor Size. Branch circuit conductors supplying loads other than cooking appliances as covered in (c) above and as listed in Section 210-2 shall have an ampacity sufficient for the loads served and shall not be smaller than No. 14.

Exception No. 1: Fixture Tap Conductors. Tap conductors for recessed lighting fixtures as permitted in Section 410-67(c), shall have an ampacity of not less than 15 ampere [Table 210-24].

Note. Branch circuit tap conductors are not permitted for receptacle outlets; see Exception No. 1 to Section 210-19(d) for details.

210-20 Overcurrent Protection

(a) Continuous Loads. Branch circuit overcurrent protection devices must be sized no less than 125 percent of the continuous loads, plus 100 percent of the noncontinuous loads; see Section 210-19(a) for branch circuit conductor size for continuous loads.

Exception. Where the assembly and the overcurrent protection device is listed for 100 percent continuous load operation, the branch circuit load

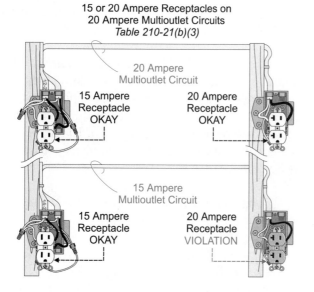

**15 or 20 Ampere Receptacles on
20 Ampere Multioutlet Circuits**
Table 210-21(b)(3)

20 Ampere
Multioutlet Circuit

15 Ampere
Receptacle
OKAY

20 Ampere
Receptacle
OKAY

15 Ampere
Multioutlet Circuit

15 Ampere
Receptacle
OKAY

20 Ampere
Receptacle
VIOLATION

Fig. 5-26 15 or 20 Ampere Receptacles on
20 Ampere Multioutlet Circuits

can be sized at 100 percent of the continuous load. Equipment suitable for 100 percent continuous loading is generally not available under 400 ampere.

(b) Branch Circuit Protection. Branch circuit conductors must be protected against **overcurrent** in accordance with Section 240-3. Branch circuit equipment must be protected in accordance with Section 240-2 and outlet (receptacles and lighting) in accordance with Section 210-21.

Exception No. 1: Fixture Tap Conductors. Tap conductors for recessed lighting fixtures are considered protected by the branch circuit protection device when in compliance with Table 210-24. See Sections 210-19(d) Ex. 1 and 410-67(c).

Exception No. 2: Fixture Wire and Cords. Fixture wires and cords must be protected in accordance with Section 240-4.

210-21 Outlet Device Rating

Receptacle and lighting outlet devices shall have an ampere rating not less than the load and shall comply with (a) and (b).

(a) Lampholder Ratings. Lampholders connected to a branch circuit with a rating over 20 ampere shall be of the heavy-duty type.

> **WARNING:**
> Lampholders used for fluorescent lamps are not rated heavy duty; therefore, fluorescent lighting fixtures cannot be installed on circuits rated over 20 ampere.

(b) Receptacle Ratings and Loadings.

(1) Single Receptacle. A single receptacle installed on an **individual branch circuit** shall have an ampacity not less than the rating of the overcurrent protection device. A single receptacle is one receptacle on a strap or yoke. A duplex receptacle is a multioutlet receptacle; see Article 100.

(2) Multiple Receptacle Loading. Where connected to a branch circuit supplying two or more receptacles or outlets, receptacles must have the load limited to 80 percent of the receptacle rating, according Table 210-21(b)(2).

(3) Multiple Receptacle Rating. Where connected to a branch circuit supplying two or more receptacles or outlets, receptacles must be rated and installed on circuits according to Table 210-21(b)(3), Fig. 5-26.

210-23 Permissible Loads

An individual branch circuit can supply any load for which it is rated. A multioutlet branch circuit must supply loads in accordance with (a) below.

(a) 15 and 20 Ampere Branch Circuit. 125 volt, 15 and 20 ampere branch circuits can supply lighting loads, utilization equipment, or any combination of both. This means that receptacles and lights are permitted on the same circuit, except for temporary power circuits [305-4(d)].

Portable Equipment. Cord- and plug-connected utilization equipment shall not exceed 80 percent of the branch circuit rating.

Fixed Equipment. The total rating of utilization equipment fastened in place "other than lighting fixtures" shall not exceed 50 percent of the branch circuit ampere rating, where the circuit supplies

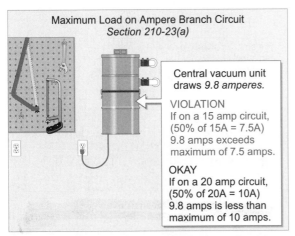

Maximum Load on Ampere Branch Circuit
Section 210-23(a)

Central vacuum unit draws *9.8 amperes.*

VIOLATION
If on a 15 amp circuit, (50% of 15A = 7.5A) 9.8 amps exceeds maximum of 7.5 amps.

OKAY
If on a 20 amp circuit, (50% of 20A = 10A) 9.8 amps is less than maximum of 10 amps.

The total rating of lighting fixtures and utilization equipment fastened in place shall not exceed 50% of the rating of a branch circuit that supplies cord- and plug-connected equipment.

Fig. 5-27 Maximum Load on Ampere Branch Circuit

lighting and receptacles for cord- and plug-connected equipment.

Question. Can a 9.8 ampere, 120 volt central vacuum be supplied by an existing 120 volt, 15 ampere receptacle circuit, Fig. 5-27?

Answer. **No**
No, because the 9.8 ampere equipment exceeds 50 percent of the 15 ampere branch circuit rating.

210-24 Multiple Outlet Branch Circuit Summary

The requirements for multiple outlet circuits are shown in Table 210-24.

Table 210-24   Multioutlet Branch Circuit Requirements

Circuit Rating	15 Amp	20 Amp	30 Amp	40 Amp	50 Amp
Minimum Circuit Wires	14	12	10	8	6
Minimum Tap Wires	14	14	14	12	12
Fixture Wires & Cords		See Section 204-4			
Receptacle Ampere Rating	15	15 or 20	30	40 or 50	50

210-25 Common Area Branch Circuits

Branch circuits in a dwelling unit shall supply only loads within or associated with the dwelling unit. Branch circuits for lighting, central alarm, signal, fire alarm, communications, or other needs for public safety shall not originate from a dwelling unit panelboard. This rule reduces the likelihood of the safety circuits being turned off by tenants, or of loss of power due to nonpayment of electric bills.

Note. This Section is not intended to prevent the dwelling unit panelboard from supplying circuits in locker, laundry, or garage facilities that are not safety-related.

PART C. REQUIRED OUTLETS

210-50 General

Receptacle outlets must be installed according to the requirements of Sections 210-52 through 210-63. Generally, receptacles are required in dwelling units, but not in commercial or industrial occupancies.

(a) Cord Pendant Receptacle Outlets. Permanently installed cord pendants receptacles are considered receptacle outlets. This is common in commercial and industrial buildings, such as when a cord with strain relief is dropped from the ceiling for equipment in the work area [400-8 Exception], Fig. 5-28.

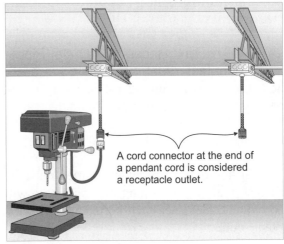

Cord Connector At End Of Cord Pendant
Section 210-50(a)

A cord connector at the end of a pendant cord is considered a receptacle outlet.

Fig. 5-28 Cord Connector at End of Cord Pendant

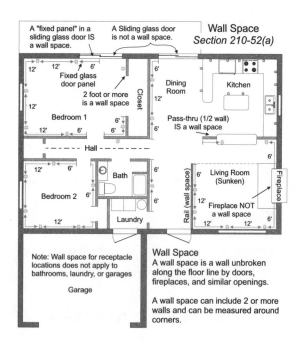

Fig. 5-29 Wall Space

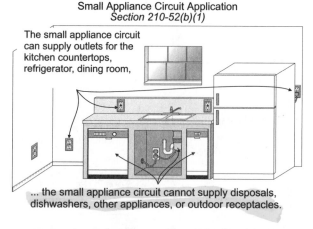

Fig. 5-30 Small Appliance Circuit Application

The ampacity of the cord [Table 400-5] shall not be less than the ampacity of the receptacle, and must be protected according to Sections 240-4 and 400-5(b). Boxes can be supported by a cord when installed according to Section 370-23(h).

(c) Appliance Receptacle Outlet Location. Receptacle outlets installed for specific appliances, such as washers, dryers, ranges, and refrigerators, must be within 6 feet of the intended location of the appliance.

210-52 Dwelling Unit Receptacle Outlet Requirements

(a) Dwelling Unit Receptacle Outlet Placement. A receptacle shall be installed in every kitchen, family room, dining room, living room, parlor, libraries, den, sun room, bedroom, recreation rooms, and other similar rooms or areas according to (1) through (3), Fig. 5-29.

(1) Spacing. A receptacle outlet must be installed so that no point along the wall space will be more than 6 feet (measured horizontally) from a receptacle outlet.

(2) Wall Space. Wall space is considered walls, fixed exterior glass, bar-counter and railing

that are at least 2 feet long, unbroken along the floor line by doors or fireplaces. Sliding glass doors on exterior walls are not considered wall space.

> **Note.** The purpose of this rule in the placement of receptacles is to avoid the use of extension cords across openings such as doors.

Spacing of Receptacles. Receptacle outlets must be equally spaced where practical.

Floor Outlets. Floor outlets located within 18 inches of the wall are considered to meet the requirements of this Section.

Receptacles Not Counted. Receptacle outlets that are part of a lighting fixture or appliance, located inside cabinets, or located over 5 feet above the floor, shall not be used to meet the requirements of this Section.

Baseboard Heaters. Listed electric baseboard heaters that have permanently installed receptacles are considered to meet the requirements of this Section [424-9].

> FPN: Not all baseboard heaters are listed for this purpose, so be sure to carefully read the equipment instructions [110-3(b)].

(b) Small Appliances.

(1) Dwelling Unit Small Appliance Outlets. Two or more small appliance branch circuits shall be required to supplied receptacle outlets in the

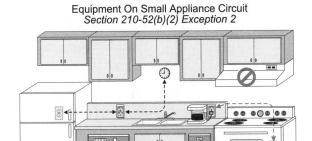

Equipment On Small Appliance Circuit
Section 210-52(b)(2) Exception 2

Gas Range

Receptacles that provide power to supplemental
equipment and lighting on gas ranges, ovens, or
cooktops are permitted on the small appliance circuit.

See Section 210-52(b)(2) Exception 1 for clock
outlet permitted on small appliance circuit.

Fig. 5-31 Equipment on Small Appliance Circuit

kitchen, pantry, breakfast room, dining room area. However a receptacle for refrigeration equipment can be on the small appliance branch circuit, Fig. 5-30.

Appliances such as disposals, dishwashers, hood fans, or lighting outlets cannot be supplied by the small appliance circuit.

Note. The *Code* does not require more than two small appliance circuits.

Exception No. 1: Switched Receptacle for Lighting. A switched receptacle for lighting is permitted in the dining room, breakfast room, or other areas, except the bathroom and kitchen. The switched receptacle for lighting cannot be connected to the 20 ampere small appliance circuit.

Exception No. 2: Refrigeration Equipment. Refrigeration equipment can be supplied from an individual 15 or 20 ampere branch circuit.

(2) **Not Supply Other Outlets.** The two 20 ampere small appliance circuits required in Section 210-11(c)(1) for these areas cannot supply any other outlet(s). This means that the kitchen light cannot be connected to the small appliance receptacle circuit.

Exception No. 1: Clock Outlet. A receptacle solely for an electric clock can be connected to the small appliance circuit, Fig. 5-31.

Exception No. 2: Gas Fired Appliances. Receptacles for supplemental equipment and lighting on gas-fired ranges, ovens, or counter-mounted cooking units can be connected to the small appliance circuit, Fig. 5-31.

CAUTION: The use of switched receptacles for lighting outlets does not apply to hallways, stairways, attached garages, or detached garages.

(3) **Kitchen Countertop Receptacles. Two Circuits Required.** 125 volt, 15 or 20 ampere receptacles used for countertop surface appliances in a dwelling unit kitchen must be supplied by at least two 20 ampere circuits [210-11(c)(1) and 220-16(a)]. These small appliance branch circuits can supply receptacles in the kitchen as well as the pantry, breakfast room, dining room, or other similar areas where food is likely to be served.

The two 20 ampere small appliance circuits cannot supply more than one kitchen of a dwelling. In effect, if a dwelling unit has two kitchens then each kitchen requires two small appliance circuits (four circuits). This rule does not apply to a dwelling unit that has multiple dining rooms.

Note. The *Code* does not define a kitchen. What if a stove is located in a basement, garage, or remote building, are these areas considered to be kitchens? Check with your inspector.

(c) **Kitchen and Dining Countertop Receptacle Location.** In kitchens and dining rooms of dwelling units, receptacle outlets for counter spaces shall be installed according to (1) through (5) below. GFCI protection is required for all 125 volt, 15 and 20 ampere receptacles that supply kitchen countertop surfaces [210-8(a)(6)].

(1) **Wall Counter Space.** A receptacle outlet must be installed for every kitchen and dining area counter wall space 12 inches or wider. Receptacles must be installed so that no point along the counter wall space is more than 2 feet (measured horizontally) from a receptacle outlet, Fig. 5-32.

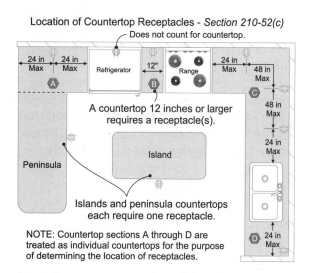

Fig. 5-32 Location of Countertop Receptacles

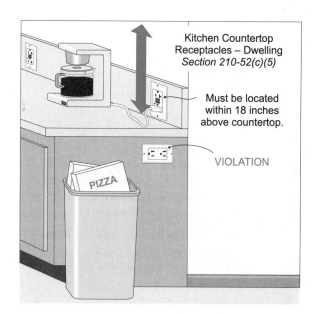

Fig. 5-33   Kitchen Countertop Receptacles - Dwelling

(2) Island Countertop Space. This Section mandates only one receptacle outlet to be installed at each island countertop that has a long dimension of 24 inches or greater, and a short dimension of 12 inches or greater.

(3) Peninsular Countertop Space. This Section mandates only one receptacle outlet to be installed at each peninsular countertop that has a long dimension of 24 inches or greater, and a short dimension of 12 inches or greater.

(4) Separate Spaces. When breaks occur in countertop spaces (ranges, refrigerators, sinks, etc.), each countertop surface is considered a separate counter for determining receptacle placement.

(5) Receptacle Outlet Location. Receptacle outlets shall be located above the countertop, but not more than 18 inches from the countertop surface. Receptacles shall not be installed in a face-up position in the work surfaces or countertops and they must not be located on the sides of cabinets, Fig. 5-33.

Receptacles that are rendered not readily accessible by appliances fastened in place, or appliances in dedicated spaces (dishwasher, microwave, etc.), are not considered meeting the receptacle outlets as required by this Section.

Exception. The receptacle outlet required for the countertop can be installed below the countertop

where necessary for the physically impaired or where there is no wall space above the island or peninsular counter. The receptacle must be located not more than 12 inches below the countertop surface and it cannot extend more than 6 inches measured horizontally from the counter's edge, Fig. 5-34.

(d) Dwelling Unit Bathroom Receptacles. At least one receptacle outlet must be installed within 36 inches of the outside edge of each basin and the receptacle shall be located on a wall that is adjacent to the basin location. This rule insures that a receptacle is located next to each basin to keep equipment cords from being draped across a sink, Fig. 5-35.

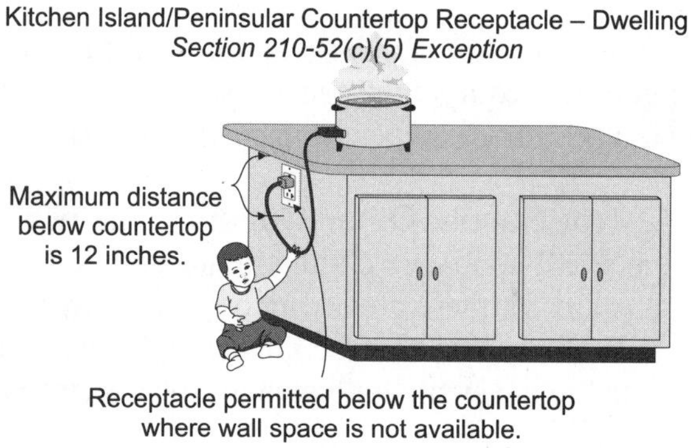

Kitchen Island/Peninsular Countertop Receptacle – Dwelling
Section 210-52(c)(5) Exception

Maximum distance below countertop is 12 inches.

Receptacle permitted below the countertop where wall space is not available.

Fig. 5-34 Kitchen Island/Peninsular Countertop Receptacle - Dwelling

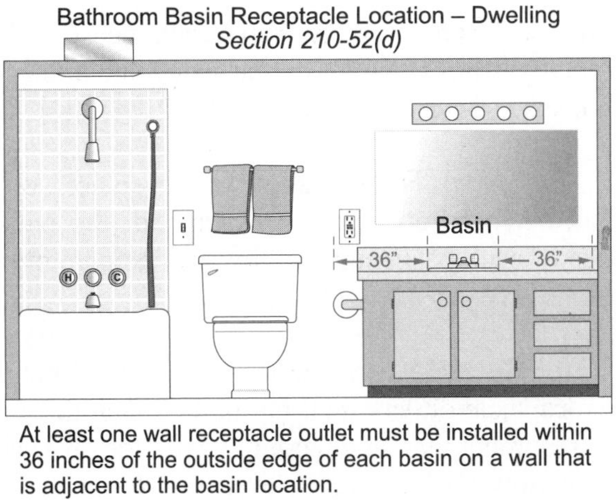

Bathroom Basin Receptacle Location – Dwelling
Section 210-52(d)

Basin

⟵ 36" ⟶   ⟵ 36" ⟶

At least one wall receptacle outlet must be installed within 36 inches of the outside edge of each basin on a wall that is adjacent to the basin location.

Fig. 5-35 Bathroom Basin Receptacle Location - Dwelling

Basement Receptacle Outlet – Dwelling
Section 210-52(g)

Unfinished Basement

In a one-family dwelling, where a portion of the basement is finished, at least one receptacle outlet is required for the unfinished area of the basement.

Fig. 5-36 Basement Receptacle Outlet - Dwelling

Note. One receptacle outlet can be used between two basins, if the receptacle is located within 36 inches of the outside edge of each basin.

(e) Dwelling Unit Outside Receptacles.

One-Family Dwelling Unit. Two receptacle outlets shall be installed not more than 6½ feet above grade, one at the front and one at the back of the **one-family dwelling.** GFCI protection is required for these receptacles [210-8(a)(3)].

Two-Family Dwelling Unit. Each grade level unit of a **two-family dwelling** must have two receptacles outlets installed not more than 6 feet, 6 inches above grade. One at the front and one at the back of each dwelling unit at grade level. GFCI protection is required for these receptacles as well [210-8(a)(3)].

Note. This Section does not require a receptacle outlet to be installed for dwelling units in a multifamily dwelling building. However, if a receptacle outlet is installed outdoors of a dwelling unit in a multifamily dwelling it must be GFCI protected [210-8(a)(3)].

(f) Dwelling Unit Laundry Area Receptacles.
All dwelling units shall have at least one 20 ampere receptacle for the laundry area receptacles. The *Code* does not require a separate circuit for the washing machine but it does require a separate 20 ampere circuit for the laundry receptacle outlet(s) [210-11(c)(2)].

Exception No. 1: Multifamily Building. If laundry facilities are available to all building occupants in a multifamily building a laundry circuit is not required in the dwelling units.

(g) Dwelling Unit Basement and Garage Receptacles.

Basement Receptacles. For a one-family dwelling, at least one 125 volt, 15 or 20 ampere receptacle outlet, in addition to any provided for laundry equipment, shall be installed in each basement. Where a portion of the basement is finished into a habitable room(s), a 125 volt, 15 or 20 ampere receptacle outlet shall be installed in the unfinished portion of the basement, Fig. 5-36.

Note. Section 210-8(a)(5) requires 125 volt, 15 or 20 ampere receptacles installed in the area of a basement not intended as a habitable room to be GFCI protected. The combination of Section 210-52(g) and 210-8(a)(5) insures that the unfinished portion of a basement has one GFCI protected 125 volt, 15 or 20 ampere receptacle outlet, even if a portion of the basement is finished into a habitable room(s). This prevents an extension cord from a non-GFCI protected receptacle to supply power to loads in the unfinished portion of the basement.

Garage Receptacles. At least one receptacle outlet must be installed at each dwelling unit garage with electric power. Section 210-8(a)(2) requires 125 volt, 15 or 20 ampere receptacles installed in the garage to be GFCI protected.

Note. The laundry outlet required in Section 210-52(f) is not permitted to satisfy the receptacle outlet requirement for either the basement or garage.

(h) Dwelling Unit Hallway Receptacle. One receptacle outlet shall be installed for each hallway.

Hallway. A hallway is a corridor that is at least 10 feet long measured along the centerline of the hall without passing through a doorway.

210-60   Receptacles in Guestroom for Hotels and Motels

(a) Receptacle Requirements. Guest rooms in hotels, motels, and similar occupancies shall have receptacle outlets installed in accordance with Section 210-52, such as:

Number of Receptacles. Receptacle outlets shall be installed so that no point along the floor line in any wall space is more than 6 feet, measured horizontally, from an outlet in that space, including any wall space 2 feet or more in width [210-52(a)].

Bathroom Receptacles. A receptacle outlet shall be installed within 36 inches of the outside edge of each basin supplied by a 20 ampere branch circuit [210-52(d)].

(b) Receptacle Placement. The total number of receptacle outlets shall not be less than the minimum number that would comply with the provisions of Section 210-52(a). However, the receptacle outlets can be located so as to be convenient for permanent furniture layout, but at least two receptacle outlets shall be readily accessible to eliminate the need for extension cords by guests for ironing, computers, refrigerators, etc. Receptacles installed behind a bed shall be located so as to prevent the bed from contacting an **attachment plug**, or the receptacle shall be provided with a suitable guard.

210-63   Receptacles For Rooftop, Attic, and Crawl Spaces

Heating, air-conditioning, refrigeration equipment located on roof tops, or in attics and crawl spaces, shall have a 125 volt, 15 or 20 ampere receptacle outlet for servicing the equipment. The receptacle shall be installed at an accessible location within 25 feet of the equipment. Receptacles located on the roof must be GFCI protected [210-8(b)(2)] and must not be connected to the *load side* of the equipment disconnect. See Section 210-70 for the requirement of a lighting outlet for equipment requiring service.

Exception. Receptacle outlets are not required for rooftop equipment in one- and two-family dwellings.

210-70   Lighting Outlets Requirements

Lighting outlets must be installed as required in subsection (a) through (c), Fig. 5-37.

(a) Dwelling Unit Lighting Outlet Requirements.

(1) Habitable Rooms. At least one wall switch-controlled **lighting outlet** shall be installed in every habitable room and bathroom.

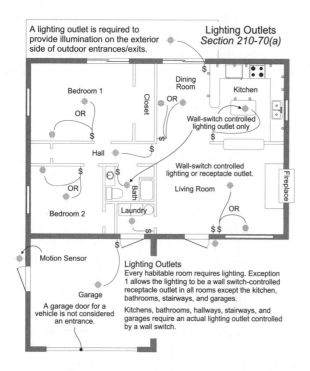

Fig. 5-37 Lighting Outlets

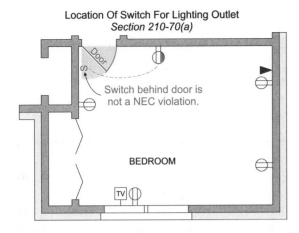

Fig. 5-38 Location of Switch for Lighting Outlet

Exception No. 1: Switched Receptacles.   In other than kitchens and bathrooms, a receptacle controlled by a wall switch can be used instead of a lighting outlet.

Exception No. 2: Occupancy Sensors.   Lighting outlets can be controlled by occupancy sensors integral with the wall switch, if a *manual* override is included.

(2) Other Locations.   At least one wall-switch controlled **lighting outlet** shall be installed in bathrooms, hallways, stairways, attached garages, detached garages with electric power, and illumination must be provided on the exterior side of outdoor entrances or exits that have grade level access. A vehicle door (garage door) in an attached garage is not considered an outdoor entrance.

Note. The *Code* contains the requirement for the location of the "lighting outlet," but does not specify the switch location. Naturally, you would not want to install the switch intentionally behind a door or other inconvenient location. But the *Code* does not require that you relocate the switch to suit the swing of the door, Fig. 5-38.

Interior Stairways.   When the difference between floor levels is six steps or more, 3-way and 4-way wall switches are required to control the lighting outlet for the interior stairways. Long pull chains won't meet *Code* requirements.

Note. It is not the intent to require 3-way switches for basements that are dead-ended when you must return to the same location you entered.

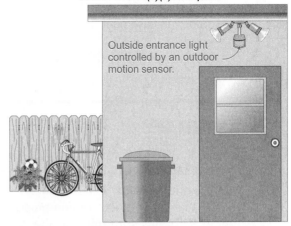

At outdoor entrances of a dwelling unit, remote, central, or automatic control of lighting is permitted in lieu of a switch.

Fig. 5-39 Control of Outdoor Entrance Lighting Outlet - Dwelling

Exception: Remote or Automatic Controls.   In hallways, stairways, and at outdoor entrances, remote, central, or automatic control of the lighting outlet is permitted, Fig. 5-39.

(3) Storage and Equipment Rooms.   Areas used for storage, or containing equipment requiring servicing, such as attics, underfloor spaces, utility rooms, and basements, must have a lighting outlet located near the equipment. The lighting outlet can

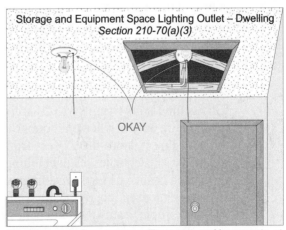

For attics, underfloor spaces, utility rooms, and basements, at least one lighting outlet containing a switch or controlled by a wall switch must be installed where these spaces are used for storage or contain equipment for servicing.

Fig. 5-40 Storage and Equipment Space Lighting Outlet - Dwelling

contain a switch or be controlled by a wall switch located at the point of entry to the attic, underfloor space, utility room, and basement, Fig. 5-40.

(c) **Commercial Attics and Underfloor Spaces.** Attics and underfloor spaces containing equipment requiring servicing must have a lighting outlet located near the equipment. The lighting outlet must be controlled by a wall switch installed at or near the point of entrance to the attic or underfloor space. See Section 210-63 for the requirement of receptacle outlets.

SUMMARY

Part A.   General Provisions   ☐ Article 210 applies to most branch circuits. The rating or setting of the overcurrent device determines the rating of the branch circuit.   ☐ The classification for multioutlet branch circuits is 15, 20, 30, 40, and 50 ampere.   ☐ Multiwire branch circuits in dwelling units that feed devices or equipment on the same yoke require a two-pole breaker.   ☐ Multiwire branch circuits are limited to line-to-neutral loads, see exceptions.   ☐ Grounded conductors of a multiwire branch circuit must be pigtailed when terminated on a device.   ☐ If a building has more than one voltage system, multiwire branch circuit ungrounded conductors must be identified by phase and system.   ☐ Only grounding-type receptacles are permitted on 15 and 20 ampere branch circuits, see exceptions.   ☐ Receptacles and cord caps that have grounding contacts must have those contacts grounded to the circuit equipment grounding conductor, see exceptions.   ☐ GFCI protection is required in dwelling unit bathrooms, garages, accessory buildings outside receptacles with direct grade access, unfinished basements, grade-level crawl spaces, kitchen countertop and countertop receptacles within 6 feet of a wet bar, as well as on outside receptacles for boathouses, docks and sea walls.   ☐ The minimum number of branch circuits required for general lighting for dwelling units shall be determined from the total general lighting load [Table 220-3(a)], and size or rating of the circuits used.   ☐ For a dwelling unit, two 20 ampere small appliance branch circuits are required in kitchens, dining rooms, breakfast rooms, pantries, or similar dining areas.   ☐ One 20 ampere branch circuit is required for the laundry outlet(s).

Part B.   Branch Circuit Ratings   ☐ The ampacity of the branch circuit conductors shall not be less than the maximum load of the circuit.   ☐ The rating of the branch circuit is determined by the overcurrent device.   ☐ The recommended voltage drop is 3 percent for branch circuits, 3 percent maximum for feeders, and 5 percent for feeders and branch circuits combined.   ☐ Circuits that supply inductive lighting loads require conductors to be sized to the ampere rating of the ballast.   ☐ An individual branch circuit can supply any load for which it is rated.

Part C.   Required Outlet Dwelling Unit Requirements   ☐ A receptacle outlet must be installed so that no point on the wall space will be more than 6 feet from a receptacle outlet.   ☐ Receptacle outlets must be equally spaced where practical and must not be spaced farther apart than 12 feet.   ☐ Dwelling units require two 20 ampere small appliance circuits.   ☐ Kitchen and dining countertops require outlets when one foot or wider, and a receptacle must be located so that no point along the wall line is more than 2 feet from a receptacle outlet.   ☐ Island and peninsular countertops require one receptacle.   ☐ Dwelling unit bathrooms require a receptacle located within 36 inches from each basin.   ☐ Each dwelling unit must have a receptacle for the laundry area.   ☐ Hallways 10 feet or more in length must have a receptacle outlet.   ☐ Lighting outlets are required in every habitable room, bathroom, kitchen, hallway, stairway, garage, and outdoor entrances (except in some cases a switched receptacle is permitted).

Commercial   ☐ A receptacle outlet is required within 25 feet of heating, air-conditioning, and refrigeration equipment on rooftops, attics, and crawl spaces, and the rooftop receptacle must be GFCI protected.   ☐ A wall switch-controlled lighting outlet is required at or near heating, air-conditioning, and refrigeration (HACR) equipment installed in attics or underfloor spaces.

REVIEW QUESTIONS

1. Explain the advantages of multiwire branch circuits.

2. Explain the disadvantages and dangers of multiwire branch circuits.

3. Explain the rules that apply to the replacement of nongrounding-type receptacles.

4. Explain how a GFCI device works.

5. Explain the GFCI protection rules that apply to dwelling-units.

6. Explain the GFCI protection rules that apply to commercial occupancies.

7. What is the maximum number of duplex receptacles on a two-wire 20 ampere circuit in a residence (dwelling unit)?

8. Explain the rules that apply to circuits that have a single receptacle.

9. Explain the rules that apply to circuits that have multioutlet receptacles.

10. Explain the rules associated with kitchen countertops for dwelling units.

11. Explain the rules that apply to receptacles outside dwelling units.

12. Explain the lighting and receptacle requirement that apply for heating, air-conditioning, and refrigeration (HACR) equipment as it relates to commercial occupancies.

13. Explain the NEC rules that apply to dwelling unit lighting and switching and habitable rooms.

14. In your own words explain the meaning of the following Article 100 definitions:

> Accessible
> Bathroom
> Continuous load
> Dwelling unit
> Feeder
> Lighting outlet
> Outlet
> Overcurrent
> Panelboard
> Readily accessible
> Receptacle
> Receptacle outlet

15. In your own words explain the meaning of the following Glossary terms:

> Arcing-fault
> Fault-current
> Harmonic Current
> High-leg
> Neutral Conductor
> Voltage Drop

Notes:

Unit 6

Article 215
Feeders

OBJECTIVES

After studying this unit, the student should be able to understand:
- what a feeder is and how it relates to service and branch circuit conductors.
- how to size feeder conductors to continuous loads.
- how to size feeder conductors for a dwelling unit or mobile home
- how to size feeder overcurrent protection devices to continuous loads.
- the identification requirements for high-leg conductors.

Definitions and Glossary Terms

To better understand the *NEC* rules contained in this unit, review the following:

Definitions Article 100 - Unit 2
Ampacity
Continuous load
Overcurrent
Grounding conductors

Glossary Terms
High-leg
Derating factors

215-1 **Scope**

Article 215 covers the installation and minimum conductor sizing requirements for feeders, Fig. 6-1.

Note. Feeder conductors that supply motor loads must be sized and protected according to Parts B and D of Article 430.

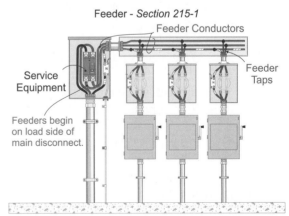

Feeder - *Section 215-1*
Feeder Conductors
Service Equipment
Feeder Taps
Feeders begin on load side of main disconnect.

Feeder: (Article 100 Definition) All circuit conductors between the service equipment or the source of a separately derived system and the final branch-circuit overcurrent device.

Fig. 6-1 Feeder Conductor Begins at the Load Side of Service Disconnect

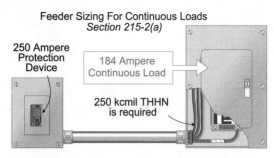

Feeder Sizing For Continuous Loads
Section 215-2(a)

250 Ampere Protection Device

184 Ampere Continuous Load

250 kcmil THHN is required

A 250 kcmil rated 232 amperes (after derating) would be protected by the 250 ampere overcurrent protection device.

Fig. 6-2 Feeder Conductors for Continuous Loads Sized at 125 Percent

215-2 Minimum Rating and Size

Feeders must have an **ampacity** not less than the load, as calculated in Article 220, Parts B, C, and D.

(a) Continuous Load Computation. Feeder conductors for **continuous loads** are sized no less than 125 percent of the continuous loads, plus 100 percent of the noncontinuous loads, based on the ampacities as listed on Table 310-16 (before any ampacity *derating factors*); see Section 110-14(c) for terminal ratings. In addition, the conductor must have an ampacity, after ampacity derating of not less than 100 percent of the total load.

> **Note.** Be sure that the conductor is protected at its derated ampacity according to Section 240-3.

> *Question.* What size conductor and protection device is required for a feeder that supplies a 184-ampere continuous load where the raceway contains four current-carrying conductors, Fig. 6-2?

> *Answer.*   250 kcmils, 250 amperes

> *Step 1.* Continuous Load
> 184 ampere × 1.25 = 230 ampere

> *Step 2.* No. 4/0 THHN at 75°C is rated 230 ampere and meets this requirement [Table 310-16].

No. 4/0 THHN has an ampacity after derating of 260 ampere × 0.8 = 208 ampere.

No. 4/0 THHN, rated 208 ampere cannot be protected by a 250 ampere overcurrent protection device because the next size up rule does not apply [240-3(b)].

(b) Minimum Feeder. Feeder conductors shall not be less than 30 ampere for any of the following:

(1) Two or more 2-wire branch circuits supplied by a 2-wire feeder;

(2) Three or more 2-wire branch circuits supplied by a 3-wire feeder;

(3) Two or more 3-wire branch circuits supplied by a 3-wire feeder; or

(4) Three or more 4-wire branch circuits supplied by a 3-phase, 4-wire feeder.

(d) Dwelling Unit and Mobile Home Feeder Sizing.   Feeder conductors are not required to be larger than the service conductors. Dwelling unit or mobile home feeder conductors, rated not over 400 ampere, can be sized according to Section 310-15(b)(6). This rule permits the feeder conductors to be sized smaller than the requirements of Table 310-16.

> FPN No. 1: See Examples D1 through D10 in Appendix D for feeder load calculations.

> FPN No. 2: Conductor voltage drop should be considered when sizing the feeder conductor. See the glossary for more details. The *NEC* suggests that conductor voltage drop not exceed 3 percent on branch circuits, 3 percent on feeders and not more than 5 percent for both branch and feeders

215-3 Overcurrent Protection

Feeder conductors must be protected against **overcurrent**, according to Section 240-3, specifically subsections (b) and (c).

Continuous Loads. Feeder overcurrent protection devices must be sized no less than 125 percent of the continuous loads, plus 100 percent of

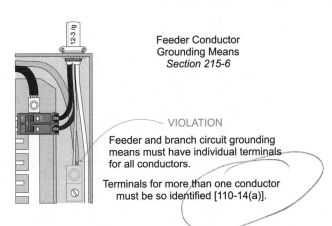

Feeder Conductor
Grounding Means
Section 215-6

VIOLATION

Feeder and branch circuit grounding
means must have individual terminals
for all conductors.

Terminals for more than one conductor
must be so identified [110-14(a)].

Fig. 6-3 Feeder and Branch Grounding Conductors
Must Terminate to Individual Terminals

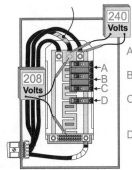

Identification Of Feeder High-Leg - *Section 215-8*

Orange identification of the feeder high-leg is required at

240 Volts

208 Volts

A: 2-pole 240 volt breaker is
   usually okay in high-leg.
B: 2-pole 120/240 volt breaker is
   okay on L1 and L3 only.
C: 1-pole 120 volt breaker can
   never be on L2 high-leg.
   Voltage is 208 volts to ground.
D: 1-pole 120 volt breaker is
   okay on L1 and L3 anytime.

Fig. 6-4 Feeder High-Leg Must Be Identified with
the Color Orange or Other Effective Means

the noncontinuous loads; see Section 215-2(a) for branch circuit conductor size for continuous loads.

Exception. Where the assembly and the overcurrent protection device is listed for 100 percent continuous load operation, the branch circuit load can be sized at 100 percent of the continuous load. Equipment suitable for 100 percent continuous loading is generally not available under 400 ampere.

215-6   Feeder Conductor Grounding Means

Panelboards containing branch circuit equipment **grounding conductors** shall have a grounding means for both the feeder and branch circuit equipment grounding conductors, see Sections 250-118, 250-119, and 384-20, Fig. 6-3.

WARNING:

When replacing disconnects, panelboards, meters, switches, or any equipment that contains the high-leg conductor, care must be taken to replace the high-leg to the proper location. Failure to install the high-leg properly can result in 120 volt circuits connected to the 208 volt high-leg, with disastrous results. See Section 384-3(f) for details on the termination of the high-leg, Fig. 6-4.

215-8   Feeder High-Leg Conductor Identification

When a grounded conductor (what is often called a neutral) is present with a *high-leg* feeder conductor, the high-leg conductor must be identified at each point where a connection is made. The identification of the high-leg conductor shall be by using orange phase tape, painting the conductor finish orange, tagging, or other effective means.

Identification of the high-leg conductor for services is covered in Section 230-56 and for panelboards in Section 384-3(e), Fig. 6-4.

215-10   Ground-Fault Protection of Equipment

Ground-fault protection of equipment (GFPE, not GFCI) is required for solidly grounded 480Y/277 volt system feeder disconnects rated 1,000 ampere or more.

Exception. Where ground-fault protection is provided on the supply side of the feeder disconnect in accordance to the requirement of Section 230-95, additional feeder ground-fault protection is not required.

Note. GFPE protection can be anywhere from 30 milliampere to up to hundreds of ampere. However, GFCI protection is limited to 4 to 6 milliampere.

SUMMARY

Article 215 covers the rules for feeders that supply branch circuits. ☐ Feeders are conductors on the load side of the service equipment, but before the branch circuit conductors. ☐ Feeder conductors must have an ampacity not less than the load, as calculated in Article 220, Parts B, C, and D. ☐ Feeder conductors for continuous loads are sized no less than 125 percent of the continuous loads, plus 100 percent of the noncontinuous loads. ☐ Feeder conductors for a dwelling unit or mobile home are not required to be larger than the service-entrance conductors, as sized in Section 310-15(b)(6). ☐ Feeder conductors must be protected against overcurrent, and overcurrent protection devices for continuous loads are sized no less than 125 percent of the continuous loads, plus 100 percent of the noncontinuous loads. ☐ The high-leg conductor must be identified with the color "orange" where a connection is made if the grounded (neutral) conductor is also present.

REVIEW QUESTIONS

1. Explain the NEC recommendations as it relates to voltage drop for both feeders and branch circuits.

2. In your own words explain the meaning of the following Article 100 definitions:

    Ampacity
    Continuous load
    Overcurrent
    Grounding conductors

3. In your own words explain the meaning of the following Glossary terms:

    High-leg
    Derating factors

Unit 7

Article 220

Branch Circuit, Feeder and Service Calculations

OBJECTIVES

After studying this unit, the student should be able to understand:
- branch circuit, feeder, and service calculations.
- how to size the service overcurrent protection device for continuous and noncontinuous loads.
- the number of duplex receptacles required in a commercial building.

Definitions and Glossary Terms

To better understand the *NEC* rules contained in this unit, review the following:

Definitions Article 100 - Unit 2

Dwelling unit	Feeder
Service	Demand factor
Garage	Nominal voltage
Nonlinear load	

Glossary Terms
kVA
Lamp
Reactance
Three-phase

PART A. GENERAL

220-1 Scope

This Article provides the requirement for sizing branch circuits, feeders, and services; and for determining the number of receptacles on a circuit, and the number of branch circuits required.

Additional Code Articles. When sizing conductors and overcurrent protection for branch circuits, feeders, and services, other *Code* Articles must be considered such as:

Air Conditioning – 440
Appliances – 422
Branch Circuits – 210
Conductors – 310
Electric Space Heating Equipment – 424
Feeders – 215
Motors – 430
Overcurrent Protection Rules – 240
Services – 230

220-2 Voltages

(a) Voltage Value to Be Used. Unless other voltages are specified, branch circuit, feeder, and service loads shall be computed using the following **nominal voltages**: 120, 120/240, 208Y/120, 240, 347, 480Y/277, 480, 600Y/347 and 600 volt, Fig. 7-1.

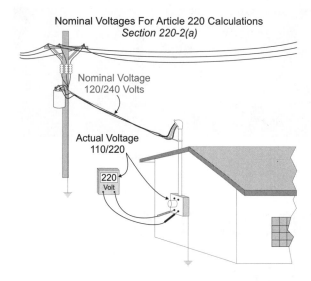

Fig. 7-1 Nominal Voltage Used for Load Calculations

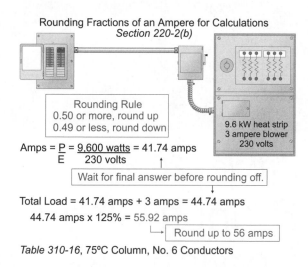

Fig. 7-2 Rounding Fractions of an Ampere

Note. The Canadian Electrical Code, Part II permits *three-phase*, 600Y/347 and single-phase 347 volt systems. This equipment can also be used and installed in the United States.

(b) Fractions of an Ampere. Where the computations result in less than 0.50 of an ampere such fractions may be dropped. When do you round? After each calculation, or at the final calculation? The *NEC* is not specific on this issue, but I guess it all depends on the answer you want to see!

Question. What size THHN conductor is required to supply a 9.6 kW, 230 volt, single-phase fixed space heater that has a 3 ampere blower motor [424-3(b)]? Terminal rating of 75°C, Figure 7-2.

Answer.   No. 6
According to Section 424-3(b) the conductors and overcurrent protection device to electric space heating equipment shall be sized no less than 125 percent of the total load (heat plus motors).

Step 1. Space Heater Ampere = Watts/Volt
9,600 watts/230 volt = 41.74 ampere

Step 2. Total Load (heat + motor)
41.74 ampere + 3 ampere = 44.74 ampere

Step 3. Conductor Size
44.74 ampere × 1.25 = 55.92 ampere

Round up to 56 ampere
No. 6, rated 65 ampere at 75°C [Table 310-16]

220-3 **Computation of Loads**

(a) General Lighting. The general lighting load of 3 VA per square foot specified in Table 220-3(a) shall be computed from the outside dimensions of the building or area involved, Fig. 7-3.

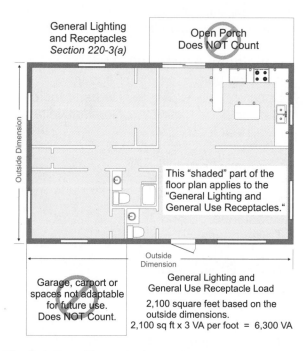

Fig. 7-3 Dwelling - General Lighting and Receptacle Demand Load

Bank - General Lighting And Receptacle Demand Load
Section 220-3(a) Note

Bank - 10,000 Square Feet
Number of Receptacles Unknown

Example: Determine demand load for lighting and receptacles.

10,000 square feet x 3½ VA per foot = 35,000 VA lighting load

35,000 VA lighting load x 1.25 [215-2(a)] = 43,750 VA lighting demand

Table 220-3(a) Note requires an additional unit load of 1 VA per square foot when the number of receptacles is unknown

43,750 VA lighting + 10,000 VA receptacle = 53,750 VA demand load

Fig. 7-4 Bank - General Lighting and Receptacle Demand Load

Dwelling Units. For **dwelling units** the computed floor area shall not include open porches, **garages**, or unfinished spaces not adaptable for future use. "The outlets specified in Section 210-52(e) and Section 210-52(g), and the outlets specified in Section 210-70(a)(2) for exterior, garage, storage, and equipment spaces, shall be permitted to be included in the general lighting load specified in Table 220-3(a)."

Note. When determining the general lighting load for a dwelling unit, you must include those spaces adaptable for future use.

For Offices and Banks. If the number of receptacles is unknown, an additional 1 VA per square foot must be included for the feeder and/or service load. See the note at the bottom of Table 220-3(a), which states that "In addition, a unit load of 1 volt-ampere per square foot shall be included for general-purpose receptacle outlets where the actual number of general-purpose receptacle outlets is unknown."

Question. What is the general lighting and receptacle load for a 10,000 square foot office building if the number of receptacles is unknown, Fig. 7-4?

Answer. 53,750 VA

Lighting 10,000 sq/ft × 3.5 VA × 1.25	= 43,750 VA
Receptacle (1 VA per square foot)	= 10,000 VA
Total	53,750 VA

Note. Section 215-2(a) and 230-42(a) requires the conductors to be sized at 125% of the continuous load.

Dwelling Unit(s). The 3 volt-ampere per square foot for general lighting includes all 125 volt, 15 and 20 ampere general use receptacles.

(b) Other Loads – Branch Circuits.

(6) Sign and Outline Lighting. Each commercial building and each commercial occupancy accessible to pedestrians shall be provided, at an accessible location outside the entrance to each tenant space, with at least one outlet for sign or outline lighting. The outlet(s) must be supplied by a 20 ampere branch circuit that supplies no other load [600-5(a)]. The minimum feeder/service VA load for the sign outlet shall be computed at a minimum of 1,200 volt-ampere.

(9) Commercial Receptacle Load. The minimum load for general use receptacle outlets shall be 180 volt-ampere per strap.

(10) Residential Occupancies. The 180 VA per receptacle strap in Section 220-3(b)(9) applies to commercial and industrial outlets, not to dwelling units. This subsection is intended to make it clear that all 125 volt, 15 and 20 ampere general purpose receptacle outlets in dwelling units are part of the 3 VA general lighting load.

Number of Receptacles on a Circuit

The NEC limits the number of receptacles on a circuit based on the 180 VA per outlet and the VA rating of the circuit.

Question. – Commercial. What is the maximum number of receptacle outlets permitted on a 20 ampere, 120 volt circuit, Fig. 7-5?

Answer. *13 receptacles*
The maximum number of single or duplex receptacles on a 20 ampere, 120 volt circuit is calculated as follows:

Step 1. Determine the circuit VA rating
120 volt × 20 ampere = 2,400 VA

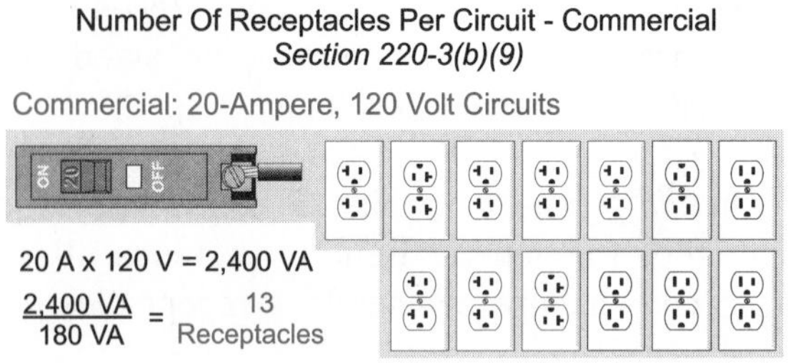

Number Of Receptacles Per Circuit - Commercial
Section 220-3(b)(9)

Commercial: 20-Ampere, 120 Volt Circuits

20 A x 120 V = 2,400 VA

$$\frac{2,400 \text{ VA}}{180 \text{ VA}} = \frac{13}{\text{Receptacles}}$$

Fig. 7-5 Commercial - Maximum of 13 Receptacles on a Circuit

Step 2. Number of receptacles permitted
2,400 VA/180 VA* = 13 receptacles
*Receptacles are not considered a continuous load.

The *NEC* is not specific on the requirements for determining the number of receptacle outlets permitted on general lighting circuits in residences. However, the *NEC* Handbook (published by the NFPA) clarifies that there is no limit as to the number of receptacle outlets on a dwelling unit circuit, because residential receptacles are lightly used, Fig. 7-6.

CAUTION: Check with your local building/ electrical code. Some local areas have a maximum number of receptacles and lighting outlets permitted on a circuit for dwelling units.

Though there is no limit on the number of lighting and/or receptacle outlets on dwelling general lighting branch circuits; the *NEC* does require a minimum number of circuits to be installed for this purpose [210-11(a)].

However, the *NEC* does require a minimum number of branch circuits [210-11(a)], and the receptacle and lighting loads must be evenly distributed among the required circuits [210-11(b)].

220-4 Maximum Loads

(a) Motors and Air Conditioners. Motor circuits shall be sized according to Article 430, specifically Sections 430-22, 24 and 430-25 and air-conditioning circuits shall be sized according to Article 440, specifically Sections 440-32, 33 and 34.

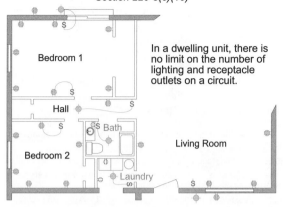

Number of General-Purpose Receptacles - Dwelling Unit
Section 220-3(c)(10)

In a dwelling unit, there is no limit on the number of lighting and receptacle outlets on a circuit.

Bedroom 1

Hall

Bath

Bedroom 2

Living Room

Laundry

Fig. 7-6 Dwelling - No Limit on the Number of General-Purpose Receptacles on a Circuit

(b) Inductive Lighting Loads. Circuits that supply inductive lighting, such as fluorescent and H.I.D. fixtures, must have the conductors sized according to the ampere rating of the ballast, not the *wattage* of the lamps. See my example in Section 220-4(b) of this book.

Inductive lighting fixtures (fluorescent, high-pressure sodium, etc.) have ballasts that operate on the principle of inductive *reactance*. Inductive reaction causes the voltage and current not to be in phase (power factor below unity, or 1.0); as a result the input VA is greater than the output watts. I know that this is getting complicated, but just remember that you do not use the *lamp* wattage rating for calculations.

Note. The maximum continuous load must not exceed 80 percent of the branch circuit rating [210-19(a)].

Question. How many 4-lamp, 34 watt fluorescent lighting fixtures (rated 1.34 ampere) are permitted on a 20 ampere circuit continuously?

Answer.   11 fixtures
20 ampere × 0.80 = 16 ampere
16 ampere/1.34 = 11 fixtures

(c) Household Cooking Appliances. Conductors for household cooking appliances can be sized according to the demand factors of Table 220-19 Note 4, see Sections 210-19(c).

PART B. FEEDER AND SERVICE CALCULATIONS

220-11 General Lighting Demand Factors

The *Code* recognizes that not all general lighting loads are on at the same time and has provided for a method to apply a **demand factor** when determining feeder and service sizes.

The dwelling unit demand factors listed in Table 220-11 apply to the 3 VA per square foot of the general lighting load [Table 220-3(a)]. The demand factors listed on Table 220-11 are also permitted to apply to the two small appliance circuits [210-11(c)(1), 220-16(a)], and the laundry circuit [210-11(c)(2), 220-16(b)].

Question. What is the general lighting demand load including the small appliance and laundry circuit for a dwelling unit of 2,700 square feet?

Answer.   6,360 VA

Step 1.

General lighting	2,700 × 3 VA	= 8,100 VA
Small appliance	1,500 × 2 VA	= 3,000 VA
Laundry circuit	1,500 × 1 VA	= 1,500 VA
Total Load		= 12,600 VA

Step 2.

Demand Factors:

First 3,000 VA at 100%	= 3,000 VA
Remainder 9,600 VA at 35%	= 3,360 VA
Total Demand Load	= 6,360 VA

See Section 210-11(a) in this book for example on calculating the number of general lighting circuits required for a dwelling unit.

220-12 Track Lighting

(b) Track Lighting. Track lighting loads in commercial occupancies shall be in addition to the general lighting loads as listed in Table 220-3(a). For other than dwelling units or the guest rooms of hotels and motels, the feeder and service load calculation for track lighting is to be determined at 150 volt-ampere for every 2 feet of track installed.

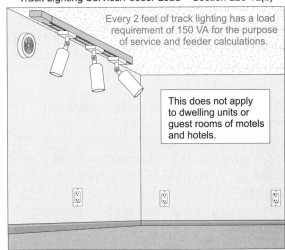

Track Lighting Service/Feeder Load – *Section 220-12(b)*

Every 2 feet of track lighting has a load requirement of 150 VA for the purpose of service and feeder calculations.

This does not apply to dwelling units or guest rooms of motels and hotels.

Note: There is no limit on the length of track on a branch circuit, but the load installed on the track is limited to the rating of the lamps and branch circuit.

Fig. 7-7 Track Lighting Service/Feeder Load at 75 VA Per Foot

Track Length. There is no maximum length of track, because the 150 VA-per-2 feet rule only applies to the feeder/service load calculation, Fig. 7-7.

Track Load. The maximum number of lighting fixtures allowed on a track is based on the wattage rating of the lamps, the voltage and ampere rating of the circuit.

Question. How many 75 watt lamps can be on a 20 ampere, 120 volt circuit continuously?

Table 220-11   Lighting Demand Factors

Type of Occupancy	Portion of Lighting Load Demand to Which Demand Factor Applies (volt-amperes)	Demand Factor %
Dwelling Units	First 3,000	100
	Next 117,000	35
	Remainder over 120,000	25
Hotels and Motels, Including Apartment Houses without Provision for Cooking by Tenants	First 20,000	50
	Next 80,000	40
	Remainder	30
Warehouses	First 12,500	100
	Remainder	50
All Others	Total VA	100

Answer.   25 fixtures
Circuit VA, 120 volt × 20 ampere = 2,400 VA
Maximum Load, 2,400 VA × 0.80 = 1,920 VA
1,920 VA/75 watts per fixtures = 25 fixtures

220-13 Commercial and Industrial Receptacle Demand Factors

Commercial and industrial receptacles are calculated at a load of 180 volt-ampere per strap [220-3(b)(9)]. The demand factors of Table 220-13 can be applied to that portion of the receptacle load that exceeds 10 kVA.

The first 10 kVA is equal to 55 receptacles: 10,000 VA/180 VA = 55 receptacles. So the first 55 receptacles are at 100 percent demand, and all additional receptacles are at 50 percent demand.

Question. What is the receptacle demand load for 150 receptacles, Fig. 7-8?

Answer.   18,500 VA
Step 1.
150 receptacles × 180 VA   = 27,000 VA

Step 2.
Demand on 27,000 VA receptacle load:
First 10,000 VA at 100%   = 10,000 VA
Remainder 17,000 VA at 50%   = 8,500 VA
Total Receptacle Demand Load   = 18,500 VA

220-14 Motors

Conductor Size for Motors and Other Loads. The feeder/service conductor must be sized no smaller than 125 percent of the largest motor ampere, plus the sum of the other motor ampere, plus the sum of the other loads as computed in this Article [430-24].

Protection Size for Motors and Other Loads. The protection device for motor feeders must not be smaller than the largest motor branch circuit protection device size, plus the sum of the other motor(s) full-load ampere, plus the sum of the other loads as computed in this Article [430-63].

Air Conditioning. Article 220 does not specify the method to compute the air-conditioning feeder and/or service load for standard load calculations in

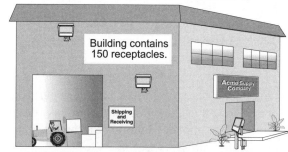

Receptacle Demand Loads - *Section 220-13*

Determine the demand load for 150 receptacles.
Section 220-3(b)(9), each receptacle = 180 VA
Table 220-13: First 10 kVA at 100% and the remainder at 50%
150 receptacles x 180 VA each = 27,000 VA
First 10,000 VA at 100% =   - 10,000 VA x 1.00 = 10,000 VA
Remainder of load at 50% =   17,000 VA x 0.50 = +8,500 VA
Receptacle Demand Load =   18,500 VA

Fig. 7-8 Commercial - Receptacle Demand Load

Article 220, Part B. However, there is a very subtle reference in Section 220-3(b)(3) that directs us to use Article 440 for Air Condition calculations. Section 440-34 specifies that the feeder conductor must be sized no smaller than 125 percent of the largest air-conditioner current, plus the sum of the other loads as computed according to Article 220.

220-15 Fixed Electric Space Heating

The feeder and service load for fixed electric space heating equipment must be computed at no less than 100 percent of the total connected load. However, the feeder load for fixed electric space heating equipment shall not be less than the rating of the largest branch circuit supplied.

220-16 Dwelling Unit Small Appliance and Laundry Loads

(a) Dwelling Unit Small Appliance Circuit Load. Each dwelling unit shall have a minimum of two 20 ampere small appliance branch circuits [210-11(c)(1)]. Each small appliance circuit shall have a feeder and service load of 1,500 volt-ampere and the demand factors of Table 220-11 can be applied to this load.

Receptacle Rating. The small appliance branch circuits are permitted to use 125 volt, 15 or

20 ampere rated receptacles [Table 210-21(b)(3) and Table 210-24].

Other Related Code Sections.
- Receptacles for small appliance circuits, [210-52(b)]
- Receptacles required for kitchen countertops, [210-52(c)]

(b) Dwelling Unit Laundry Circuit Load. Each dwelling unit shall have a 20 ampere laundry circuit [210-11(c)(2)]. The feeder and service loads 1,500 volt-ampere and the demand factors of Table 220-11 can be applied to this load.

Other Related Code rules.
- Laundry outlet required within 6 feet, [210-50(c)]
- Laundry outlet requirements, [210-52(f)]

220-17   Dwelling Unit Appliance Demand Factors

A demand factor of 75 percent can be applied to the total connected load of four or more appliances, other than electric space-heating equipment [220-15], electric clothes dryers [220-18], electric ranges [220-19], electric air-conditioning equipment [Article 440, Part D], or electric motors [220-14].

Question. Dwelling Unit. What is the feeder appliance demand load for a dwelling unit that has one 1,000 VA disposal, one 1,500 VA dishwasher, and one 4,500 VA water heater, and Fig. 7-9?

Answer.   7,000 VA, No demand

Question. Multifamily Dwelling. What is the feeder appliance demand load for a 12-unit multi-family dwelling (each unit 7,000 VA of appliances)?

Answer.   63,000 VA
Demand factor, 36 appliances is 75 percent
7,000 VA × 12 units × 0.75 = 63,000 VA

220-18   Dwelling Unit Electric Clothes Dryer Demand Factor

The load for electric clothes dryers located in a

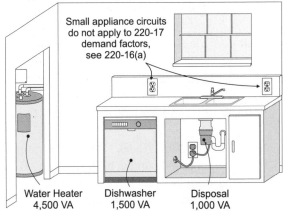

Appliance Demand Load For Service - *Section 220-17*

Small appliance circuits do not apply to 220-17 demand factors, see 220-16(a)

Water Heater 4,500 VA   Dishwasher 1,500 VA   Disposal 1,000 VA

No demand factor for 3 appliances. There must be 4 or more appliances to apply Section 220-17 demand factors.

Fig. 7-9 Dwelling - Appliance Demand Load for Service and Feeder Calculations

dwelling unit shall not be less than 5,000 volt-ampere, or the nameplate rating if greater than 5,000 volt-ampere.

Note. A dryer load is not required if the dwelling unit does not have a dryer outlet, or if it has a gas dryer.

Demand Factor. When a building contains 5 or more dryers, it shall be permissible to apply the demand factors listed in Table 220-18 to the total connected load of the dryers.

Neutral Size. The service and feeder neutral can be sized at 70 percent of the dryer demand load [220-22], as calculated according to Table 220-18.

Question. What is the demand load for a 12-unit multifamily building that contains a 5 kW dryer in each unit.

Answer   27,000 VA
12 units × 5,000 VA × 0.45 = 27,000 VA

220-19   Dwelling Unit Electric Range Demand Factor

Household cooking appliances individually rated over 1¾ kW can have the branch circuit, feeder, and service loads, calculated according to Table 220-19. See Examples No. D5(a) and D(6) in Appendix D of the *National Electrical Code.*

For Household Cooking Appliances Rated Over 1¾ kW Through 8¾ kW.

Step 1. Calculate the total connected load.

Step 2. Apply the column demand factor.
1.75 kW through 3.49 kW, use Column B.
3.50 kW through 8.75 kW, use Column C.

Note 3 of Table 220-19. Where the rating of cooking appliances fall under both columns B and C, apply the demand factor percentage for those units within their respective columns and then add that sum together.

Question. What is the demand load for ten 3 kW ovens and eight 6 kW cooktops?

Answer. 31.98 kW
Step 1. Column B: Ten 3 kW ovens
10 units × 3 kW = 30 kW × 0.49 = 14.70 kW

Step 2. Column C: Eight 6 kW cooktops
8 units × 6 kW = 48 kW × 0.36 = 17.28 kW

Step 3. 14.7 kW plus 17.28 kW = 31.98 kW

For Cooking Appliances over 8¾ kW through 12.49 kW.
The feeder and service demand load shall be calculated according to Table 220-19 Column A based on the number of appliances.

Question. What is the demand load for five 12.3 kW ranges?

Answer. 20 kW
Column A demand (5 ranges) = 20 kW

For Cooking Appliances of Equal Rating 12 kW through 27.49 kW.

Step 1. The feeder and service demand loads shall be calculated according to Table 220-19 Column A based on the number of appliances.

Step 2. The value from Column A shall be increased 5 percent for each kW, or major fraction (0.5 kW), that the average range exceeds 12 kW.

Question. What is the demand load for three ranges rated 16 kW.

Answer. 16.8 kW
Step 1. Column A demand (3 units) = 14 kW

Step 2. The average range (16 kW) exceeds 12 kW by 4 kW, increase Column A (14 kW) by 20%, 14 kW × 1.2 = 16.8 kW

For Cooking Appliances of Unequal Rating, 12.5 kW and Above.
The feeder and service demand loads shall be calculated by:

Step 1. Calculate the total connected load (using 12 kW for any range less than 12 kW).

Step 2. Calculate the average rating of the unequal ranges by dividing the total connected load by the total number of units.

Step 3. Calculate the feeder and service demand load according to Table 220-19 Column A based on the number of appliances.

Step 4. Increase Column A value 5 percent for each kW, or major fraction (0.5 kW), that the average range exceeds 12 kW.

Question. What is the demand load for 3 ranges rated 9 kW and 3 ranges rated 14 kW.

Answer. 22.05 kW
Step 1. Total connected load:
3 units × 12 kW   = 36 kW (minimum 12 kW)
3 units × 14 kW   = 42 kW
Total load   = 78 kW

Step 2. Average range rating
78 kW(total connected)/6 (units) = 13 kW

Step 3. Column A (6 units) = 21 kW

Step 4. The average exceeds 12 kW by 1 kW; therefore, increase Column A by 5 percent.
21 kW × 1.05 kW = 22.05 kW

Branch Circuit Sizing, 220-19, Note 4

To determine the load for sizing branch circuit and tap conductors be sure to apply the requirement of Note 4 to Table 220-19:

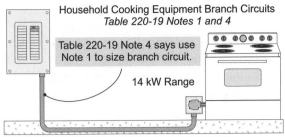

Household Cooking Equipment Branch Circuits
Table 220-19 Notes 1 and 4

Table 220-19 Note 4 says use Note 1 to size branch circuit.

14 kW Range

Example: Size of the branch circuit for a 14 kW Range.

Branch Circuit Conductor: Table 220-19 Note 4, use Note 1
Step 1: Column A, one unit = 8 kW demand
Step 2: Increase Column A answer (8 kW) by 10% because the average range (14 kW) exceeds 12 kW by 2 kW.
8 kW x 1.10 = 8.8 kW demand
I = P/E = 8,800 VA/240 volts = 37 amperes
Table 310-16, Use No. 8 conductor, see Section 210-19(c).

Fig. 7-10 Branch Circuit Calculation - Household Cooking Equipment

One Range. Calculate the branch circuit demand load, according to Table 220-19.

Question. What size branch circuit is required for a 14 kW range, Fig. 7-10?

Answer.   37 ampere
Step 1. Table 220-19 Column A, 8 kW

Step 2. Increase Column A (8 kW) 10%
8,000 VA × 1.10 = 8,800 VA
I = P/E = 8,800 VA/240 volt = 37 ampere

One Wall-Mounted Oven or Counter-Mounted Cooking Unit. The branch circuit load for one wall-mounted oven or one counter mounted cooking unit is based on the nameplate rating of the appliance.

Question. What size tap conductors are required for a 6 kW wall-mounted oven [210-19(c) Ex. 1]?

Answer.   No. 10
I = 6,000 VA/240 volt = 25 ampere, No. 10

One Counter-Mounted Cooking Unit and Up to Two Wall-Mounted Ovens. The branch circuit load for one counter-mounted cooking unit and up to two wall-mounted ovens is determined by combine the cooking unit nameplates together and treating this as a single range. Then calculate the

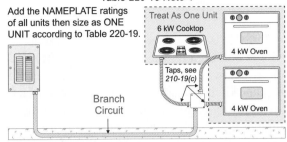

Branch Circuit For Two Ovens and One Cooktop
Table 220-19 Note 4

Add the NAMEPLATE ratings of all units then size as ONE UNIT according to Table 220-19.

Treat As One Unit
6 kW Cooktop
4 kW Oven
Taps, see 210-19(c)
4 kW Oven
Branch Circuit

Example: Branch circuit for a 6 kW cooktop and 2- 4 kW ovens.

Branch Circuit Conductor: Table 220-19 Note 4, use Note 1
Step 1: Total nameplate of all three units, 6 kW + 4 kW + 4 kW = 14 kW
Step 2: Treat as one 14 kW range. Column A for one unit is 8 kW. Increase Column A answer (8 kW) by 10% because the average range (14 kW) exceeds 12 kW by 2 kW, 8 kW x 1.10 = 8.8 kW demand
I = P/E = 8,800 VA/240 volts = 37 amperes
Table 310-16, Use No. 8 conductor, see Section 210-19(c).

Fig. 7-11 Branch Circuit Calculation - Two Ovens and One Cooktop

branch circuit demand load as one range in accordance with Table 220-19.

Question 1. What is the demand load for one counter-mounted cooking unit rated 6 kW, and one wall-mounted oven rated 3 kW on the same circuit.

Answer.   No. 8
Step 1. 6 kW + 4 kW = 9 kW

Step 2. Treat as one 9 kW range
Table 220-19, Column A = 8,000 watts
I = P/E = 8,000 watts/240 volt = 33 ampere

Question 2. What is the demand load for one 6 kW counter-mounted cooking unit and two 4 kW wall-mounted ovens on the same circuit, Fig. 7-11?

Answer.   No. 8
Step 1. 6 kW + 4 kW + 4 kW = 14 kW

Step 2. Treat as one 14 kW range.
Table 220-19 Column A for one unit is 8 kW, increase Column A (8 kW) 10% because the average range (14 kW) exceeds 12 kW by 2 kW.
8,000 VA × 1.10 = 8,800 VA,
I = P/E = 8,800 VA/240 volt = 37 ampere, No. 8

Note. See Sections 210-19(c), 220-4(c), and 422-32(b) for other rules as they relate to cooking appliances.

220-20   Commercial Kitchen Equipment Demand Factors

Table 220-20 is used to calculate the demand load for commercial electric cooking equipment, such as dishwasher booster heaters, water heaters, and other kitchen loads. The Table demand factors shall apply only to thermostat controlled electric equipment, or electric equipment that is used intermittently. Table 220-20 does not apply to space heating, ventilating, or air-conditioning equipment.

To determine the demand load, calculate the total connected nameplate load, then apply the demand factor from Table 220-20.

CAUTION: The feeder demand load shall not be less than the sum of the two largest kitchen equipment loads.

Question. What is the demand load for one 15 kW booster water heater, one 15 kW water heater, one 3 kW oven, and one 2 kW deep fryer in a commercial kitchen, Fig. 7-12?

Answer. 30 kW*
Step 1. Total connected load
15 kW* + 15 kW* + 3 kW + 2 kW = 35 kW

Step 2. Table 220-20 demand factor, 80%
35 kW × 0.8 = 28 kW
*Not less than the two largest appliances, or 30 kW.

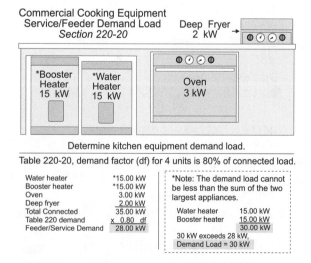

Commercial Cooking Equipment
Service/Feeder Demand Load
Section 220-20

Determine kitchen equipment demand load.

Table 220-20, demand factor (df) for 4 units is 80% of connected load.

Water heater	*15.00 kW	
Booster heater	*15.00 kW	
Oven	3.00 kW	
Deep fryer	2.00 kW	
Total Connected	35.00 kW	
Table 220 demand	x 0.80 df	
Feeder/Service Demand	28.00 kW	

*Note: The demand load cannot be less than the sum of the two largest appliances.

Water heater	15.00 kW
Booster heater	15.00 kW
	30.00 kW

30 kW exceeds 28 kW,
Demand Load = 30 kW

Fig. 7-12 Commercial - Cooking Equipment

220-21   Noncoincident Loads

Where it is unlikely that two or more noncoincident loads will be in use simultaneously, it shall be permissible to use only the largest load(s) that will be used at one time, in computing the total load of a feeder.

Note. The *Code* does not have a similar rule for branch circuits.

Question. What is the demand load for a 5 hp (230 volt, 28 ampere single-phase) air-conditioner [440-32] versus 7.5 kW heating[220-15]?

Answer.   8,050 VA
Step 1. Air-conditioner, 5 hp VA
230 volt × 28 ampere × 1.25 = 8,050 VA

Step 2. Heat
7,500 at 100% = 7,500, Omit

220-22   Feeder and Service Neutral Load

The feeder and service neutral load shall be the maximum unbalanced demand load between the neutral and any one ungrounded (hot) conductor as determined by Article 220, Part B (standard calculations). 240 volt loads are not connected to the **grounded (neutral) conductor**, therefore they are not considered when sizing the feeder or service neutral conductor.

WARNING:
The grounded (neutral) service conductor must be brought to each service disconnect and must be sized no less than required in Section 250-24(b).

Question. What size grounded (neutral) conductor is required for a feeder load of 200 ampere, that has a maximum unbalanced load is 100 ampere? System voltage – 208Y/120 volt, 3-phase, 4-wire, Fig. 7-13.

Answer.   No. 3
No. 3 has the ampacity to carry the 100 ampere unbalanced load, based on 75°C termination rating [110-14(c) and Table 310-16], but the service

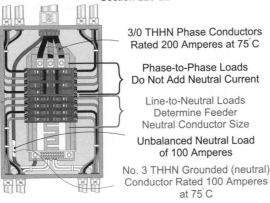

Feeder Neutral Size Based On Unbalanced Load
Section 220-22

3/0 THHN Phase Conductors
Rated 200 Amperes at 75°C

Phase-to-Phase Loads
Do Not Add Neutral Current

Line-to-Neutral Loads
Determine Feeder
Neutral Conductor Size

Unbalanced Neutral Load
of 100 Amperes

No. 3 THHN Grounded (neutral)
Conductor Rated 100 Amperes
at 75°C

Fig. 7-13 Feeder Neutral Size Based on the Maximum Unbalanced Load

grounded (neutral) conductor must be sized no smaller than No. 2 [250-24(b) and Table 250-66].

Dwelling Unit Range and Cooking Appliances Neutral Load.

The feeder and service neutral demand load for household electric ranges, wall-mounted ovens, or counter-mounted cooking units shall be calculated at 70 percent of the demand load as determined by Table 220-19.

Question. What is the neutral load for ten 12 kW ranges?

Answer.   17.5 kW
Step 1. Table 220-19 Column A
Ten units = 25 kW

Step 2. The neutral load is at 70%
25 kW × 0.7 = 17.5 kW

Dwelling Unit Dryer Neutral Load.

The feeder and service neutral demand load for electric dryers shall be calculated at 70 percent of the demand load as determined by Table 220-18.

Question. What is the neutral demand load for fifteen 5 kW dryers?

Answer. 21 kW
Step 1. Demand load
15 units × 5 kW × 0.4 = 30 kW

Step 2. Neutral demand load at 70%
30 kW × 0.7 = 21 kW

Over 200 Ampere Reduction.

For 3-wire *single-phase* (120/240 or 240/480) or 4-wire, 3-phase (208Y/120 or 480Y/277) systems, the feeder and service neutral demand can be reduced an additional 70 percent for that portion of the unbalanced load over 200 ampere.

Question. What is the neutral demand load for a feeder that totals 600 ampere of which 100 ampere is rated 240 volt, 100 ampere is ranges rated 120/240 volt, 50 ampere is dryers rated 120/240 volt, and the remaining 350 ampere is other 120 volt loads.

Answer.   379 ampere
Step 1. Total Load

100 ampere at 240 Volt	=	0 ampere
100 ampere, Ranges × 0.7	=	70 ampere
50 ampere, Dryers × 0.7	=	35 ampere
350 ampere, Other 120 v	=	350 ampere
Feeder Neutral Load	=	455 ampere

Step 2. Neutral Demand

First 200 ampere at 100%	=	200 ampere
Remainder 255 ampere × 0.70	=	179 ampere
Total neutral demand load		379 ampere

Harmonic (Nonlinear) Loads.

The feeder and service neutral demand load cannot be reduced for **nonlinear loads** supplied from a 4-wire, wye-connected, 3-phase system.

> FPN No. 2: 120 and 277 volt nonlinear loads can generate *harmonic currents* that add on the neutral conductor (instead of canceling), which can cause excessive current on the neutral conductor. To prevent overheating of the neutral conductor, the neutral can be sized 2 times larger than the calculated unbalanced load (not a *Code* rule).

No Reduction Permitted for Unbalanced Wye Systems.

The feeder and service neutral demand load shall not be permitted to be reduced for 3-wire circuits consisting of two phase wires and a neutral supplied from a 4-wire, *wye-connected*, 3-phase system, Fig. 7-14.

Note. This is because the neutral of a 3-wire circuit connected to a 4-wire, 3-phase system carries approximately the

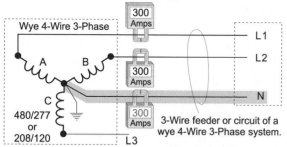

Grounded Conductor Sizing 200 Amp Reduction Not Permitted
Section 220-22

The 70% reduction of the unbalanced load over 200 amperes does *NOT* apply to *wye 3-wire 1-phase circuits* of a *wye 4-wire 3-phase system* because the neutral carries about the same current as the phase conductors.

Proof: Determine neutral current of 300 Ampere 3-Wire Feeder.

$$I \text{ of Neutral } = \sqrt{(A^2 + B^2) - (AB)}$$
$$= \sqrt{(300^2 + 300^2) - (300 \times 300)}$$
$$= \sqrt{(90,000 + 90,000) - (90,000)}$$
$$= \sqrt{180,000 - 90,000}$$
$$= \sqrt{90,000}$$
$$I \text{ of Neutral } = \ \ 300 \text{ Amperes}$$

Fig. 7-14 Grounded (Neutral) Conductor Sizing

same current as the phase conductors [310-15(b)(4)(c)].

PART C. OPTIONAL CALCULATIONS FOR COMPUTING FEEDER AND SERVICE LOADS

220-30 Optional Calculation-Dwelling Unit

(a) Feeder and Service Load. For a dwelling unit having an ampacity of 100 ampere or more, it is permissible to compute the feeder and service loads in accordance with the optional method listed in Table 220-30, instead of the standard method specified in Part B of Article 220.

(b) The calculated demand load shall not be less than 100 percent of the first 10 kVA plus 40 percent of the remainder of the following loads.

(1) Small Appliance and Laundry Branch Circuits. 1,500 volt-ampere for each 20 ampere small appliance and laundry branch circuit. Since we are required to have two small appliance circuits and a laundry circuit, the minimum will be 4,500 VA.

(2) General Lighting. 3 volt-ampere per square foot for general lighting and general use receptacles.

(3) Appliance Nameplate. The nameplate rating of all appliances that are fastened in place, permanently connected, or located to be on a specific circuit.

(4) Motor Nameplate VA. The nameplate VA rating of all motors. (simply multiply motor volt by the motor ampere).

(c) Heating and Air Conditioning. Include the largest of the following:

(1) 100 percent of the nameplate rating of the air-conditioning.

(2) 100 percent of the nameplate rating of the heat pump and supplemental heating.

(3) 100 percent of the nameplate rating of thermal storage heating. Thermal storage heating is the process of continuously heating bricks or water at night when the electric rates are lower. Then during the day, the building uses the thermally stored heat.

(4) 65 percent of nameplate rating of central space heating.

(5) 65 percent of nameplate rating of three or fewer separately controlled space heating units.

(6) 40 percent of nameplate rating of four or more separately controlled space heating units.

Question. What size service is required for a 1,500 square foot dwelling unit that contains the following loads:

System voltage is 115/230 single-phase.

Dishwasher	= 1,200 VA
Water heater	= 4,500 VA
Disposal	= 900 VA
Dryer	= 4,000 VA
Cooktop	= 6,000 VA
Ovens (two) each	= 3,000 VA
A/C	= 5 hp
Heat	= 7.5 kW

*Answer. **109 ampere***

Step 1. Determine total connected load:

(1) Small appliance $1,500 \times 2$	= 3,000 VA
Laundry	= 1,500 VA

(2) General lighting $1,500 \times 3$	= 4,500 VA

(3) Dishwasher	= 1,200 VA
Water heater	= 4,500 VA
Disposal	= 900 VA
Dryer	= 4,000 VA
Cooktop	= 6,000 VA
Ovens $3,000 \times 2$	= 6,000 VA
	= 31,600 VA

Step 2. Apply demand factor:

First 10,000 at 100%	=10,000 VA
Remainder at 40% ($21,600 \times 0.4$)	= 8,640 VA
	=18,640 VA

Step 3. Largest of A/C vs. heat:

A/C: 5 hp = 28 ampere $\times$ 230 =
6,440 VA at 100%	= 6,440 VA
Heat: 7,500 at 65%	= 4,875 VA

Step 4. Add 2 and 3:

18,640 VA + 6,440 VA	=25,080 VA

I = P/E, I = 25,080 VA/230 volt = 109 ampere

Section 215-2(d) permits feeder and service conductors for dwelling units to be sized according to the values listed in to Table 310-15(b)(6), No. 3.

SUMMARY

Part A.   General   ☐ When a calculation results in 0.5 or larger round up to the next ampere. ☐ Table 220-3(a) specifies the general lighting load per square foot of floor area. ☐ The floor area shall be computed from the outside dimensions of the building or area. ☐ In dwelling units the floor area does not include open porches, garages, or unused or unfinished spaces, not adaptable for future use, but it does include basement spaces. ☐ Recessed inductive lighting fixture loads shall be based on the VA rating of the fixture, not the wattage of the lamps.

Part B.   Feeders and Service Calculations   ☐ The demand factor for general lighting load is listed in Table 220-11. ☐ Demand factors for nondwelling receptacle loads are given in Table 220-13. ☐ Fixed electric space heating loads shall be computed at 100 percent of the connected load. ☐ Dwelling unit appliance demand load for one to three units is 100 percent, and for four or more units the demand factor is 75 percent. ☐ Feeder and service demand factors for household electric clothes dryers are listed in Table 220-18. ☐ Demand loads for household electric ranges, wall-mounted ovens, counter-mounted cooking units, etc. are listed in Table 220-19. ☐ To calculate the demand load for commercial electric cooking equipment, dishwasher booster heaters, water heaters, and other kitchen equipment, see Table 220-20. ☐ When sizing feeders and services, you may omit the smaller of two noncoincidental loads. ☐ The feeder and service neutral load shall be the maximum unbalanced demand load between the neutral and any one ungrounded conductor. ☐ The neutral can be reduced to 70 percent for household ranges, dryers and ovens. ☐ If the total neutral load exceeds 200 ampere, the neutral can be further reduced in some cases.

Part C.   Optional Calculations   ☐ Optional calculations for dwelling units are covered in Section 220-30 and Table 220-30.

REVIEW QUESTIONS

1.   Unless other voltages are specified, when computing branch circuit, feeder, and service calculations, a nominal system voltage of _____ shall be used.

2. For commercial lighting other than general illumination, receptacle outlets shall be rated as _____ VA.

3. Explain the rules that apply to dwelling branch circuits in Article 220.

4. In your own words explain the meaning of the following Article 100 definitions:

   Dwelling unit
   Feeder
   Service
   Demand factor
   Garage
   Nominal voltage
   Nonlinear load

5. In your own words explain the meaning of the following Glossary terms:

   kVA
   Lamp
   Reactance
   Three-phase

Unit 8

Article 225
Outside Branch Circuits and Feeders

OBJECTIVES

After studying this unit, the student should be able to understand:
- disconnect requirements for second buildings.
- grounding conductor requirements for a second building if the feed to the second building is run overhead or underground in a rigid nonmetallic conduit.
- the requirements for the overhead conductor, such as weatherheads and drip loops, point of attachment, and clearances above roofs.
- the requirements for underground rigid nonmetallic conduits, such as minimum burial depths, backfill, and raceway seals.
- grounding electrode requirements at the second building.

Definitions and Glossary Terms

To better understand the *NEC* rules contained in this unit, review the following:

Definitions Article 100 - Unit 2

Approved	Building
Disconnect	Feeder
Readily accessible	Service
Wet location	Within sight

Glossary Terms
Drip loop
Flex
Manually

PART A. GENERAL REQUIREMENTS

225-1 SCOPE

Article 225 covers installation requirements for equipment, including conductors located outside on or between **buildings,** poles, and other structures on the premises, Fig. 8-1.

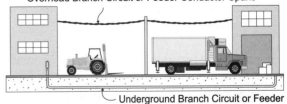

Outside Branch Circuits And Feeders
Article 225 Scope
Overhead Branch Circuit or Feeder Conductor Spans

Underground Branch Circuit or Feeder

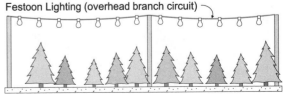

Festoon Lighting (overhead branch circuit)

Article 225 applies to outside branch circuits and feeders run on or between buildings, structures, or poles, and electric equipment and wiring for utilization equipment located outside.

Fig. 8-1 Outside Branch Circuits and Feeders

225-2   OTHER ARTICLES

Other important Articles that apply to outside branch circuits and feeders include:

Article 200	Grounded (neutral) Conductor
Article 210	Branch circuits
Article 215	Feeders
Article 220	Load Calculations
Article 230	Overhead Conductors
Article 240	Conductor Overcurrent Protection
Article 250	Grounding
Article 300	Wiring Methods
Article 305	Temporary Wiring
Article 310	Conductors for General Wiring
Article 320	Open Wiring on Insulators
Article 321	Messenger Supported Wiring
Article 410	Lighting Fixture

225-4   CONDUCTOR COVERING

Conductors installed in **wet locations** shall be TW, THW, THWN, XHHW, or other types listed for wet locations. Section 310-8 contains similar requirements for conductors installed in wet locations. Table 310-13 lists which conductors can be installed in a wet location.

225-6   MINIMUM SIZE CONDUCTORS

(a) Overhead Conductor Spans. The minimum size individual overhead conductor is dependent on the length of the conductor span. No. 10 copper or No. 8 aluminum is required for spans up to 50 feet, and No. 8 Copper or No. 6 aluminum for spans over 50 feet.

Exception: Messenger Wire. Overhead conductors supported by messenger wire, when installed according to Article 321, can be smaller.

(b) Festoon Lighting. The minimum size conductor for festoon lighting shall not be smaller than No. 12 copper.

Definition of Festoon Lighting. Festoon lighting is a string of outdoor lights suspended between two points. This type of wiring is commonly installed at used car lots and for temporary lighting. See Section 225-24 for installation requirement of outdoor lampholders.

225-7   LIGHTING EQUIPMENT INSTALLED OUTDOORS

(c) 277 Volt to Ground. Lighting fixtures rated 277 volt to ground are permitted for outdoor areas of commercial and industrial establishments. Lighting fixtures rated and connected at 277 volt shall not be installed within 3 feet of windows that open, platforms, fire escapes, and so on, Fig. 8-2.

See Section 210-6 for the types of lighting fixtures permitted on branch circuits rated more than 120 volt nominal.

225-10   WIRING ON BUILDINGS

Wiring methods permitted for outside wiring are covered in the following *Code* Articles:

Article 318	Cable Tray
Article 320	Open Wiring on Insulators
Article 321	Messenger Supported Wiring
Article 330	Mineral Insulated Cable
Article 334	Metal Clad Cable
Article 345	Intermediate Metal Conduit
Article 346	Galvanized Rigid Metal Conduit
Article 347	Rigid Nonmetallic Conduit
Article 348	Electrical Metal Tubing
Article 350	Flexible Metal Conduit
Article 351	Liquidtight Flexible Conduit
Article 362	Wireways
Article 364	Busways
Article 365	Cablebus
Article 374	Auxiliary Gutters

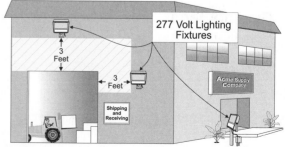

277 Volt Outdoor Lighting Equipment
Section 225-7(c)

277 Volt Lighting Fixtures

3 Feet

3 Feet

Acme Supply Company

Shipping and Receiving

Lighting fixtures supplied by voltages over 120 volts but not more than 277 volts to ground must be kept 3 feet from doors, windows, platforms, and similar locations.

Fig. 8-2 Outdoor 277 Volt Lighting Limitations

225-16 POINT OF ATTACHMENT

The point of attachment for overhead conductors shall not be less than 10 feet above the finish grade. The conductors, including the *drip loop* must be installed to meet the clearance requirements of Section 225-18.

> CAUTION: Overhead conductors might need to have the point of attachment raised to an acceptable height so that the conductors' final *sag* complies with the clearances listed in Section 225-18.

225-18 CLEARANCES

Overhead conductor spans not over 600 volt, nominal, shall conform to the clearance requirement as listed in the following summary table.

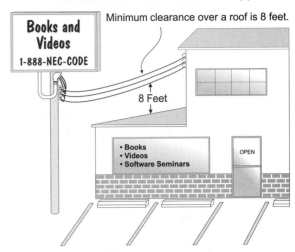

Clearance Above Roofs - *Section 225-19(a)*

Minimum clearance over a roof is 8 feet.

Books and Videos 1-888-NEC-CODE

8 Feet

• Books
• Videos
• Software Seminars

OPEN

Fig. 8-3 Overhead Conductor Roof Clearance

Table 225-18 Overhead Conductor Clearances Location	From Final Grade or Other Accessible Surfaces
0-150 Volts to Ground: Conductors at entrance equipment, drip loops, and over areas or sidewalks accessible only to pedestrians.	10 Feet
151-300 Volts to Ground: Conductors over residential property and driveways, and over commercial areas not subject to truck traffic.	12 Feet
301-600 Volts to Ground: Conductors over residential property and driveways, and over commercial areas not subject to truck traffic.	15 Feet
Truck Traffic (any voltage): Conductors over public streets, alleys, roads, parking areas subject to truck traffic, commercial driveways, and other areas traveled by large vehicles, such as forests or orchards.	18 Feet

225-19 CLEARANCES FROM BUILDING

(a) Above Roofs. Overhead conductors passing over a roof require a minimum clearance of 8 feet above the surface of the roof. This clearance is required for a minimum distance of 3 feet in all directions from the edge of the roof, Fig. 8-3.

Exception No. 1: Parking Garage Roofs. Where pedestrians or vehicles are normally on the roof, such as a parking garage, overhead conductors must have a clearance according to Section 225-18.

Exception No. 2: Steeply Sloped Roofs. Where the voltage does not exceed 300 volt between conductors, overhead conductor clearances from the roof can be reduced from 8 feet to 3 feet, if the slope of the roof exceeds 4 inches in 12 inches.

Note. The danger of persons contacting overhead conductors is lessened when there is reduced voltage and the roofs have a slope or angle that makes them difficult to walk upon.

Exception No. 3: Overhang Portion Only. If the voltage between conductors does not exceed 300 volt, the conductor clearance over the roof overhang can be reduced from 8 feet to 18 inches This is only permitted if no more than 6 feet of overhead conductors pass over no more than 4 feet of roof overhang, and the conductors terminate at a through-the-roof raceway or approved support.

Exception No. 4: Point of Attachment. The 3 foot vertical clearance that extends from the roof shall not apply when the point of attachment is on the side of the building below the roof.

(b) From Non-Building or Non-Bridge Structures. Overhead conductors not over 600 volt, nominal, shall maintain vertical, diagonal, and horizontal clearance of not less than 3 feet from signs, chimneys, radio and television antennas, tanks, and other nonbuilding or nonbridge structures.

(d) Final Spans and Clearance from Building Openings. Overhead conductor to a building shall maintain a clearance of 3 feet from windows that are designed to be opened, doors, porches, balconies, ladders, stairs, fire escapes, or similar locations, Fig. 8-4.

In addition overhead conductors shall maintain a vertical clearance of 10 feet above platforms, projections or surfaces from which they might be reached. This vertical clearance shall extend 3 feet measured horizontally from the platform, projections or surfaces from which they might be reached.

Exception: Conductors Above Windows. Overhead conductors run above a window are not required to maintain the 3 foot distance.

Below Opening. Conductors cannot be installed under an opening through which materials might pass. Overhead conductors shall not be run or installed where they will obstruct entrance to these building openings. For example, the upper opening in a barn loft is often used to move hay in, or out, of the loft storage area.

Note. Overhead conductors are permitted above pools, diving structures, observation stands, towers, or platforms. See Section 680-8 for complete details.

225-20 MECHANICAL PROTECTION

Outside wiring conductors that are subject to physical damage, or subject to contact with objects, shall be protected according to Section 230-50. This Section requires conductors to be protected by rigid metal conduit, intermediate metal conduit, rigid

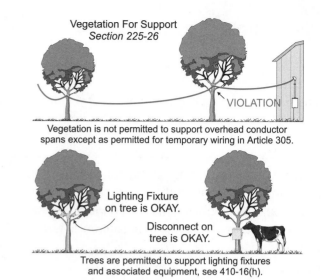

Vegetation is not permitted to support overhead conductor spans except as permitted for temporary wiring in Article 305.

Trees are permitted to support lighting fixtures and associated equipment, see 410-16(h).

Fig. 8-5 Vegetation Not Permitted to Support Overhead Conductors

nonmetallic conduit, electrical metallic tubing, or other **approved** means [110-27].

225-22 RACEWAYS RAINTIGHT AND ARRANGED TO DRAIN

Raceways installed outdoors shall be **raintight** and arranged to drain. Check with your inspector for an interpretation of arranged to drain. Some inspectors permit or even require that the raceway have a small hole drilled at the lowest point.

Exception. Flexible metal conduit can be installed outdoors [225-10 and 350-5(1)] and it is not required to be raintight. But the *flex* must be installed so that water will not enter the enclosure.

Note. Many electricians, contractors, and engineers, as well as inspectors are surprised to learn that flexible metal conduit can be installed outdoors in a wet location.

225-26 TREES FOR CONDUCTOR SUPPORT

Trees or other vegetation cannot be used for the support of overhead conductor spans. However, vegetation can be used to support wiring and electrical equipment as well as lighting fixtures [410-16(h)], Fig. 8-5.

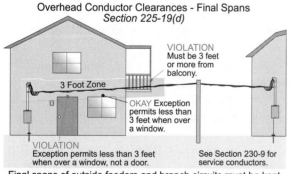

Final spans of outside feeders and branch circuits must be kept a minimum of 3 feet from windows that open, doors, porches, balconies, ladders, stairs, fire escapes, or similar locations.

Fig. 8-4 Overhead Conductor Clearance from Accessible Locations

Exception: Temporary Wiring. Vegetation can be used for the support of conductors for temporary wiring, according to Article 305.

PART B. MORE THAN ONE BUILDING OR STRUCTURE

225-30 Number of Supplies

A building or structure can be served by only one supply, except as permitted in the following:

(a) Special Conditions.
(1) Fire Pumps
(2) Emergency Power
(3) Legally required stand-by power
(4) Optional stand-by power
(5) Parallel power production systems.

(b) Special Occupancy. An additional supply is permitted by **special permission**, for:

(1) Multiple-Occupancy Building. Where there is no space available for the disconnecting means accessible to all the occupants.

(2) Large Building. A single building or other sufficiently large structure that requires two or more supplies.

(c) Capacity Requirements. Two or more supplies shall be permitted where the calculated load according to Article 220 exceeds 2,000 ampere for system supply voltage of 600 volt or less.

(d) Different Characteristics. An additional supply is permitted for a building having different voltages, frequencies, number of phases, or for different uses.

225-31 Disconnecting Means

A disconnect must be provided to disconnect all conductors that enter or pass through a building or structure. The construction of the feeder **disconnecting means** is contained in Section 225-38. A *pushbutton* that activates the electromagnetic *coil* of a *shunt-trip circuit-breaker* is not a disconnect.

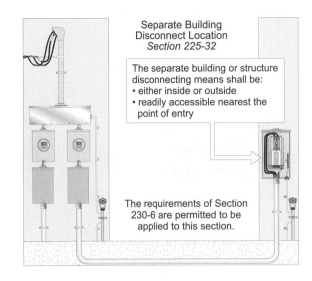

Fig. 8-6 Separate Building Disconnect Location

225-32 Disconnect Location

The **feeder** disconnecting means for a building or structure shall be installed either inside or outside of the building or structure served, or where the conductors pass through the building or structure. The disconnecting means shall be at a **readily accessible** location "nearest" the point of entrance of the conductors, Fig. 8-6.

The intent of this rule is to insure that feeders do not enter or pass through a building without having the ability to disconnect them. Naturally the conductors have the option of passing under or around the building without the required disconnect.

Note. If the disconnect is located outdoors, the *NEC* does not require it to be located on the building or structure. How far away can the disconnect be located from the building? This is up to the electrical inspector.

Feeder conductors are considered outside of a building or other structure where encased or installed under not less than 2 inches of concrete or brick, when installed in accordance with the requirements of Section 230-6, Fig. 8-7.

Exception No. 1: Safe Switching Practices. Disconnecting means can be located elsewhere on the premises where documented safe switching

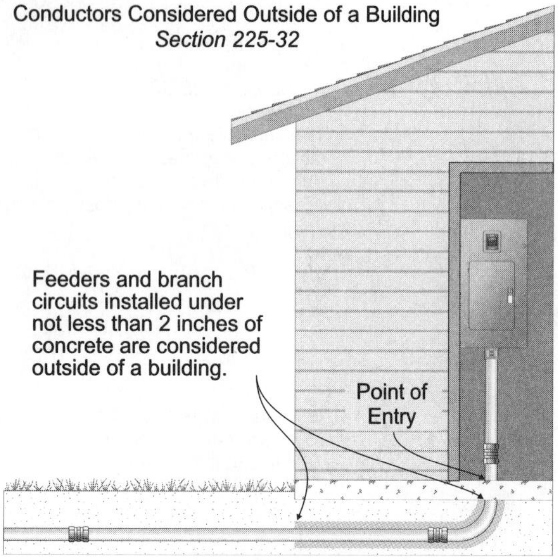

Conductors Considered Outside of a Building
Section 225-32

Feeders and branch circuits installed under not less than 2 inches of concrete are considered outside of a building.

Point of Entry

Fig. 8-7 Conductors are Considered Outside of a Building When Under 2 inches of Concrete

practices are maintained such as at universities and other types of similar occupancies.

Exception No. 3: Poles for Lighting Standards. Disconnecting means for poles used as lighting standards may be located remote from the pole, Fig. 8-8.

Exception No. 4: Signs. The disconnecting means for a sign structure is not required to be readily accessible, because it must comply with the disconnect requirements for signs located in Section 600-6.

Note. Section 600-6 requires each sign to be controlled by an externally operable switch or circuit breaker that opens all ungrounded conductors of the sign. The sign disconnecting means shall be **within sight** of the sign, or the disconnecting means shall be capable of being locked in the open position.

225-33 Maximum Number of Disconnects

(a) Maximum of Six Per Building.   There shall be no more than six disconnects consisting of up to six switches or six circuit breakers mounted in a single enclosure, or separate enclosures for each feeder as permitted in Section 225-30.

225-34 Grouping of Disconnects

(a) Two to Six Disconnects.   The building or structure disconnection means shall be grouped. Each disconnect shall be marked to indicate the load served.

(b) Fire Pump, Emergency, and Stand-By Power.   The disconnect for fire pumps, or stand-by power as permitted in Section 225-30(a), shall be located remotely away from the two to six disconnects for normal power. The reason these disconnects are located away from the normal power disconnect(s) is to minimize the accidental interruption of the emergency power supply.

225-35 Access to Occupants

In a multiple occupancy building, each occupant shall have access to his or her disconnecting means.

Exception. In multiple occupancy buildings where electrical maintenance is provided by continuous building management, the disconnecting means can be accessible only to building management personnel.

225-36 Identified As Suitable for Service Equipment

The disconnecting means must be marked to identify it as suitable for use as service equipment. This means that the disconnect is supplied with a

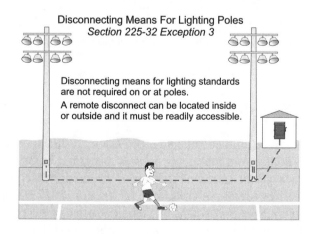

Disconnecting Means For Lighting Poles
Section 225-32 Exception 3

Disconnecting means for lighting standards are not required on or at poles.
A remote disconnect can be located inside or outside and it must be readily accessible.

Fig. 8-8 Disconnecting Means for Lighting Poles

main bonding jumper so that a neutral-to-ground connection can be made as permitted in Section 250-32(b)(2), Fig. 8-9.

225-37 Identification

Where a building or structure is supplied by more than one **service**, or a combination of branch circuits, feeders, and services, a permanent plaque or directory shall be installed at each service, feeder, or branch circuit disconnect location denoting all other services, feeders, and branch circuits supplying that building or structure and the area served by each.

Note. When additional feeders are installed at a building or structure, they are not required to be grouped together, see 225-33(a).

225-38 Manual or Power-Operated Circuit Breakers

The disconnecting means can consist of either a *manually*-operable switch or circuit breaker, or a power-operated switch or circuit breaker. If power-operated, the switch or circuit breaker must also be capable of being operated by hand.

225-39 Rating of Disconnect

The disconnecting means shall have an ampere rating not less than the computed load according to Article 220. In no case shall the rating be lower than specified in (a), (b), (c), or (d) below.

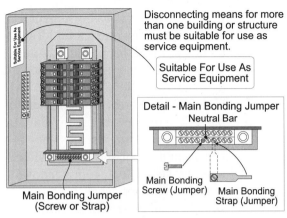

Fig. 8-9 Disconnecting for Separate Building Must Be Rated "Suitable for Service Equipment"

(a) One-Circuit Installation. For installations that supply loads limited to a single branch circuit, the disconnecting means shall have a rating of not less than 15 ampere.

(b) Two-Circuit Installations. For installations consisting of not more than two, 2-wire branch circuits, the disconnecting means shall have a rating of not less than 30 ampere.

(c) One-Family Dwelling. For a one-family dwelling, the disconnecting means shall have a rating of not less than 100 amperes, 3-wire.

(d) All Others. For all other installations, the disconnecting means shall have a rating of not less than 60 ampere.

SUMMARY

Part A.   General Requirements   ☐ Article 225 covers the installation requirements for equipment, including conductors, located outside buildings, between buildings, poles, and structures. ☐ In wet locations, conductors must be TW, THW, THWN, XHHW, or other types listed for wet locations.   ☐ Overhead conductor spans up to 50 feet require No. 10 CU or No. 8 AL; spans over 50 feet require No. 8 CU or No. 6 AL.   ☐ Festoon lighting minimum size conductor is No. 12 CU. ☐ Lighting fixtures are permitted for outdoor areas of commercial and industrial establishments, but 277 and 480 volt lighting fixtures must not be installed closer than 3 feet from window openings, platforms, fire escapes, etc.   ☐ Conductors must have a vertical clearance of not less than 8 feet from the roof surface and maintained in all directions for a distance of 3 feet.   ☐ Conductors must be kept 3 feet from windows that open, doors, fire escapes, platforms, etc.   ☐ Raceways installed outdoors shall be raintight and arranged to drain.   ☐ Underground circuits shall be installed and protected, according to the requirements of Section 300-5.   ☐ Vegetation cannot be used for the support of overhead conductor spans.

REVIEW QUESTIONS

Part A.   General Requirements

1. The point of attachment for overhead conductors shall not be less than _____ feet above the finish grade.

2. Overhead conductors shall maintain vertical, diagonal, and horizontal clearance of not less than _____ feet from signs, chimneys, radio and television antennas, or other structures.

3. Conductors cannot be installed _____ an opening through which materials might pass.

4. Can overhead conductors be above pools?

5. When can a tree be used to support overhead conductor spans?

Part B. More than One Building or Structure

6. Building or structure can be served by _____ supply.

7. Is a disconnecting means required to disconnect all conductors that enter or pass through a building or structure?

8. What are the rules for building feeder disconnect?

9. How many disconnects are permitted for each feeder?

10. Does the building/structure disconnection means have to be grouped and marked to indicate the load served?

11. In your own words explain the meaning of the following Article 100 definitions:

   Approved
   Building
   Disconnect
   Feeder
   Readily accessible
   Service
   Wet location
   Within sight

12. In your own words explain the meaning of the following Glossary terms:

   Drip loop
   Flex
   Manually

Unit 9

Article 230

Services

OBJECTIVES

After studying this unit, the student should be able to understand:

• when additional services can be added without receiving special permission from the inspector.

• the rules that apply to drip loop conductors.

• the rules that apply to service cables, as covered in this book.

• the rules that apply to the high-leg conductor in meter equipment, disconnects, panelboards, and branch circuits.

• service disconnect requirements such as location, number, grouping, and access by occupants.

Definitions and Glossary Terms

To better understand the *NEC* rules contained in this unit, review the following:

Definitions Article 100 - Unit 2

Ampacity	Bathroom
Bonding	Bonding jumper
Building	Continuous load
Disconnect	Raceway
Readily accessible	Service drop
Service entrance	Service lateral
Service point	Weatherproof

Glossary Terms

Bus	Drip loop
Guy wires	High-leg
Load side	Weather head

Introduction

Over the years, the electrical industry has developed "rules of thumb" which do not accurately reflect actual code requirements, such as "no more than six mains per building." This is not true; the *NEC* does at times permit more than six mains per building, see Section 230-72. Likewise, many people believe that "line and load conductors cannot be in the same raceway." This is not true; line and load conductors can be installed in the same raceway, see Section 230-7.

These two examples illustrate how careful we must be to understand what current *Code* rules actually say. For example, try to keep an open mind when reviewing Article 230. It will be a big test of your ability to overcome your natural tendency to believe what you have always thought.

Other *NEC* Articles that are important for understanding services include: Article 220 – Service Calculations, sizing the grounded and ungrounded service conductors, and Article 240 – Overcurrent Protection, protecting conductors against overcurrent.

Services have special grounding and **bonding** requirements, which are specified in Article 250, in particular Sections 250-92 and 250-94. The general rules for the installation of conductors and wiring methods covered in Articles 110, 300, and 310 are also very important.

GENERAL STRUCTURE OF ARTICLE 230 **SERVICES**

Because of its size, Article 230 is divided into eight parts.

Part A.	General
Part B.	Overhead Service – Drop
Part C.	Underground Service – Lateral
Part D.	Service-Entrance Conductors
Part E.	Service Equipment – General
Part F.	Service Disconnect
Part G.	Overcurrent Protection

PART A. **GENERAL**

230-1 **Scope**

Article 230 covers the installation requirements for service conductors and equipment. It is very important to know where the service begins and where the service ends when applying the rule of Articles 230 and 250.

According to Article 100 definitions, the service would include the conductors and equipment for delivering electric energy from the serving utility to the wiring system of the premises served. What this means is that the service begins at the **service point** (the connection of the premises wiring to the utility power conductors) and ends at the service disconnecting means, Fig. 9-1.

Conductors and equipment supplied from a battery uninterruptible power supply system, a solar voltaic system, generator, transformer, or phase converters are not considered service conductors, they are feeder conductors.

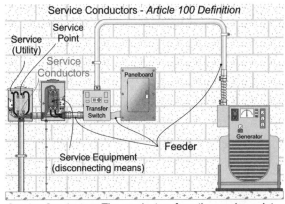

Service Conductors: The conductors from the service point to the service disconnecting means. *Feeder:* All circuit conductors between the service equipment, the source of a separately derived system, or other power supply source and the final branch circuit overcurrent device.

Fig. 9-1 Service Conductors Terminate at the Service Disconnect

230-2 **Number of Services**

A building or structure can be served by only one service (service-drop or service-lateral), except as permitted in (a) through (d) that follows.

(a) Special Conditions.
(1) Fire Pumps
(2) Emergency Power
(3) Legally required stand-by power
(4) Optional stand-by power
(5) Parallel power production systems

Note. To avoid a disruption of fire pump equipment, emergency, legally required stand-by, optional stand-by, or parallel power production systems, Section 230-72(b) requires that the disconnect for these systems be remote from the normal service disconnects.

CAUTION: A single building with a 2-hour fire wall between two (or more) occupancies is considered by the building code as separate buildings. Therefore a separate service is permitted for each building, Fig. 9-2.

(b) Special Occupancy. An additional supply is permitted by special permission, for:

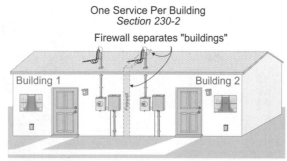

Fig. 9-2 Only One Service Is Permitted for Each Building

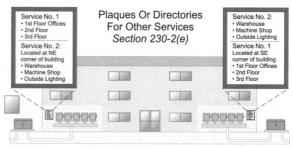

Fig. 9-3 Multiple Services Require Plaques or Directories

(1) Multiple-Occupancy Building. An additional service is permitted in multiple-occupancy building where there is no space available for the disconnecting means accessible to all the occupants.

(2) Large Building. Where a single building or other large structure requires two or more supplies. This would include a building such as automobile manufacturing plants of great size in which it would be impractical to have one service.

(c) Capacity Requirements. Two or more supplies shall be permitted where:

(1) Over 2,000 Amperes. Where the calculated load (according to Article 220) exceeds 2,000 ampere for system supply voltage of 600 volt or less. Since the type of occupancy is not specified, it can be residential, commercial, industrial, single occupancy, or multiple occupancy. This rule permits a separate service for demand loads over 2,000 ampere at 600 volt or less without special permission from the electrical inspector.

(2) Exceed Utility Capacity. An additional service is permitted where the load requirements of a single-phase installation exceed the utilities normal capacity.

(3) Special Permission. If the conditions of (c)(1) or (c)(2) cannot be met, then special written permission is required.

(d) Different Characteristics. An additional service is permitted for different characteristics, voltages, frequencies, number of phases, or for different uses, such as for different electricity rate schedules.

(e) Identification of Multiple Services. Where a building or structure is supplied by more than one service, or a combination of branch circuits, feeders, and services, a permanent plaque or directory shall be installed at each service, feeder, or branch circuit disconnect location denoting all other services, feeders, and branch circuits supplying that building or structure and the area served by each, Fig. 9-3.

> **Note.** When additional services (**service-drops** or **service-laterals**) are installed at a building or structure, they are not required to be grouped together.

230-3 One Building or Other Structure Not to Be Supplied through Another

Service conductors cannot pass inside a **building** to serve another building or other structure.

> **Note.** Service conductors passing through or under the building with 2 inches of concrete cover are considered outside the building, according to Section 230-6.

230-6 Conductors Considered Outside a Building

Conductors in a raceway are considered outside

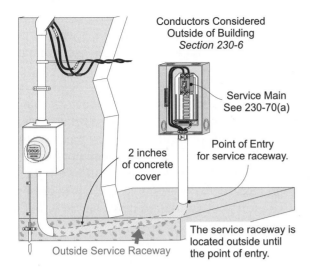

Conductors Considered
Outside of Building
Section 230-6

Service Main
See 230-70(a)

Point of Entry
for service raceway.

2 inches
of concrete
cover

The service raceway is
located outside until
the point of entry.

Outside Service Raceway

Fig. 9-4 Service Conductors Considered Outside of
Building If Covered with 2 Inches of Concrete

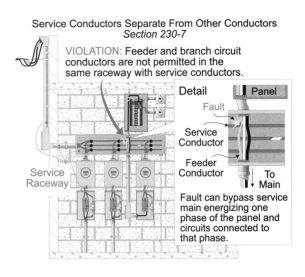

Service Conductors Separate From Other Conductors
Section 230-7

VIOLATION: Feeder and branch circuit
conductors are not permitted in the
same raceway with service conductors.

Detail   Panel

Fault

Service
Conductor

Feeder
Conductor

To
Main

Service
Raceway

Fault can bypass service
main energizing one
phase of the panel and
circuits connected to
that phase.

Fig. 9-5 Feeder and Branch Not Permitted in
Same Raceway with Service Conductors

a building when they are under the building, encased within 2 inches or more of concrete or brick, or installed in a transformer vault [Article 450, Part C], Fig. 9-4.

> **WARNING:**
> Overcurrent required by Section 230-90 could be bypassed if we mix service conductors with other conductors. Under this condition one phase would always be energized, even if the service disconnect is open, placing the electrical worker in danger, Fig. 9-5.

230-7 Service Conductors Separate from Other Conductors

Conductors other than service conductors shall not be installed in the same **raceway** or cable. Feeder and branch circuit conductors cannot be installed in the same raceway with service conductors, but the rule does not limit the mixing of service conductors with branch circuit and feeder conductors in the same enclosure.

Exception No. 1: Grounding and Bonding Conductors. Grounding conductors and **bonding jumpers** are permitted in the same raceway with service conductors.

Note. This requirement may be the root of the misconception that line and load conductors are not permitted in the same raceway. It is true that service conductors are not permitted in the same raceway with feeder or branch circuit conductors. However, line and load conductors of feeders and branch circuits can all be in the same raceway, cable, or enclosure, Fig. 9-6.

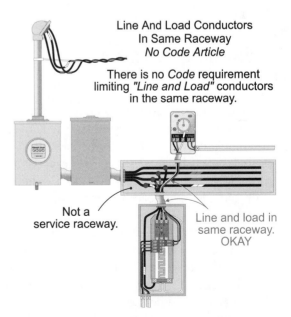

Line And Load Conductors
In Same Raceway
No Code Article

There is no *Code* requirement
limiting *"Line and Load"* conductors
in the same raceway.

Not a
service raceway.

Line and load in
same raceway.
OKAY

Fig. 9-6 Line and Load Conductors Can Be
Installed within the Same Raceway or Enclosure

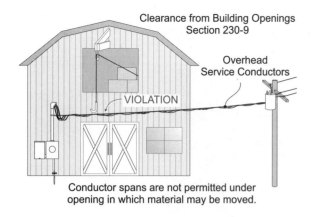

Clearance from Building Openings
Section 230-9

Overhead
Service Conductors

VIOLATION

Conductor spans are not permitted under
opening in which material may be moved.

Fig. 9-7 Service Drop Conductors not Permitted under Building Openings

230-8 Raceway Seals

Used or unused underground raceways shall be sealed according to Section 300-5(g). The purpose of the seal is to prevent water, resulting from condensation, from entering the service equipment. This can be accomplished with the use of a putty-like material called duct seal or a fitting identified for the purpose.

Note. A seal of the types required in Chapter 5, specifically Section 501-5 of the *NEC* for hazardous (classified) locations is not required in this Section.

230-9 Clearance from Building Openings

Service conductors installed as open conductors shall have a clearance of not less than 3 feet from windows that are designed to be opened, doors, porches, balconies, ladders, stairs, fire escapes or similar locations. See 225-19(d) for similar requirements for outside overhead feeders and branch circuit conductors.

Overhead service conductors to a building shall maintain a vertical clearance of at least 10 feet above platforms, projections or surfaces from which they might be reached. This vertical clearance shall extend 3 feet (measured horizontally) from the platform, projections or surfaces from which they might be reached.

Exception: Service Conductors Above Win-

dows. Service conductors run above a window are not required to maintain the 3 feet distance from the window.

Below Opening. Service conductors cannot be installed under an opening through which materials might pass. Overhead service-drop conductors and **service-entrance** conductors shall not be run or installed where they will obstruct entrance to these building openings. For example, the upper opening in a barn loft is often used to move hay in or out of the loft storage area, Fig. 9-7.

PART B. OVERHEAD SERVICE-DROP CONDUCTORS

Author's Comment: Overhead service-drop conductors from the electric utility must be installed in accordance with the National Electric Safety Code (NESC), and are exempted from the requirements of the *National Electrical Code*, [90-2(b)(5)].

230-23 Size and Rating

(a) Ampacity of Service-Drop Conductors. Service-drop conductors shall have adequate mechanical strength and sufficient ampacity for the loads served, as computed according to Article 220.

(b) Ungrounded (Hot) Conductor Size. Service-drop conductors shall not be smaller than No. 8 copper, No. 6 aluminum or copper-clad aluminum.

Exception: Limited Loads. Service-drop conductors can be as small as No. 12 for limited load installations.

(c) Grounded (Neutral) Conductor Size. Service-drop grounded (neutral) conductors shall be sized according to the maximum unbalanced load, according to Section 220-22. In addition, the grounded (neutral) service conductor shall not be smaller than as required by Section 250-24(b).

230-24 Clearances

Service-drop conductors shall not be readily

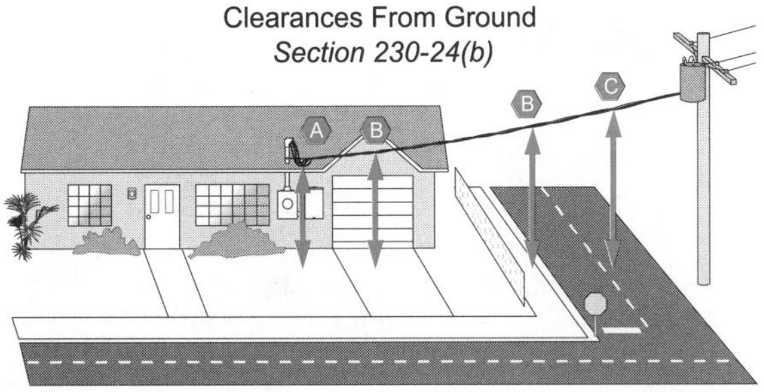

Clearances From Ground
Section 230-24(b)

Fig. 9-8 Service Drop Conductor Clearances Must
Be Maintained

accessible, and must comply with the clearances as
specified in (a) through (d) of this Section.

(a) **Above Roofs.** Service-drop conductors
passing over all roofs require a minimum clearance
of 8 feet above the surface of the roof. This clearance is required for a minimum distance of 3 feet
in all directions from the edge of the roof. This
prevents the overhead conductors from being
contacted.

Exception No. 1: Parking Garage Roofs. If
pedestrians or vehicles are normally on the roof
(such as parking garages), service-drop conductors
must have a clearance complying with Section 230-
24(b).

Exception No. 2: Steeply Sloped Roofs. Where
the voltage does not exceed 300 volt between conductors, conductor clearances can be reduced from
8 feet to 3 feet, where the roof slopes greater than 4
inches in 12 inches.

Note. The danger of persons contacting
overhead conductors is lessened when there
is reduced voltage and the roofs have a
slope or angle that makes walking difficult.

Exception No. 3: Overhang Portion Only. If
the voltage between conductors does not exceed
300 volt, the conductor clearance over the roof
overhang can be reduced from 8 feet, to 18 inches.
This is only permitted if no more than 6 feet of service-drop conductors pass over no more than a four
foot roof overhang, and the conductors terminate at
a through-the-roof raceway or approved support.

Exception No. 4: Point of Attachment. The ver-

tical clearance that extends 3 feet from the roof shall
not apply to the final conductor span where the
point of attachment is on the side of the building
(below the roof).

(b) **Service-Drop Vertical Clearance from
Ground.** Service-drop conductors and the lowest
point of the *drip loop*, where not in excess of 600
volt, nominal, shall have a minimum clearance of
10 feet from final grade. The ten-foot vertical clearance above grade for service-drop conductors also
applies to the lowest point of the service drip loop
conductors.

Section 230-24(b) Vertical Clearance Summary Location	From Final Grade or Other Accessible Surfaces
0-150 Volts to Ground: Service drop conductors at service entrance equipment and the lowest point of the drip loop conductors and over areas or sidewalks accessible only to pedestrians.	10 Feet
151-300 Volts to Ground: Service drop conductors and the lowest point of the drip loop conductors over residential property/driveways and over commercial areas not subject to truck traffic.	12 Feet
301-600 Volts to Ground: Service drop conductors and the lowest point of the drip loop conductors over residential property/driveways and commercial areas not subjected to truck traffic.	15 Feet

In addition the minimum height above final
grade for conductors 600 volt or less, cannot be less
than listed in the following table, Fig. 9-8:

(d) **Conductor Clearance – Swimming
Pools.** Service conductors are permitted above
pools, diving structures, observation stands, towers,
or platforms. See Section 680-8 for details.

230-26 **Point of Attachment**

The point of attachment for service-drop conductors shall not be less than 10 feet above the
finish grade and must be located so that the minimum service conductor clearance required by
Section 230-24(b), can be maintained, Fig. 9-9.

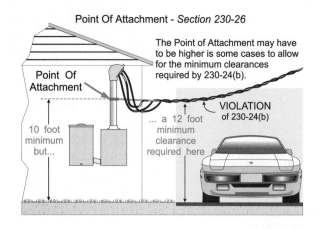

Fig. 9-9 Point of Attachment Might Need to Be Elevated to Accommodate Clearance Requirement

> CAUTION: Conductors might need to have the point of attachment raised higher so that the conductor's sag complies with the clearance requirements of Section 230-24, Fig. 9-9.

230-28 Service Masts Used As Supports

The service mast must have adequate mechanical strength or be supported by braces or *guy wires* to withstand the strain caused by the service-drop conductors. Some local building codes require a minimum 2 inch rigid conduit to be used for the service mast. Also many electric utilities contain specific guidelines.

Only power service-drop conductors can be attached to a service mast. Sections 810-12 and 820-10(c) specify that aerial cables for radio, TV, or CATV cannot be attached to the electric service mast, and Section 810-12 prohibits antennas from being attached to the service mast, Fig. 9-10.

PART C. UNDERGROUND SERVICE LATERAL CONDUCTORS

230-31 Size and Rating

(a) Service-Laterals. Service-lateral conductors must have adequate mechanical strength and sufficient ampacity for the loads, as computed according to Article 220.

(b) Ungrounded (Hot) Conductor Size.
Service-lateral conductors shall not be smaller than No. 8 copper, No. 6 aluminum, or copper-clad aluminum.

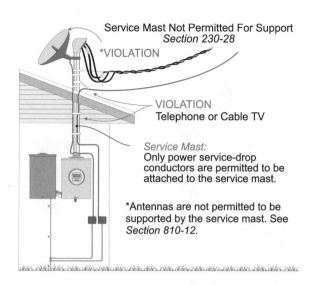

Fig. 9-10 Only Service Conductors Permitted to Be Attached to the Service Mast

Exception: Limited Loads. Service-lateral conductors can be as small as No. 12 for limited load installations.

(c) Grounded (Neutral) Conductor Size. Service-lateral grounded (neutral) conductors shall be sized according to the maximum unbalanced load, according to Section 220-22. In addition, the grounded (neutral) service conductor shall not be smaller than as required in Section 250-24(b).

230-32 Underground Protection against Damage

Underground service conductors must be protected against physical damage according to the requirements of Section 300-5. The cable or raceway must be buried according to the minimum burial depths listed in Table 300-5.

Under Building. Section 300-5(c). Cables installed under a building must be installed in a raceway.

Physical Damage. Section 300-5(d). Protection Against Physical Damage. Emerging direct buried conductors must be protected by enclosures or raceways to a point of 8 feet above grade.

Entering Building. Where service-lateral conductors enter a building, they must be encased in 2 inches of concrete or brick, according to

Section 230-6, or installed in a raceway as identified in Section 230-43.

PART D. SERVICE-ENTRANCE CONDUCTORS

230-40 Number of Service-Entrance Conductor Sets

Each service-drop or lateral shall supply only one set of service-entrance conductors.

Exception No. 1: Multiple Occupancy Building. Buildings with more than one occupancy can have one set of service-entrance conductors run to each occupancy.

Exception No. 2: Two to Six Disconnecting Means. One set of service-entrance conductors can supply two to six service disconnecting means as permitted in Section 230-71(a).

Exception No. 3: One-Family Dwelling Unit. A single-family dwelling unit and a separate structure can have one set of service-entrance conductors run to each from a single service-drop or lateral, Fig. 9-11.

Note. No more than 6 mains at each location [230-71].

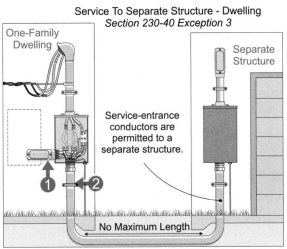

Service To Separate Structure - Dwelling
Section 230-40 Exception 3

One-Family Dwelling

Separate Structure

Service-entrance conductors are permitted to a separate structure.

No Maximum Length

A single-family dwelling and a separate structure are permitted to have service-entrance conductors run to each from a single service drop or lateral.

Fig. 9-11 Dwelling - Service Entrance Conductors Permitted to Separate Structure

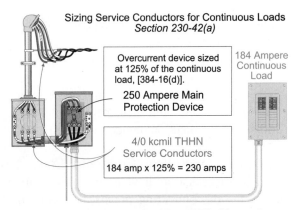

Sizing Service Conductors for Continuous Loads
Section 230-42(a)

Overcurrent device sized at 125% of the continuous load, [384-16(d)].

250 Ampere Main Protection Device

184 Ampere Continuous Load

4/0 kcmil THHN Service Conductors
184 amp x 125% = 230 amps

The service conductor size is based on 125% of the continuous load according to conductor ampacity before derating.

Fig. 9-12 Service Conductors for Continuous Loads Sized at Not Less than 125 Percent

230-42 Size and Rating

(a) Load Calculations. Service-entrance conductors must have sufficient ampacity for the loads they serve, as computed in Article 220.

Continuous Load. Service conductors that supply **continuous loads** must be sized no less than 125 percent of the continuous loads, plus 100 percent of the noncontinuous loads. This calculation is based on the conductor ampacities listed in Table 310-16 before any ampacity adjustment in accordance with the terminal rating [110-14(c)].

> **Note.** In addition, conductors shall have the **ampacity**, after any ampacity adjustment, to carry the load, and the conductors shall have proper overcurrent protection in accordance with Section 240-3 based on the adjusted ampacity.

Question. What size THHN conductor and overcurrent protection device is required for a service that supplies 184 ampere of continuous load? The nipple contains four current carrying conductors, Fig. 9-12.

Answer. No. 4/0, 250 ampere protection
Step 1. Determine Protection Device Size [230-42(a), 240-6(a) and 384-16a]
184 ampere × 1.25 = 230 ampere
250 ampere protection device

Step 2. Size Conductor [110-14(c), 230-42(a)]
184 ampere × 1.25 = 230 ampere
No. 4/0 rated 230 ampere at 75°C, Table 310-16

No. 4/0 THHN is rated 230 ampere can carry the 184 ampere load, and it properly protected by the 250 ampere protection device [240-3(b)].

(b) Ungrounded (Hot) Conductor Size. Service conductors must not have a rating less than the rating of the disconnecting means as specified in Section 230-79.

(c) Grounded (Neutral) Conductor Size. The grounded (neutral) conductor shall be sized to the maximum unbalanced load, according to Section 220-22, and shall not be smaller than required in Section 250-24(b).

230-43 Wiring Methods Not Over 600 Volt Nominal

Service conductors must be installed in the following wiring methods according to the *Code* requirements that pertain to each type:

Wiring Method	Article
Auxiliary Gutters	374
Busways	364
Cablebus	365
Electrical Metallic Conduit	348
Electrical Nonmetallic Tubing	331
Flexible Metal Conduit	350
Intermediate Metal Conduit	345
Liquidtight Nonmetallic Conduit	351
Liquidtight Metallic Conduit	351
Mineral-insulated Cable	330
Open Wiring on Insulators	320
Rigid Metal Conduit	346
Rigid Nonmetallic Conduit	347
Service-Entrance Cables	338
Type MC Cable	334
Wireways	362

Note. There is some concern in the electrical industry about the ability of some of these wiring methods to withstand ground-fault current.

230-46 Splicing Service Entrance Conductors

Service-entrance conductors can be spliced or tapped by clamped or bolted connections at anytime as long as:

The free ends of conductors are covered with an insulation that is equivalent to that of the conductors or with an insulating device identified for the purpose [110-14(b)].

Wire connectors or other splicing means installed on conductors that are buried in the earth shall be listed for direct burial [110-14(b)]; splices and taps shall be done in an enclosure or with a listed underground splice kit [300-15]. No splice can be made within a raceway [300-13].

230-50 Protection against Physical Damage-Above Ground

(a) Service-Entrance Conductors Above Ground. Service-entrance cable service conductors or cables shall be protected against physical damage. Cables subject to physical damage shall be protected with rigid metal conduit, intermediate metal conduit, rigid nonmetallic conduit suitable for the location, electrical metallic tubing, or other approved means.

230-51 Service Cable Supports

(a) Service Cable Supports. Service-entrance cable must be supported every 30 inches and within 12 inches of service heads and raceway connections.

230-53 Raceways Arranged to Drain

Raceways enclosing service-entrance conductors outdoors shall be raintight and arranged to drain when exposed to the weather.

Raceways may accumulate moisture through condensation so arrangements must be made for raceways to drain. Check with your inspector on how to arrange the raceway for drainage.

230-54 Connections at Service Head

(a) Weatherhead – Raintight. Raceways for overhead service-drops are required to have a raintight service head (*weatherhead*) at the point of attachment [230-26].

(c) Weatherhead Above the Point of Attachment. Service heads must be located above the point of attachment of the service-drop.

Exception: Below Point of Attachment. Where it is impractical to locate the service head above the point of attachment, the service head can be located no more than 24 inches from the point of attachment.

(e) Opposite Polarity through Separately Bushed Holes. Service heads must provide a bushed opening, and phase conductors must be in separate openings. Most weatherheads have a plastic plate with knockouts to accomplish this.

(f) Drip Loops. Drip loop conductors shall be below the service head or below the termination of the service-entrance cable sheath.

(g) Arranged So that Water Will Not Enter. Service-drops and service-entrance conductors must be arranged to prevent water from entering the service equipment. This is accomplished by installing the point of attachment below the weatherhead, and drip loops shall be formed on the individual conductors.

230-56 High-Leg Identification

The *high-leg* service conductor of a 4-wire, 3-phase delta-connected service where the midpoint of one winding is grounded must be permanently marked orange, or identified by other effective means. Paint or marking tape are some acceptable methods of identifying the high-leg. See Sections 215-8 for feeders, and 384-3(e) for panelboard high-leg identification requirements.

High Leg Termination. The American National Standards Institute (ANSI) requires that the metering equipment be designed for the high-leg conductor to terminate on Line 3 or the "C" phase.

Section 384-3(f) requires the high-leg conductor to terminate on the "B" or center phase of panelboards and switchboards. There are no *NEC* requirements for the location of the high-leg in disconnects and switches.

PART E. SERVICE EQUIPMENT – GENERAL

Article 250 contains special rules for grounding and bonding of services, see Sections 250-24, 250-92, and 250-94.

230-66 Identified As Suitable for Service Equipment

Service equipment rated at 600 volt or less must be marked to identify it as suitable for use as service equipment. This does not apply to meter socket enclosures because they are not considered service equipment. Review the definition of service equipment in Article 100.

PART F. SERVICE EQUIPMENT – DISCONNECTING MEANS

230-70 General

The service disconnect must disconnect all service-entrance conductors from the building or structure premises wiring.

Note. Section 230-76 requires the service disconnecting means to consist of either a *manually* operable switch or circuit breaker, or a power-operated switch or circuit breaker, provided it can be opened by hand in the event of a power supply failure. The pushbutton that operates an electromagnetic relay that opens a circuit breaker is not considered a disconnect because it does not open the main power supply conductors; it only activates an electromagnetic relay to open the circuit breaker.

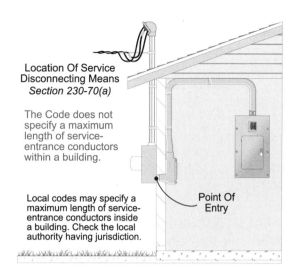

Location Of Service
Disconnecting Means
Section 230-70(a)

The Code does not
specify a maximum
length of service-
entrance conductors
within a building.

Local codes may specify a
maximum length of service-
entrance conductors inside
a building. Check the local
authority having jurisdiction.

Point Of
Entry

Fig. 9-13 Service Disconnect Must Be Readily
Accessible, Inside or Outside the Building

(a) Location. The service disconnect may be
at any **readily accessible** location outside the building or structure, or at any readily accessible indoor
location nearest the point of entry of the service
conductors. When the disconnect is installed
indoors, it cannot be installed in any **bathroom**,
Fig. 9-13.

WARNING:

Service conductor fault-current can produce tremendous heat which can ignite
combustible materials; therefore, service
conductors should be limited in length
when installed inside a building. Some
local jurisdictions have established, within
their local electrical code, a specific service
conductor length that is permitted within a
building, Fig. 9-13.

Note. If the service disconnect is located
outdoors, the *NEC* does not require that the
disconnect be located on the building or
structure. How far away can the disconnect
be located from the building? This is up to
the electrical inspector.

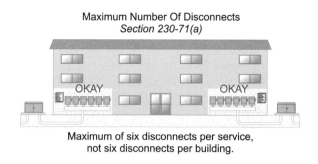

Maximum Number Of Disconnects
Section 230-71(a)

OKAY   OKAY

Maximum of six disconnects per service,
not six disconnects per building.

Fig. 9-14 Maximum of 6 Disconnects Per Service

(b) Disconnect Identification. Each service
disconnect must be permanently marked to identify
them as service disconnecting means, and Section
230-72(a) requires that all disconnects be marked to
indicate the load served.

Note. Section 110-22 specifies that each
disconnect required by this *Code* shall be
legibly marked to indicate its purpose, and
the marking shall be of sufficient durability
to withstand the environment involved.

(c) Suitable for the Conditions. All service
disconnects must be suitable for the prevailing conditions. For instance, if installed outdoors, it must
be **weatherproof**, or if in a hazardous (classified)
location it must comply with the requirements contained in Articles 500 through 517.

230-71 **Number of Disconnects**

(a) Maximum of Six Per Service. There shall
be no more than six disconnects for each service as
permitted by Section 230-2, or for each set of service-entrance conductors permitted as permitted by
Exceptions No. 1 and No. 3 to Section 230-40.

CAUTION: There is no Code rule limiting six
disconnects to a building. The rule is six disconnects for each service, not each building.
If the building is permitted to have two services, there can be a total of twelve
disconnects; two groups of a maximum of six
disconnects each, Fig. 9-14.

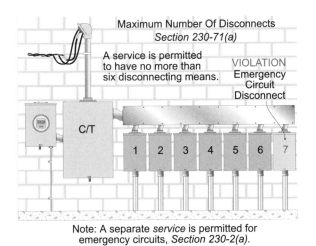

Maximum Number Of Disconnects
Section 230-71(a)

A service is permitted
to have no more than
six disconnecting means.

VIOLATION
Emergency
Circuit
Disconnect

C/T

1 2 3 4 5 6 7

Note: A separate *service* is permitted for
emergency circuits, *Section 230-2(a)*.

Fig. 9-15 No More than 6 Disconnects for Each Service

The service disconnecting means shall consist of up to six switches or six circuit breakers mounted in a single enclosure, or up to six separate enclosures located in a group, or up to six switches or six circuit breakers on a switchboard, Fig. 9-15.

230-72 Grouping of Disconnects

(a) Two to Six Disconnects. Services are not required to be grouped, but the two to six disconnecting means for each service must be grouped.

Note. The *Code* does not require a shunt trip controlled circuit breaker to disconnect all service disconnects when there is more than one service.

When a building or structure has two or more services, feeders, or branch circuits, a plaque is required at each service location to show the location of the other service, see Sections 225-37 and 230-2(b). Each service disconnect must be marked to indicate the load it serves, see Sections 110-22, 230-70(b).

(b) Fire Pump, Emergency, and Stand-By Power Service. The disconnect for fire pumps, or stand-by services as permitted in Section 230-2 Exceptions 1 and 2, shall be located remotely away from the two to six disconnects for normal service. The reason these disconnects are located away from the normal power disconnect(s) is to minimize the accidental interruption of the emergency power supply, Fig. 9-15.

Note: The power for emergency circuit cannot be supplied by the normal power and a separate service is required [700-12(d)], See the text in this book in Section 230-82(4).

(c) Access to Occupants. In a multiple occupancy building, each occupant shall have access to his or her disconnecting means.

Exception. In multiple occupancy buildings where electrical maintenance is provided by continuous building management, the service disconnecting means can be accessible only to building management personnel.

230-75 Disconnecting of Grounded Conductor

Means shall be provided at the service equipment to disconnect the grounded (neutral) conductor. This is normally accomplished by a terminal, or *bus*, on which all grounded (neutral) conductors terminate. See Section 250-24(b) for the requirements of terminating the grounded (neutral) conductor to the service disconnect enclosure.

Note. You cannot run a grounded (neutral) conductor unbroken through the service disconnect because the service grounded (neutral) conductor must be bonded to the service disconnecting means [250-24(b)].

230-76 Manual or Power-Operated Circuit Breakers

The disconnecting means can consist of either a manually-operable switch or circuit breaker, or a power-operated switch or circuit breaker. If power-operated, the switch or circuit breaker must also be capable of being operated by hand.

CAUTION: A pushbutton that controls a shunt trip circuit breaker is not considered a disconnect because it does not meet the requirements of Sections 230-76 and 230-79, Fig. 9-16.

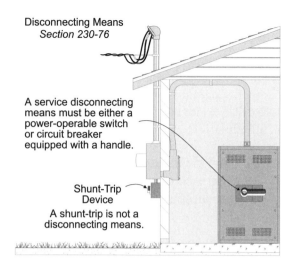

Disconnecting Means
Section 230-76

A service disconnecting
means must be either a
power-operable switch
or circuit breaker
equipped with a handle.

Shunt-Trip
Device
A shunt-trip is not a
disconnecting means.

Fig. 9-16 Shunt-Trip Device Does Not Meet
the Service Disconnect Requirements

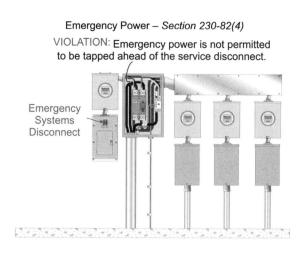

Emergency Power – *Section 230-82(4)*
VIOLATION: Emergency power is not permitted
to be tapped ahead of the service disconnect.

Emergency
Systems
Disconnect

Fig. 9-17 Emergency Power Not Permitted to
be Tapped Ahead of Service Equipment

230-79 Rating of Disconnect

The service disconnecting means shall have an ampere rating not less than the computed load according to Article 220. In no case shall the rating be lower than specified in (a), (b), (c), or (d) below.

(a) One-Circuit Installation. For installations that supply loads limited to a single branch circuit, the service disconnecting means shall have a rating of not less than 15 ampere.

(b) Two-Circuit Installations. For installations consisting of not more than two, 2-wire branch circuits, the service disconnecting means shall have a rating of not less than 30 ampere.

(c) One-Family Dwelling. For a one-family dwelling, the service disconnecting means shall have a rating of not less than 100 amperes, 3-wire.

(d) All Others. For all other installations, the service disconnecting means shall have a rating of not less than 60 ampere.

230-82 Equipment Connected to the Supply Side

Equipment shall not be connected to the supply side of the service disconnect enclosure. Individual meter socket enclosures should not be considered service equipment. See Article 100 definition of service equipment.

(2) Meters. Meters rated not in excess of 600 volt and properly grounded according with Article 250 can be ahead of the service disconnect.

(4) Tap Conductors. Tap conductors for stand-by systems, fire pump equipment, fire and sprinkler alarms, and load (energy) management devices can be connected to the supply side of the service disconnect.

Note. The permission to connect ahead of the service main for emergency power was deleted from the 1996 *Code*, Fig. 9-17.

PART E. SERVICE EQUIPMENT OVER-CURRENT PROTECTION

230-90 Overload Protection Required

Each ungrounded service conductor must have overload protection. Section 240-21(d) recognizes that it is not possible to have the protection device at the point where the service conductors receive their supply.

(a) Overcurrent Protection Size. The rating or setting of the overcurrent protection device shall not be greater than the ampacity of the conductors, provided the calculated load per Article 220 does not exceed the ampacity of the service conductors.

Exception No. 2: Next Size Up Rule. Where the ampacity of the conductors does not correspond with the standard ampere rating of overcurrent

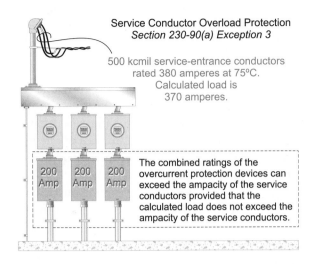

Service Conductor Overload Protection
Section 230-90(a) Exception 3

500 kcmil service-entrance conductors
rated 380 amperes at 75°C.
Calculated load is
370 amperes.

The combined ratings of the overcurrent protection devices can exceed the ampacity of the service conductors provided that the calculated load does not exceed the ampacity of the service conductors.

Fig. 9-18 Combined Rating of Overcurrent Devices Can Exceed the Service Conductor Ampacity

devices, as listed in Section 240-6(a), it shall be permissible to use the next higher overcurrent device, but only if the next size up device does not exceed 800 ampere [240-3(b)].

For example, two sets of 300,000 THHN kcmil conductors (each rated 285 ampere at 75 degrees C) can be protected by a 600 ampere overcurrent protection device, Fig. 9-18.

Exception No. 3: Service Disconnecting Means. Up to six circuit breakers or six sets of fuses of any size shall be permitted as the overcurrent device. In addition, the sum of the ratings of the two to six circuit breakers or fuses "can exceed" the ampacity of the service conductors, provided the calculated load does not exceed the ampacity of the service conductors.

230-91 Overcurrent Protection Device

(a) General. The service overcurrent protection device must be an integral part of the service disconnecting means, or located immediately next to the disconnect. Generally, the overcurrent protection is part of the service disconnecting means.

230-92 Locked Service Overcurrent Devices

Where the service overcurrent devices are locked or sealed, or not readily accessible to the occupant, branch-circuit overcurrent devices shall be installed on the load side, shall be mounted in a readily accessible location, and shall be of a lower ampere rating than the service overcurrent device.

SUMMARY

Part A. General ☐ Article 230 covers the installation requirements for service conductors and equipment. ☐ The general rule is that a building or structure is permitted only one service-drop or service-lateral. ☐ Two or more services are permitted if the ampacity is over 2,000 ampere or where the load requirements of a single-phase installation exceed the utilities normal capacity, or by special permission. ☐ Service conductors cannot pass through the interior of one building to another. ☐ Service conductors are considered outside the building if installed under a building with 2 inches of concrete cover, or encased within 2 inches of concrete or brick, or installed in a transformer vault. ☐ Service conductors cannot be installed with feeder and/or branch circuit conductors in the same raceway, but they can be mixed in the same enclosure, such as the service equipment disconnect or wireway.

Part B. Overhead Service-Drop Conductors ☐ The minimum size service-drop conductor is No. 8 copper, or No. 6 aluminum. ☐ The grounded service conductors must not be smaller than the grounding-electrode conductor as sized in Table 250-66 and not smaller than as required by Section 250-24(b). ☐ Service-drop conductors shall not be readily accessible. ☐ Service conductors passing over roofs require a minimum 8 feet clearance, with a 3 foot clearance from the edge of the roof. ☐ Service-drop clearances [230-24(b)]. ☐ Service-drop conductors are permitted above pools, diving structures, etc. ☐ The point of attachment cannot be less than 10 feet above finished grade. ☐ The service mast must have adequate mechanical strength.

Part C. Underground Service-Lateral Conductors ☐ Underground service conductors must be protected against damage.

Part D. Service-Entrance Conductors ☐ A service-drop or lateral can supply only one set of service-entrance conductors. ☐ For wiring methods permitted [230-43]. ☐ Service-entrance cable that is subject to physical damage must be protected. ☐ Service cables must be supported every 30 inches and within 12 inches of service heads, goosenecks, and raceway connections. ☐ Service raceways where exposed to the weather must be raintight and arranged to drain. ☐ Raceways for overhead service-drops are required to have a raintight service head. ☐ Service heads must be located above the point of attachment. ☐ Service heads must provide bushed openings. ☐ Drip loops must be below the service head. ☐ The high-leg service conductor must be permanently marked orange.

Part E. Service Equipment ☐ All metal parts of service equipment must be grounded.

Part F. Disconnecting Means ☐ The disconnecting means must disconnect all service-entrance conductors from the premises wiring system. ☐ Service disconnecting means must be permanently marked to identify it as a service disconnect. ☐ No more than six service disconnects are permitted per service [230-71]. ☐ Service disconnects must be grouped. ☐ Each occupant must have access to his or her disconnecting means in a multiple occupancy building. ☐ The service disconnecting means must disconnect all ungrounded service conductors simultaneously. ☐ Means shall be provided at the service equipment to disconnect the grounded (neutral) conductor. ☐ Service disconnecting means are required to indicate plainly whether they are in the open or closed position. ☐ Generally, equipment shall not be connected to the supply side of the service disconnecting means.

Part G. Overcurrent Protection ☐ Each ungrounded service conductor must have overload protection.

REVIEW QUESTIONS

1. When can additional services be added without requiring special permission from the inspector?

2. Provide a summary of all the rules that apply to drip loop conductors.

3. Provide a summary of all the rules that apply to clearance of service conductors over pools.

4. What are the rules that apply to the high-leg conductor in meter equipment, disconnects, and panelboards?

5. Give a summary of the service disconnect requirements, such as location, number, grouping, and access by occupants.

6. In your own words explain the meaning of the following Article 100 definitions:

Ampacity
Bathroom
Bonding
Bonding jumper
Building
Continuous load
Disconnect
Raceway
Readily accessible
Service drop
Service entrance
Service lateral
Service point
Weatherproof

7. In your own words explain the meaning of the following Glossary terms:

Bus
Drip loop
Guy wires
High-leg
Load side
Weather head

Unit 10

Article 240
Overcurrent Protection

OBJECTIVES

After studying this unit, the student should be able to understand:

- feeder tap rules not over 10 feet.
- feeder tap rules not over 25 feet.
- feeder tap rules not over 100 feet.
- dual ratings of circuit breakers and fuses.
- the dangers of Edison base-type fuses and circuit breakers.
- the possible dangers associated with replacing current-limiting fuses with noncurrent-limiting fuses.

Definitions and Glossary Terms

To better understand the *NEC* rules contained in this unit, review the following:

Definitions Article 100 - Unit 2

Ampacity	Approved
Bathroom	Identified
Interrupting rating	Overcurrent
Overload	Qualified person
Readily accessible	Weatherproof

Glossary Terms

AIC	Current rating
Cutout box	Fuse
Horsepower	Inverse time
Primary	Secondary

PART A. GENERAL

240-1 Scope

Article 240 is divided into nine parts:

Part A.	General
Part B.	Location
Part C.	Enclosures
Part D.	Disconnecting and Grounding
Part E.	Plug Fuses, Fuseholders, and Adapters
Part F.	Cartridge Fuses and Fuseholders
Part G.	Circuit Breakers
Part H.	Supervised Industrial Installations
Part I.	Overcurrent Protection over 600 Volts

Article 240 covers the general requirements for **overcurrent** protection and the installation requirements of overcurrent protection devices.

Overcurrent - *Section 240-1 and FPN*

20 A CB and Conductor

25 Ampere Load "OVERLOAD"

Overload

Short Circuits

Ground Fault

Phase to phase short

Phase to neutral short

Phase to ground short

Overcurrent: (Article 100 Definition) Any current in excess of the rated current of equipment or materials. Causes of overcurrent are overloads, short-circuits, and ground-faults.

Fig. 10-1 Overcurrent Includes Overload, Short-Circuit and Ground-Fault

Overcurrent is a condition where the current exceeds the rating of conductors or equipment due to **overload,** short-circuit or ground-fault, Fig. 10-1.

FPN: Overcurrent protection is provided to open the circuit if the current reaches a value that will cause an excessive or dangerous temperature in conductors or conductor insulation.

In addition to overcurrent protection we must comply with Section 110-9 for requirements of interrupting rating and Section 110-10 for protection against fault-currents.

240-2 Protection of Equipment

Equipment and conductors must be protected against overcurrent according to the Article covering that type of equipment.

Equipment	Section
Air-conditioning	440-22
Branch Circuit	210-20
Feeder Conductors	215-3
Flexible Cords	240-4
Fixed Electric Heating	424-3(b)
Fixture Wire	240-4
Panelboards	384-16(a)
Service Conductors	230-90(a)
Temporary Conductors	305-4
Transformers	450-3

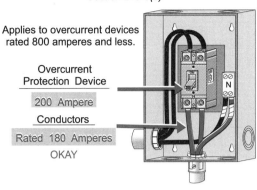

Overcurrent Protection - Next Size Up
Section 240-3(b)

Applies to overcurrent devices rated 800 amperes and less.

Overcurrent Protection Device

200 Ampere

Conductors

Rated 180 Amperes

OKAY

Fig. 10-2 Overcurrent Protection - Next Size Up

240-3 Protection of Conductors

There are many different rules for sizing and protecting conductors and equipment. It is not simply No. 12 wire and a 20 ampere breaker. The general rule is that conductors must be protected according to their **ampacity** as listed in Table 310-16.

Other methods of protection are permitted or required as listed in subsections (b) through (m) of this Section.

(b) Next Higher Overcurrent Device Rating. The next higher protection device is permitted if all of the following conditions are met, Fig. 10-2:

(1) Conductors do not supply multioutlet receptacle branch circuits for portable cord- and plug-connected loads.

(2) The ampacity of a conductor does not correspond with the standard ampere rating of a fuse or circuit breaker as listed in Section 240-6(a).

(3) The next size up breaker or fuse does not exceed 800 ampere.

(c) Circuits with Overcurrent Protection over 800 Ampere. If the circuit overcurrent protection device exceeds 800 ampere, the circuit conductor ampacity must not be less than the rating of the overcurrent protection device as listed in Section 240-6(a), Fig. 10-3.

(d) Small Conductors. Unless specifically permitted in 240-3(e) through 240-3(g), overcurrent

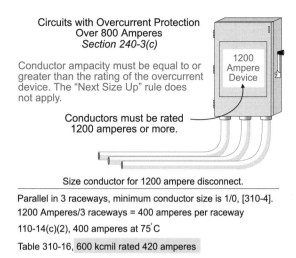

Size conductor for 1200 ampere disconnect.

Parallel in 3 raceways, minimum conductor size is 1/0, [310-4].

1200 Amperes/3 raceways = 400 amperes per raceway

110-14(c)(2), 400 amperes at 75°C

Table 310-16, 600 kcmil rated 420 amperes

Fig. 10-3 Overcurrent Protection Over 800 Amperes - "Next Size Up" Rule Not Permitted

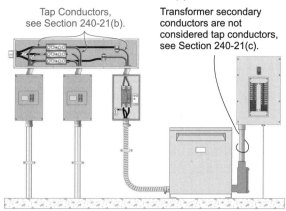

A tap conductor is any conductor (except service conductors) that has overcurrent protection ahead of its point of supply that exceeds the value permitted.

Fig. 10-5 Tap Conductor Definition

protection shall not exceed 15 ampere for No. 14, 20 ampere for No. 12, and 30 ampere for No. 10 copper, or 15 ampere for No. 12, and 25 ampere for No. 10 aluminum and copper-clad aluminum after ampacity correction, Fig. 10-4.

(e) Tap Conductors. For the purpose of this Article, a tap conductor is defined as a conductor having a *current rating* less than the rating or setting of the overcurrent device that protects the conductors from which the tap is derived, Fig. 10-5.

The overcurrent protection requirements for tap conductor are contained in Sections 210-19(d), 240-21(b).

(f) Transformer Secondary Conductors. 3- or 4-wire *secondary* conductors are not considered protected by the *primary* transformer

protection device. 2-wire transformer secondary conductors can be protected by the primary protection device sized according to Section 450-3(b).

Note. Conductors on the *load side* of a transformer (secondary) are not considered to be tap conductors and must have overcurrent protection in accordance with the requirements of Section 240-21(c), Fig. 10-5.

(g) Overcurrent for Specific Applications

Air Conditioning. Air conditioning equipment circuit conductors are protected according to Section 440-22.

Question. What size conductor and protection device is required for an air-conditioner when the nameplate indicates that the motor-compressor is 18 amperes, the minimum circuit ampacity is 23 ampere and the maximum fuse size is 40 ampere, Fig. 10-6.

Answer.   No. 12, 40 ampere protection
Step 1. Conductor [440-32] This Section requires the conductor be sized at 125 percent of the motor-compressor rated-load current.
18 ampere × 1.25 = 22.5 ampere
No. 12 THHN is rated at 25 ampere at 60°C

Step 2. Overcurrent Protection [440-22(a)] The protection device shall not be greater than 225

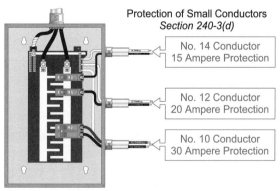

Unless specifically permitted in (e) through (g), the overcurrent protection shall not exceed 15 amperes for No. 14, 20 amperes for No. 12, and 30 amperes for No. 10 copper.

Fig. 10-4 Protection of No. 14, No. 12, and No. 10 Conductors

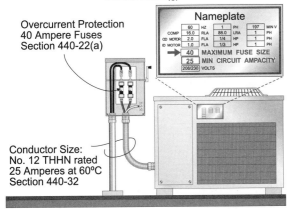

Fig. 10-6 Overcurrent Protection for A/C Must Comply with Article 440, Not Article 240

percent of the motor-compressor rated-load current.

18 ampere × 2.25 = 40.5 ampere

Next size down, 40 ampere, see 240-6(a)

Fire Alarm Circuits. Fire alarm circuit conductors are protected according to Sections 760-23 and 760-24.

Motors. Motor circuit conductors are protected according to Section 430-31, 430-52, and 62.

Question. What size motor branch circuit conductor and protection device (circuit breaker) is required for a 7½ hp (22 ampere), 230 volt, 3-phase motor, Fig. 10-7?

Answer. No. 10, 60 ampere breaker

Step 1. Conductor must be sized at 125 percent of the motor full-load current based on the 75°C column of Table 310-16 [110-14(c)].

22 ampere × 1.25 = 27.5 ampere

No. 10 THHN rated 30 ampere at 75°C

Step 2. Short-circuit ground-fault protection sized to 250 percent of FLC in Table 430-150 [430-52(c)(1) Ex.1 and Table 430-152].

22 ampere × 2.5 = 55 ampere

Next size up, 60 ampere device [240-6(a)]

Motor Control. Motor and motor control circuit conductors are protected according to Section 430-72.

Motor operated appliances are protected

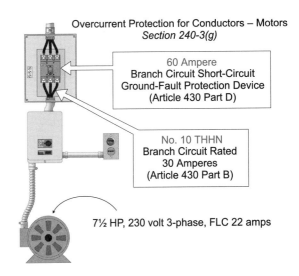

Fig. 10-7 Overcurrent Protection for Motors Must Comply with Article 430, Not Article 240

according to Article 422 Parts B and D.

Refrigeration. Air-conditioning and refrigeration equipment circuit conductors are protected according to Article 440, Parts C and F.

Remote Control and Signaling. Remote control, signaling, and power-limited circuit conductors shall be protected against overcurrent according to Sections 725-23 and 725-41.

240-4 Protection of Flexible Cords and Fixture Wires

(a) Ampacity. Flexible cords shall be protected against overcurrent at a setting not greater than the cord ampacity, as listed in Sections 400-5(a) and 400-5(b). Supplementary overcurrent protection devices can be used for the protection [240-10].

(b) Branch Circuit Overcurrent Protection

(1) Listed Appliances or Lamps. Overcurrent protection is not required for flexible cords that are approved for use with specific listed appliances or portable lamps.

(2) Fixture Wire. Fixture wires taps in accordance with Section 210-19(d) Exception No. 1 must follow the following:

- No. 18 (up to 50 feet runs), maximum 20 ampere protected circuit
- No. 16 (up to 100 feet runs), maximum 20 ampere protected circuit
- No. 14 and larger, maximum 20 or 30 ampere protected circuit
- No. 12 and larger, maximum 40 or 50 ampere protected circuit

240-6   Standard Ampere Ratings

(a) Standard Ratings of Circuit Breaker and Fuses. The standard ampere ratings for overcurrent protection devices (fuses and inverse time circuit breakers) are as follows: 1, 3, 6, 10, 15, 20, 25, 30, 35, 40, 45, 50, 60, 70, 80, 90, 100, 110, 125, 150, 175, 200, 225, 250, 300, 350, 400, 450, 500, 600, 601, 700, 800, 1000, 1200, 1600, 2000, 2500, 3000, 4000, 5000, and 6000.

Note. Fuses rated less than 15 ampere are sometimes required for the protection of fractional *horsepower* motor circuits [430-52], and for the protection of motor control circuits [430-72 and 725-12].

(b) Adjustable Circuit Breakers. The ampere rating of an adjustable circuit breaker shall be the highest possible setting of the external long-time pickup adjustment. The long-time pickup adjustment is an overload setting, not a short-circuit or ground-fault.

(c) Accessible to Qualified Persons Only. If the adjustment feature of an adjustable longtime pickup circuit breaker is located behind a removable cover, bolted equipment, or locked door accessible only to **qualified persons**, then the ampacity shall be at the adjusted setting.

240-10   Supplementary Overcurrent Protection

Supplementary overcurrent protection devices add a further degree of protection in addition to the branch circuit protection device and are often used for lighting fixtures, appliances, and the protection of internal circuits and components of equipment. However, supplementary overcurrent protection devices cannot be used as a substitute for branch circuit protection.

240-11   Current-Limiting Devices

A current-limiting overcurrent protective device has such a high speed of response that it cuts off a short-circuit before it can build up to its full peak value, Fig. 10-8.

240-13   Ground Fault Protection of Equipment

The general rule is that ground-fault protection of equipment shall be provided for solidly grounded 480/277 wye electrical systems for each building or structure main disconnecting means rated 1,000 ampere or more. This rule does not apply to the feeder if the service is protected with ground-fault protection in accordance with Section 230-95.

PART B.   LOCATION

240-20   Ungrounded Conductors

(a) Overcurrent Device Required. Overcurrent protection devices must be installed in series with each ungrounded conductor (hot, phase wire).

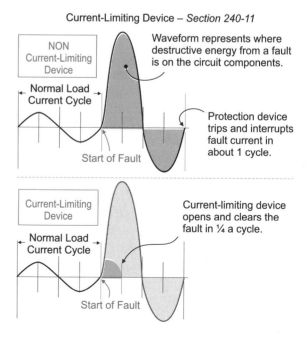

Fig. 10-8 Current-Limiting Device Open Faster than Noncurrent-Limiting Device

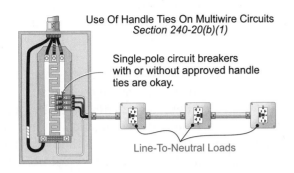

Use Of Handle Ties On Multiwire Circuits
Section 240-20(b)(1)

Single-pole circuit breakers with or without approved handle ties are okay.

Line-To-Neutral Loads

Fig. 10-9 Handle Ties Permitted for Multiwire Circuits Supplying Line-to-Neutral Loads

(b) Circuit Breaker As an Overcurrent Device. Circuit breakers shall open all ungrounded conductors of the circuit unless otherwise permitted in (1), (2), or (3) below:

(1) Multiwire Branch Circuit. Individual single-pole breakers are permitted for each ungrounded conductor of a multiwire branch circuit that supplies only single-phase line-to-neutral loads, except as required in dwelling units [210-4(b)], Fig. 10-9.

(2) Single-Phase Line-to-Line Loads. Individual single-pole circuit breakers with **approved** handle ties can be used for each ungrounded conductor of a branch circuit that supplies only single-phase line-to-line loads, Fig. 10-10.

(3) Three-Phase Line-to-Line Loads. Individual single-pole breakers with approved handle ties are permitted for each ungrounded conductor of

a branch circuit that serves only 3-phase line-to-line loads.

Note. Approved handle ties must be used; nails, screws, wires or other nonconforming methods are generally not approved by the electrical inspector.

240-21 **Location in Circuit**

Overcurrent protection devices must be connected in series with each ungrounded conductor. The overcurrent protection device must be at the point where branch circuit and feeder conductors receive their power, except as permitted in (a) through (g).

In addition, no tap conductor shall supply another conductor. In other words, you can't make a tap from a tap.

(a) Branch Circuit Taps. Branch circuit conductors can be tapped in accordance with Section 210-19. Tap conductors supplying a single household electric range, are considered protected by the branch circuit overcurrent device when the requirements of Sections 210-19(c) Exception No. 1, 220-4(c), and 210-24 are met.

(b) Feeder Taps.

(1) Taps Not Over 10 Feet. Conductors can be tapped from feeders, if the tap conductors are no longer than 10 feet and have an ampacity no less than; (1) the computed load according to Article 220, and (2) the ampere rating of the termination device, such as a panelboard or overcurrent device, Fig. 10-11.

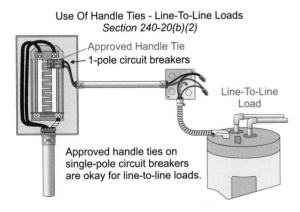

Use Of Handle Ties - Line-To-Line Loads
Section 240-20(b)(2)

Approved Handle Tie
1-pole circuit breakers

Line-To-Line Load

Approved handle ties on single-pole circuit breakers are okay for line-to-line loads.

Fig. 10-10 Approved Handle Ties Permitted for Line-to-Line Loads

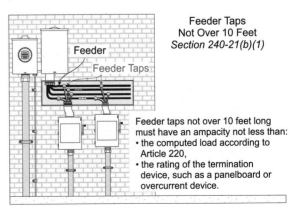

Feeder Taps Not Over 10 Feet
Section 240-21(b)(1)

Feeder

Feeder Taps

Feeder taps not over 10 feet long must have an ampacity not less than:
• the computed load according to Article 220,
• the rating of the termination device, such as a panelboard or overcurrent device.

Fig. 10-11 Feeder Taps Not Over 10 Feet

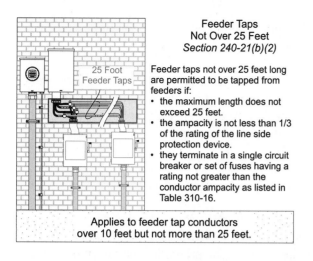

Feeder Taps Not Over 25 Feet
Section 240-21(b)(2)

Feeder taps not over 25 feet long are permitted to be tapped from feeders if:
• the maximum length does not exceed 25 feet.
• the ampacity is not less than 1/3 of the rating of the line side protection device.
• they terminate in a single circuit breaker or set of fuses having a rating not greater than the conductor ampacity as listed in Table 310-16.

Applies to feeder tap conductors over 10 feet but not more than 25 feet.

Fig. 10-12 Feeder Taps Not Over 25 Feet

The tap conductors that leave the enclosure in which the tap is made must be installed in a raceway, and the tap conductors must have an ampacity of no less than the ampacity of the feeder overcurrent protection device from which the conductors are tapped.

(2) Taps Not Exceeding 25 Feet. Conductors can be tapped from feeders, if the total length of the tap conductors is no longer than 25 feet. The tap conductors must have an ampacity not less than 1/3 the ampacity of the overcurrent protection device

protecting the feeder. The tap conductors must terminate in a single circuit breaker, or set of fuses having a rating no greater than the conductor ampacity as listed in Table 310-16, Fig. 10-12.

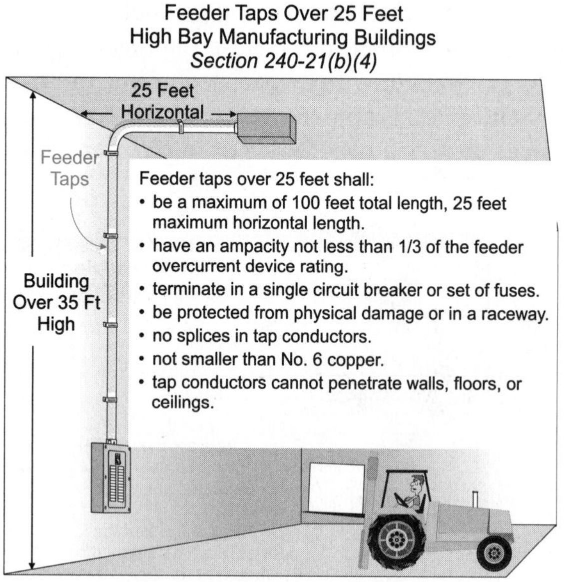

(4) Taps Over 25 Feet. In high-bay manufacturing buildings over 35 feet high at walls, tap conductors are permitted in lengths up to 100 feet. This is permitted only if the electrical system is maintained and supervised by qualified persons. The tap shall be made not less than 30 feet from the floor, be run no more than 25 feet horizontally, and have a total length not more than 100 feet without a splice.

Tap conductors 100 feet in length must terminate in a single circuit breaker, or set of fuses, not have any splice, must have an ampacity not less than the ampacity of the feeder overcurrent protection device, and be no smaller than No. 6 copper, or No. 4 aluminum. The tap conductors must be run in a raceway and must not penetrate walls, floors, or ceilings, Fig. 10-13.

(5) Outside Feeder Taps. Outside feeder tap conductors must remain outdoors, except at the point where the conductors terminate, and they must, Fig. 10-14:

(a) Be suitably protected from physical damage.

(b) Terminate in an overcurrent device that limits the load to the ampacity of the conductors. The

Feeder Taps Over 25 Feet
High Bay Manufacturing Buildings
Section 240-21(b)(4)

25 Feet Horizontal

Feeder Taps

Building Over 35 Ft High

Feeder taps over 25 feet shall:
• be a maximum of 100 feet total length, 25 feet maximum horizontal length.
• have an ampacity not less than 1/3 of the feeder overcurrent device rating.
• terminate in a single circuit breaker or set of fuses.
• be protected from physical damage or in a raceway.
• no splices in tap conductors.
• not smaller than No. 6 copper.
• tap conductors cannot penetrate walls, floors, or ceilings.

Fig. 10-13 Feeder Taps Over 25 Feet for High Bay Manufacturing Buildings

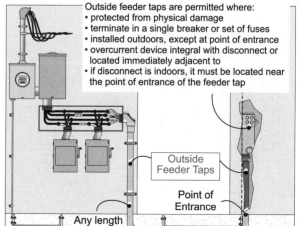

Outside Feeder Taps - *Section 240-21(b)(5)*

Outside feeder taps are permitted where:
• protected from physical damage
• terminate in a single breaker or set of fuses
• installed outdoors, except at point of entrance
• overcurrent device integral with disconnect or located immediately adjacent to
• if disconnect is indoors, it must be located near the point of entrance of the feeder tap

Outside Feeder Taps

Point of Entrance

Any length permitted

Fig. 10-14 Outside Feeders Taps

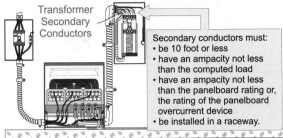

Fig. 10-15 Transformer Secondary Conductors

conductors must terminate in a single circuit breaker or a single set of fuses.

(c) Have overcurrent protection device for the conductors that is part of the disconnecting means.

(d) Must have the disconnect located at a **readily accessible** location either outside or inside the building or structure, nearest the point of conductor entrance or termination.

(c) Transformer Secondary Conductors

(1) Primary Protection. Three or four-wire secondary conductors are not considered protected by the primary transformer protection device. 2-wire transformer secondary conductors can be protected by the primary protection device sized according to Section 450-3(b).

(2) Secondary Not Exceeding 10 Feet. Secondary transformer conductors not longer than 10 feet are permitted if, Fig. 10-15:

(a) The ampacity of the secondary conductors is;
(1) Not less than the computed load according to Article 220.

(2) Not less than the ampere rating of the termination device, such as a panelboard or overcurrent device.

. (c) The secondary conductors must be installed in a raceway.

(3) Secondary Conductors Not Over 25 Feet. Secondary conductors can be longer than

10 feet; however, the secondary conductors can not be longer than 25 feet. The secondary conductors must be suitably protected from physical damage and they must terminate in a single circuit breaker or set of fuses that has an ampere rating no greater than the secondary conductor ampacity as listed in Table 310-16.

(4) Outside Secondary Conductors. Secondary conductors must remain outdoors, except at the point of termination, and they must:

(a) Be suitably protected from physical damage.

(b) The conductors terminate at a single circuit breaker or a single set of fuses that will limit the load to the ampacity of the conductors.

(c) Have the overcurrent protection device for the conductors as part of the disconnecting means.

(d) Having the disconnecting means for the conductors installed at a readily accessible location either outside of a building or structure, or inside nearest the point of entrance of the conductors.

(d) Service Conductors. Service conductors must be protected against overcurrent in accordance with Sections 230-90 and 230-91.

(f) Motor Circuit Taps. Motor tap conductors shall be protected against overcurrent in accordance with Sections 430-28 and 430-53.

240-24 Location in Premises

(a) Readily Accessible. The general requirement is that overcurrent protection devices must be readily accessible, unless one of the following applies, Fig. 10-16:

(2) Supplementary Overcurrent Protection. Supplementary overcurrent protection devices are not required to be readily accessible [240-10].

(4) Adjacent to Equipment. Overcurrent devices installed adjacent to motors, appliances, or other equipment that they supply can be accessible by portable means, Fig. 10-17.

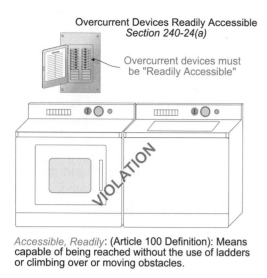

Fig. 10-16 Overcurrent Devices Must Be Readily Accessible

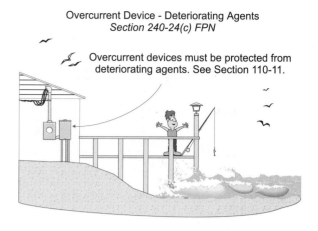

Fig. 10-18 Overcurrent Devices Must not Be Subjected to Deteriorating Agents

Note. A similar exception applies for a switch (disconnect) [380-8(a) Ex. 2].

(b) Occupant to Have Ready Access. Building occupants must have ready access to the overcurrent protection device protecting the supply conductors to their occupancy.

(1) Multiple-Occupancy Building. In a multiple-occupancy building with 24-hour management supervision, the service and feeder overcurrent protection devices can be kept locked and be accessible only to authorized management personnel.

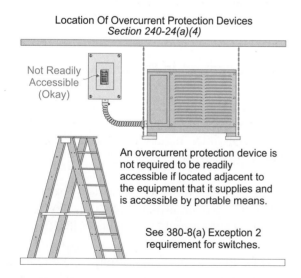

Fig. 10-17 Overcurrent Devices Permitted to Be Located Adjacent to Equipment It Controls

(2) Guest Room of Hotels and Motels. The overcurrent protection devices for guest rooms of hotels and motels intended for transient occupancy, and under continuous building management supervision, can be accessible only to authorized management personnel.

(c) Exposed to Physical Damage. Overcurrent devices must be located where they will not be exposed to physical damage [240-30]. In addition, Section 110-27(b) contains a similar requirement.

FPN: Sections 110-11 and 110-12(c) contains requirements for deteriorating agents, Fig. 10-18.

(d) Vicinity of Easily Ignitable Material. Overcurrent protection devices cannot be located near easily ignitable material, such as in clothes closets.

Note. The reference to clothes closets in this Section is an example of an area that might have easily ignitable materials. The purpose of preventing overcurrent protection devices near easily ignitable material is to prevent fires, not to keep them out of clothes closets, Fig. 10-19.

(e) Not Located in Bathrooms. Overcurrent protection devices, other than supplementary overcurrent protection, shall not be located in the **bathrooms** of dwelling units or guest rooms of hotels or motels, Fig. 10-20.

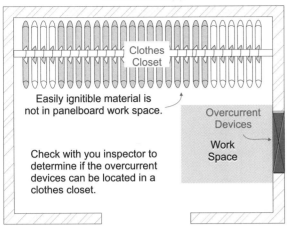

Overcurrent Devices - Easily Ignitible Material
Section 240-24(d)

Clothes
Closet

Easily ignitible material is
not in panelboard work space.

Overcurrent
Devices

Work
Space

Check with you inspector to
determine if the overcurrent
devices can be located in a
clothes closet.

Fig. 10-19 Overcurrent Protection Devices Not
Permitted to Be Near Easily Ignitible Material

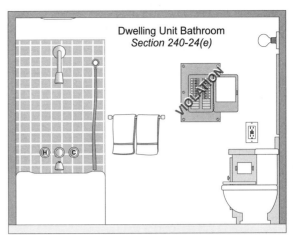

Dwelling Unit Bathroom
Section 240-24(e)

VIOLATION

Overcurrent devices are not permitted in dwelling unit bathrooms.
This includes the guest room bathrooms of hotels and motels.

Fig. 10-20 Overcurrent Protection Devices Not
Permitted in Dwelling Unit Bathrooms

Note. A GFCI receptacle is not an overcurrent protection device and can be installed in bathrooms.

PART C.   ENCLOSURES

240-30   General

Overcurrent devices must be protected from physical damage. One way of complying is to enclose the overcurrent protection devices in **cabinets** or *cutout boxes*, see Article 373.

240-32   Damp or Wet Locations

Cabinet or cutout box enclosures for overcurrent devices installed in damp or wet locations must be **identified** for use in those locations. The enclosure must prevent moisture or water from entering or accumulating and it must be listed as **weatherproof**. When the cabinet or cutout box is surface-mounted in a wet location, the enclosure must be mounted with at least a ¼ inch air space between it and the mounting surface, see Section 373-2(a).

240-33   Enclosure – Vertical Position

A cabinet or cutout box enclosure that contains

overcurrent protection devices must be mounted in a vertical position. However, circuit breakers enclosures can be installed horizontally. Where the circuit breaker is operated vertically, the "up" position of the handle must be the "on" position [240-81].

> CAUTION: The cabinet cannot be mounted sideways for panelboards that have circuit breakers mounted on both sides of the enclosure, see Section 240-81.

PART E.   PLUG FUSES, FUSEHOLDERS, AND ADAPTERS

240-50   General

(a) Maximum Voltage. Plug fuses and fuseholders shall be used on the following circuits:

(1) Not Over 125 Volt. Where the operating voltage does not exceed 125 volt between conductors.

(2) Not Over 150 Volt. Where the voltage does not exceed 150 volt to ground, such as a 3-wire 120/240 volt, single-phase, and 208Y/120 volt, three-phase systems.

Note. Plug fuses cannot be used on 240/120 volt 3-phase systems, because the voltage of one phase (high-leg) to ground is 208 volt. They cannot be used on 480Y/277 volt systems either.

240-51   Edison Base Fuse

(a) Classification. Edison base *fuses* are an old type of fuse, classified to operate at not more than 125 volt and not over 30 ampere.

(b) Replacement Only. Edison base fuses are permitted only for replacement in an existing installation, but only if there is no evidence of tampering or overfusing.

Note. Many homeowners replace blown fuses with whatever they have available. They are not aware of the dangers of overfusing the conductors. The *Code* places the responsibility to prevent this dangerous condition from continuing on the electrical trade.

WARNING:

1. Edison base screw shells have no restriction on the installation of fuses of different ampacities. A 30 ampere Edison fuse could be, and often is, installed in circuits of No. 14 or No. 12 conductors.

2. The Edison base also accepts the Edison base circuit breaker. Care must be taken when installing this type of protection device, because Edison base circuit breakers are approximately ¼ inch deeper than Edison base plug fuses. When the door is closed on the panelboard, it could press the circuit breaker reset button. The pressing of the reset button (which prevents normal circuit breaker operation) by closing the panel cover has caused fires.

240-52   Edison Base Fuseholders

When Edison base fuseholders are installed in new construction, they must be converted to accept Type S fuses by the use of adapters.

Type S Plug Fuses
Sections 240-53 and 240-54

Fig. 10-21 Type S Plug Fuses

240-53   Type S Fuses

(a) Classification of Type S Fuses. Fuses of type S shall be classified at not over 125 volt and 15, 20, and 30 ampere, Fig. 10-21.

(b) Not Interchangeable with Lower Ampere Fuses. Type S fuses shall be made so that ampere classification cannot be interchanged with any other classification. Type S fuses are designed to be used only in a Type S fuseholder, or an Edison base fuseholder with a Type S adapter.

240-54   Type S Fuses, Adapters, and Fuseholders

(a) Edison Base Fuseholders. Type S adapters are designed to fit Edison base fuseholders.

(b) Prevent Edison Base Fuses. Type S fuseholders or Type S adapters are designed for Type S fuses only.

(c) Non-Removable Adapters. Type S adapters are designed so they cannot be removed once installed.

Note. Do not destroy the Edison base fuseholder by trying to remove a Type S adapter with needle nose pliers. There are tools specifically designed for the electrician to remove a Type S adapter.

(d) Tamper-Proof. Type S fuses, fuseholders, and adapters must be designed so that tampering or shunting would be difficult (a penny will not fit behind it).

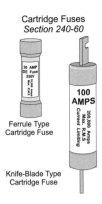

Cartridge Fuses
Section 240-60

Ferrule Type
Cartridge Fuse

Knife-Blade Type
Cartridge Fuse

Fig. 10-22 Cartridge Fuse Types

PART F. CARTRIDGE FUSES AND FUSEHOLDERS

240-60 General

There are two basic shapes of cartridge fuses, the ferrule type (maximum 60 ampere) and the knife blade type (over 60 ampere). The physical size of the fuse (length and diameter) is dependent on the fuse voltage and current rating, Fig. 10-22.

Nontime-Delay Fuses. Blow once, throw away, nontime-delay fuses are usually Class H fuses with standard 10,000 ampere **interrupting ratings** [240-60(c)]. Nontime delay (one time) fuses are general duty fuses, and can be used for most circuits. One-time fuses are less expensive than dual-element fuses and should not be used for inductive loads, such as motors, motor control circuits, and so on.

Dual-Element. This fuse has two fuse elements as the name implies. One element provides short-circuit protection, and the other provides thermal overload protection. These fuses are often used for motors and motor control circuits because of high *inrush current.* These fuses range in size from 1/10th of ampere up to 600 ampere.

Current-Limiting. A fuse that can limit the available fault let-through current. Current-limiting fuse enclosures have a special feature that keeps noncurrent-limiting fuses from being installed. See the comments following Section 240-60(b).

(a) Maximum Voltage Requirement. 300-volt type cartridge fuses and fuseholders are not permitted on circuits exceeding 300 volt between conductors.

Exception: 300 Volt to Ground. Single-phase line-to-neutral circuits, supplied from a 3-phase 4-wire, grounded neutral system, where the maximum voltage from any ungrounded conductor to ground, does not exceed 300 volt.

(b) Non-Interchangeable Fuseholders. Fuseholders must be designed to make interchanging of fuses for different voltages and current classifications difficult.

Current-Limiting. Fuseholders for current-limiting fuses must be designed to make it impossible to insert fuses that are not current limiting. Current-limiting fuses can limit the available fault-current. This Section requires that both ferrule and knife blade type fuseholders have fittings that will reject noncurrent-limiting fuses (rejection clips). See Section 240-11 for more details.

> **WARNING:**
> Care must be taken when replacing current-limiting fuses; be sure to replace the fuse with the proper AIC rating. Using the wrong interrupting current capacity fuse could literally blow the equipment off the wall [110-9 and 110-10], Fig. 10-23.

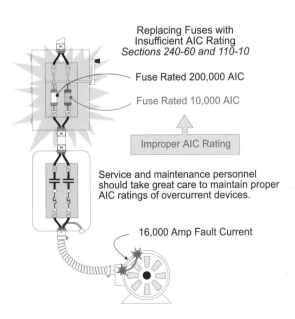

Replacing Fuses with
Insufficient AIC Rating
Sections 240-60 and 110-10

Fuse Rated 200,000 AIC

Fuse Rated 10,000 AIC

Improper AIC Rating

Service and maintenance personnel should take great care to maintain proper AIC ratings of overcurrent devices.

16,000 Amp Fault Current

Fig. 10-23 Careful When Current-Limiting Fuses

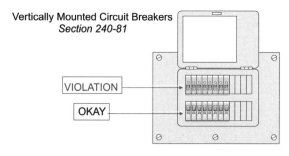

Fig. 10-24 Vertically Mounted Circuit Breakers Must Have the "Up" Position the "On" Position

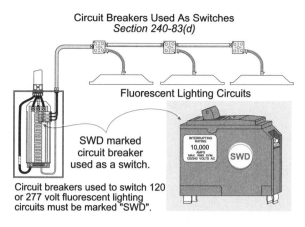

Fig. 10-25 Circuit Breakers Used as Switches for Switching Fluorescent Lighting Must Be "SWD"

(c) Marking. Cartridge fuses shall be marked with their ampere rating, voltage rating, *AIC* rating if other than 10,000 ampere, current-limiting if applicable, and the manufacturer's name or trademark.

Exception. Cartridge fuses used for supplementary protection are not required to have an interrupting rating [240-10].

PART G.   CIRCUIT BREAKERS

240-80   Method of Operation

Circuit breakers of all ratings must be capable of being opened and closed by hand. Nonmanual means of operating a circuit breaker, such as an electrical (shunt trip) or pneumatic operation, are permitted so long as the circuit breaker can also be operated manually.

Instantaneous trip circuit breakers, or motor circuit protectors (MCP), are used only for the protection of motor and motor conductors for short-circuit and ground-fault conditions. These types of circuit breakers have an adjustable trip setting that permits the protection to be adjusted from 700 percent to 1,300 percent of their basic rating. They are installed in motor control centers with coordinated overloads.

Inverse time circuit breakers are the most common types of circuit breakers, and we use them every day. They are often called molded-case circuit breakers. The term *inverse time* means that as the current level increases, the time needed for the

opening of the device decreases. This type of device is used on most circuits, but will not protect a motor against overload.

240-81   Indicating

Circuit breakers are required to show whether they are in the open "off" or closed "on," position.

Vertical Mount. Where the handle of a circuit breaker is operated vertically, the "up" position of the handle must be the "on" position [240-33 and 380-7], Fig. 10-24.

240-83   Markings

(c) Interrupting Rating. Unless marked otherwise, all circuit breakers have an interrupting rating of 5,000 ampere (AIC).

> **WARNING:**
> DO NOT replace a 14,000 ampere interrupting Capacity (AIC) circuit breaker with a 10,000 AIC circuit breaker. If circuit breakers are installed or replaced with an interrupting ampere rating insufficient to withstand the fault-current, a time bomb has been set [110-9].

(d) Circuit Breakers Used As Switches for Fluorescent Light Switching. Circuit breakers used for switching 120 or 277 volt lighting shall be listed and marked switching duty, "SWD," Fig. 10-25.

(e) Voltage Markings. Circuit breakers must be marked with a voltage rating that corresponds with their interrupting rating. See Section 110-9 for details on interrupting rating.

240-85   Use of Circuit Breakers

Straight Voltage Rating. A circuit breaker with a straight voltage rating, such as 240 or 480 volt, may be connected to a circuit where the nominal voltage between any two conductors does not exceed the circuit breaker voltage rating. They may be installed in grounded or ungrounded systems.

Slash Voltage Rating. Circuit breakers with a slash voltage rating, such as 120/240, 208/120, or 480/277 volt, can only be connected to a circuit when the nominal voltage of any conductor to ground does not exceed the lower of the two values. The nominal voltage between any two conductors cannot exceed the higher value of the circuit breaker voltage ratings. Slash voltage rated circuit breakers can be connected either line-to-neutral or line-to-line loads. This type of overcurrent protection device may be installed only on grounded systems.

240-86   Series Rated Equipment

Equipment must be marked, when suitable, as having an available fault-current higher than its marked interrupting rating, as permitted for series rated equipment. See Section 110-22 for additional rules on series rated equipment.

A new Part H titled "Supervised Industrial Installations" has been added. Specifically Sections 240-90, 91, and 92.

A new Part I titled "Overcurrent Protection Over 600 Volts" has been added. Specifically Sections 240-100 and 101.

SUMMARY

Part A.   General □ Equipment and conductors must be protected against overcurrent. □ Overcurrent includes short-circuit, ground-fault, and overload conditions. □ Fault-current is the result of short-circuits or ground-faults. □ It is generally permitted to use the next size up overcurrent protection device (800 ampere or less) if the conductor ampacity does not correspond to standard size overcurrent devices. □ Supplementary overcurrent protection is permitted for flexible cords and fixture wires. □ See Section 240-6(a) for standard ampere ratings for fuses and circuit breakers. □ Overcurrent protection devices must be connected in series with each ungrounded conductor.

Part B.   Location □ Overcurrent protection devices must be located at the point where the circuit conductors receive their power. □ A tap conductor is a conductor that does not have overcurrent protection in accordance with its ampacity. □ Overcurrent devices must be readily accessible.   □ Overcurrent protection devices must be located where they will be protected against physical damage. □ Overcurrent protection devices cannot be located near easily ignitable materials.

Part C.   Enclosures □ Overcurrent protection devices must be installed in an enclosed cabinet or cutout box. □ Enclosures that contain overcurrent protection devices must be mounted in a vertical position.

Part D.   Plug Fuses, Fuseholders, and Devices □ Edison base plug fuses are the old type of fuses. □ Plug fuses and fuseholders can only be used on circuits with not more than 125 volt between conductors. □ Ampere ratings must be marked on all fuses, fuseholders, and fuse adapters. □ Edison base fuses are permitted only for replacement in an existing installation. □ Type S fuses and fuseholders must be tamper resistant.

Part E.   Cartridge Fuses and Fuseholders ☐ A dual-element fuse has two elements. ☐ Unless otherwise marked, all fuses have a 10,000 AIC rating.

Part F.   Circuit Breakers ☐ All circuit breakers must be capable of being opened by hand. ☐ When circuit breakers are operated vertically, the "up" position must be the "on" position. ☐ Unless otherwise marked, all circuit breakers are rated 5,000 AIC. ☐ Circuit breakers used to switch 120 and 277 volt fluorescent loads must be marked switching duty, "SWD." ☐ Circuit breakers with a straight voltage rating may be connected where the nominal voltage between any two conductors does not exceed the circuit breaker voltage rating. These circuit breakers cannot be used for multiwire branch circuits or for any line-to-neutral loads. ☐ Circuit breakers with a slash voltage rating can only be connected to a circuit when the nominal voltage of any conductor to ground does not exceed the lower of the two values, and the nominal voltage between any two conductors cannot exceed the higher value.

REVIEW QUESTIONS

1. Explain the rules for overcurrent protection of circuits not over 800 ampere.

2. At what point in the circuit is overcurrent protection required?

3. Draw an example of a feeder tap not over 10 feet and indicate the *NEC* requirements.

4. Draw an example of a feeder tap not over 25 feet and indicate the *NEC* requirements.

5. Draw an example of a feeder tap not over 100 feet and indicate the *NEC* requirements.

6. Circuit breaker enclosures must be mounted in a _____ position.

7. What are the dangers of Edison base type fuses and circuit breakers?

8. When are Edison base type fuses permitted to be installed?

9. What are some of the possible dangers associated with replacing current-limiting fuses with noncurrent-limiting fuses?

10. When circuit breakers are operated vertically, the up position must be the _____ position.

11. When are circuit breakers required to be marked "SWD"?

12. Explain the rules of "straight voltage rating" and "slash voltage ratings" on circuit breakers.

13. In your own words explain the meaning of the following Article 100 definitions:

    Ampacity
    Approved
    Bathroom
    Identified
    Interrupting rating
    Overcurrent
    Overload
    Qualified person
    Readily accessible
    Weatherproof

14. In your own words explain the meaning of the following Glossary terms:

    AIC
    Current rating
    Cutout box
    Fuse
    Horsepower
    Inverse time
    Primary
    Secondary

Notes:

Unit 11

Article 250
Grounding and Bonding
Part I (Sections 250-1 through 250-70)

OBJECTIVES

After studying this unit, the student should be able to understand:
- why systems and circuit conductors are grounded.
- which conductor is solidly grounded and why.
- the purpose of the equipment grounding conductor.
- how a system is grounded.
- what a separately derived system is.
- the grounding and bonding requirements for a transformer.
- when raceways must be grounded.
- when a grounding conductor is required in a nonmetallic raceway.
- the purpose of the grounding path and how it helps clear ground-fault current.
- how many grounding electrodes are permitted when a building has two services.
- the purpose of the service main bonding jumper.
- when the neutral is permitted to serve as the equipment grounding conductor.
- the most common types of equipment grounding conductors.

Definitions and Glossary Terms

To better understand the *NEC* rules contained in this unit, review the following:

Definitions Article 100 - Unit 2

Accessible	Bonded
Bonding jumper	Effectively grounded
Equipment	Exposed
Grounding conductor	Grounding electrode
Listed	Main bonding Jumper
Pressure connectors	Qualified persons
Service	Service-entrance

Glossary Terms

Delta-connected	Electrically continuous
Electrocution	Exothermic welding
Fault-current	Ground-fault
Impedance	Load side
Low impedance	Lugs
Nonferrous	Parallel
Phase	Resistance
Spliced	Wye-connected

INTRODUCTION TO GROUNDING

Grounding is one of the most important, but one of the least understood Article in the *NEC*. This unit should help you to understand the purpose of grounding, how grounding systems work, and the terms used to describe their components. Before talking about the specific details of the grounding,

131

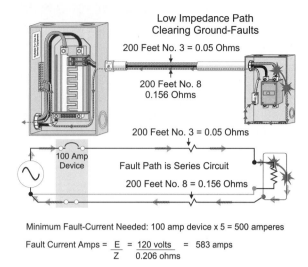

Fig. 11-1 Low Impedance Path Permits the Clearing of Ground-Faults

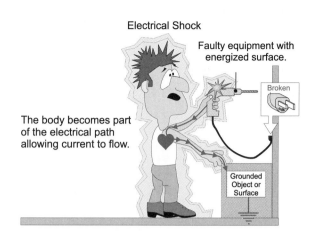

Fig. 11-2 Electrical Shock Occurs When the Body Becomes Part of the Electric Circuit

let's try to understand its purpose and what it is trying to accomplish.

Electrical systems have the potential to kill people and destroy property. Anyone associated with the installation, maintenance, or repair of an electrical system has a responsibility for the safety of others.

Case 1. An electrician accidentally separates a raceway from its coupling while working in an attic. One end of the raceway cuts through the conductor insulation and the raceway becomes energized. The electrician makes contact with the energized raceway and dies.

Case 2. The electrician fails to properly bond the service raceway. When the maintenance person closes the heavy-duty safety switch, a *ground-fault* occurs in the service equipment, resulting in an explosion. The maintenance person dies. There are many cases of overcurrent protection devices not clearing the *fault-current* due to poor grounding practices, resulting in death and fires.

The grounding system has nothing to do with how efficiently equipment operates. A washing machine will clean, a stove will cook, and lights will turn on, even if the equipment is not properly grounded. It is sad that despite today's sophisticated technology, we do not have an accessory that would show when equipment is not properly grounded.

The proper application and installation of equipment grounding will significantly reduce the hazards that exist in the use of electricity. The pur-

pose of the *National Electrical Code* is the practical safeguarding of persons and property from hazards arising from the use of electricity [90-1(a)]. Proper grounding and bonding is essential for maximum protection of life and property.

Before exploring the *Code* requirements for grounding, let's review concepts and terms that are important for us to understand Article 250.

Purpose of Grounding

The purpose of grounding electrical metal enclosures is to remove dangerous voltage (potential) that could exist during a ground-fault condition. To remove the dangerous voltage, the circuit overcurrent protection device must open to clear the ground-fault. The effectiveness of the grounding path plays a vital role in the activation of the overcurrent protection device to clear the ground-fault. To open the protection device quickly, the grounding path must be low enough to permit ground-fault current to reach a level of at least 5 times (preferably 10 times) the overcurrent protection device's rating, Fig. 11-1.

Electrical Shock Hazard

People die when voltage pushes electrons through their bodies, particularly through the heart. If a person makes contact between an object that has voltage and another object that has a different potential, current will flow between those contact points, Fig. 11-2.

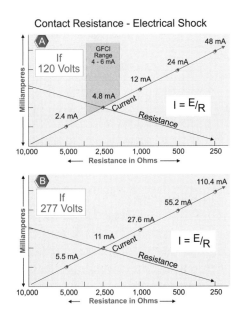

Fig. 11-3 Contact Resistance and Voltage
Determines the Level of Electrical Shock

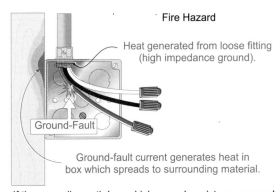

Fig. 11-4 Improper Ground - Fire Hazard·

The magnitude of the current flow through any object is dependent on two factors, contact *resistance* and the magnitude of the electromotive force (voltage). The greater the voltage, or the lower the contact resistance (wet location), the greater the current flow, (I = E/R), Fig. 11-3.

In addition, the greater the duration of time that current flows through a person, the greater the likelihood of death from electric shock.

> **Note.** Humans are susceptible to death when exposed to currents as low as 30 milliampere (.0030 ampere) for a fraction of a second.

Fire Hazard

In addition to electric shock and electrocution, electrical current can create a fire. Fire is caused by heat, and heat is a function of current squared times resistance over a period of time (Heat = Ampere² × Resistance). If the grounding path has a high resistance, the ground-fault current might not be of a sufficient magnitude to open the circuit protection device to clear the fault. This will result in dangerous voltage on all metal parts, and ground-fault current flowing (generating heat) for a period of time that could cause a fire, Fig. 11-4.

> **Note.** This happened at the MGM Grand Hotel in Las Vegas in 1980. Eighty-four people died because of a poor grounding path. There was a ground-fault, but the grounding path impedance was so high that it did not allow enough current to trip the circuit protection device. This ground-fault current continued to heat the metal raceway until it ignited nearby combustible materials.

PART A. GENERAL

250-1 Scope

Article 250 of the *NEC* covers the general rules and specific requirements of when systems must be grounded, the locations of grounding connections, the size and types of grounding conductors, bonding jumpers, and electrodes, and the methods of grounding and bonding. To better apply the *NEC* grounding rules, you must understand that there are two different methods of grounding. They are earth and safety ground and they serve different purposes.

250-2 Grounding and Bonding Requirements

(a) Earth Ground (System Grounding). The system ground is the intentional grounding of one system conductor at the power supply. The system

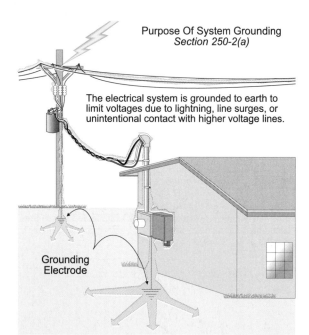

Fig. 11-5 System Grounding - Lightning Protection

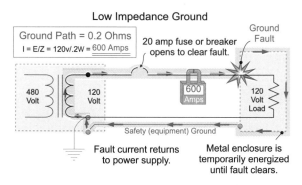

Fig. 11-7 Low Impedance Ground Path Causes Overcurrent Protection Device to Open Quickly

(b) Equipment Grounding (Safety Grounding).   The purpose of grounding electrical metal enclosures is to remove dangerous voltage (potential) that could exist during a ground-fault condition. To remove the dangerous voltage, we must clear the ground-fault by quickly opening the circuit overcurrent protection device.

To open the circuit overcurrent protection device, the safety grounding path must have sufficient *low impedance,* Fig. 11-7.

This is accomplished by bonding metal parts to each other and then bonding the metal parts to the grounded (neutral) conductor at one location in accordance with the requirements of Section 250-142.

is grounded to earth to limit voltages due to lightning, line surges, or unintentional contact with higher voltage lines. This is accomplished by bonding the **grounded (neutral) conductor** and the metal parts of the service equipment to a grounding electrode, Fig. 11-5.

In addition, systems are grounded to an electrode to stabilize the voltage to ground during normal operation, Fig. 11-6.

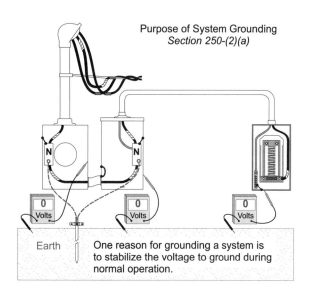

Fig. 11-6 System Grounding - Stabilizes Voltage to Ground

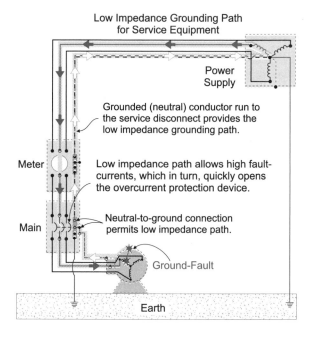

Fig. 11-8 Grounded (Neutral) Conductor Provides Low Impedance Path at Service Equipment

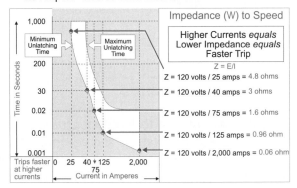

Fig. 11-9 Proper Grounding Causes the Overcurrent Protection Device to Open Quickly

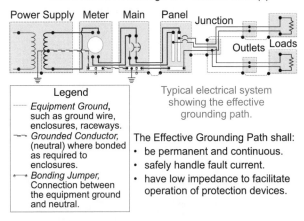

Fig. 11-10 Fault Current Grounding Path Requirements

For service equipment the grounded (neutral) conductor provides the low impedance path, when it is bonded to the service disconnecting means, Fig. 11-8.

Note. The low impedance path can consist of conductors, raceways, enclosures, boxes, housings, and other metal parts of equipment associated with an electrical system [250-118].

The time it takes for a circuit breaker or fuse to clear the fault (open) is inversely related to the magnitude of the current. This means that the greater the fault-current (low impedance path), the quicker the overcurrent device will clear the fault. For example, a 20 ampere circuit breaker will clear a fault according to the following values, Fig. 11-9:

Current	Time to Open	Impedance
25 ampere	500 seconds	4.80 ohms
40 ampere	30 seconds	3.00 ohms
125 ampere	0.01 seconds	0.96 ohms

CAUTION: The low impedance equipment safety grounding path is intended to carry sufficient fault-current to open the circuit overcurrent protection device. This is accomplished by allowing the ground-fault currents to reach a level of at least 5 times (and preferably ten times) the rating of the overcurrent protection device for 1 to 2 seconds.

(c) **Bonding Electrically Conductive Materials.** Electrically conductive materials such as metal water piping, metal gas piping and structural steel members likely to become energized shall be bonded to the system grounded (neutral) conductor. This provides a low impedance path for fault-current that will facilitate the operation of overcurrent protection device to clear the fault [250-104].

(d) **Fault Current Path.** An effective grounding path must be mechanically and *electrically continuous* and must have the capacity to safely conduct fault-currents without damage to itself. In addition, the equipment grounding path must be of sufficiently low impedance to limit dangerous voltages on metal parts by clearing the ground-fault through the opening of the circuit protection device, Fig. 11-10.

CAUTION: The equipment grounding conductor must be of sufficient size to withstand ground-fault currents [250-96(a) and Note to Table 250-122].

Note. The *NEC* requires all conductors, including the grounding conductor, to be installed in the same raceway so that the the impedance of the ground-fault path will not increase [300-3(b), 300-5(i), 300-20(a)].

Earth Not Used for Grounding. The earth cannot be used for safety grounding because the high impedance of the earth to ground-faults does

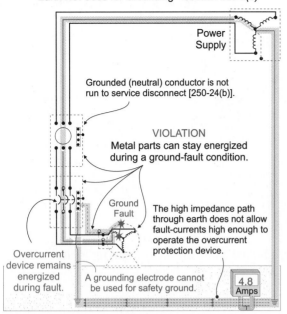

Earth Not Used for Grounding - *Section 250-2(d)*

Power Supply

Grounded (neutral) conductor is not run to service disconnect [250-24(b)].

VIOLATION
Metal parts can stay energized during a ground-fault condition.

Ground Fault

The high impedance path through earth does not allow fault-currents high enough to operate the overcurrent protection device.

Overcurrent device remains energized during fault.

A grounding electrode cannot be used for safety ground.

4.8 Amps

Fig. 11-11 Earth Cannot Be Used as a Low Impedance Ground Path

not permit sufficient current for the protection device to open [250-54]. This results in dangerous voltages remaining on the surface of metal parts, Fig. 11-11.

> CAUTION: A grounding electrode, such as a ground rod or concrete foundation steel, cannot be used for equipment grounding. The high impedance of the earth does not permit sufficient ground-fault current to clear the protection device. For example, a 25 ohm resistance ground rod would only permit 4.8 ampere of ground-fault current, Fig. 11-11.
> I = E/Z, I = 120 volt/25 ohm = 4.8 ampere

250-4 Other Code Sections

Other *Code* Sections that have additional requirements or that modify the requirements of Article 250 are:

Hazardous Locations, 501-16, 502-16, 503-16
Panelboards, 384-20
Receptacles, 210-7, 410-58 and 517-13(a)
Receptacle Cover Plates, 410-56(d)
Swimming Pools and Spas, 680-22, 680-25

Switches, 380-9(b) and 517-13(a)
Switch Cover Plates, 380-12

250-6 Objectionable Current Over Grounding Conductors

(a) Arrangement to Prevent Objectionable Current. Electrical systems and **equipment** must be grounded in a manner that will prevent objectionable neutral current from flowing through the grounding conductors or the grounding path because of an improper neutral-to-ground connections [250-142].

(b) Fixing Objectionable Currents. Objectionable neutral currents caused by improper neutral-to-ground connections must be fixed by one of the following methods:

(1) Discontinue all but one of the neutral-to-ground connections.

(2) Change the location of the grounding connections, or break the conductive path inter-connecting the grounding connections.

(3) Other methods acceptable to the authority having jurisdiction.

(c) Ground-Fault Currents Not Classified as Objectionable Currents. Ground-fault currents on the grounding conductor are not considered objectionable currents for the purposes of this Section.

(d) Computer Equipment Objectionable Currents. Currents that cause electrical noise, electromagnetic interference, or data errors in electronic equipment such as computers, magnetic tape drives, or electronic machines are not considered objectionable currents.

> **Note.** Computer/data processing equipment must be safety grounded in accordance with all the grounding requirements of Article 250. An isolated grounding conductor is permitted for computer equipment, but it must be connected to the system grounded (neutral) conductor at the **service** or separately derived system [250-146(d)].

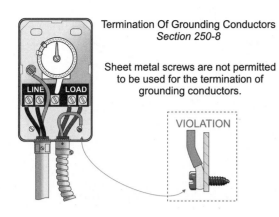

Termination Of Grounding Conductors
Section 250-8

Sheet metal screws are not permitted
to be used for the termination of
grounding conductors.

Fig. 11-12 Grounding Conductors Must Be
Terminated Properly

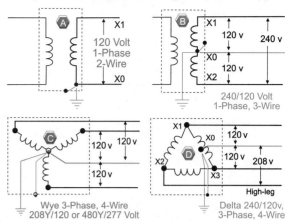

Fig. 11-13 Most Systems Are Required to
Be Grounded

250-8 Grounding and Bonding Conductor Termination

Equipment grounding conductors, grounding electrode conductors, and bonding jumpers must be terminated by *exothermic welding*, listed **pressure connectors** (set screw, compression), **listed** clamps, or other listed fitting.

Sheet-metal screws shall not be used for the termination of grounding conductors, Fig. 11-12.

250-10 Protection of Ground Clamps

Ground clamps and fittings shall be protected from physical damage by being enclosed in metal, wood, or its equivalent. For all practical purposes, the easiest way to comply with this requirement outdoors is to terminate the ground clamp on a buried electrode [250-52(c)(3), 250-68(a) Ex., 250-70].

250-12 Clean Surface

The contact surface for grounding fittings must be clean so it will provide proper electrical continuity. Paint, tarnish, or any other substance such as plaster, must be removed terminating the grounding conductors [250-52, 250-96(a)].

PART B. SYSTEM GROUNDING

250-20 System Circuit Grounding

(b) Systems Required to Be Grounded. Alternating current systems of the following types

must be earth grounded and have a neutral-to-ground connection installed, Fig. 11-13.

(1) Single-phase systems in which the maximum voltage to ground does not exceed 150, such as a 120 volt, 2-wire or 120/240 volt, 3-wire system.

(2) Three-phase, 4-wire system *wye-connected* system, such as 480Y/277 volt or 208Y/120 volt.

(3) Three-phase, 4-wire *delta-connected* system, such as 120/240 volt with high-leg.

(d) Separately Derived Systems. Wiring for **separately derived systems** whose power is derived from a generator, transformer, or converter windings, if required to be grounded as in (a) above, shall be grounded as specified in Section 250-30.

> FPN No. 1: A generator is not a separately derived system if the neutral is solidly interconnected to a service-supplied system neutral.

> **Note.** The earth ground and neutral-to-ground connection must be installed in accordance to Section 250-24 for services.

250-24 Service System Grounding

(a) Methods of Grounding System. Alternating current systems that must be grounded, as specified by Section 250-20(b), shall have a **grounding electrode conductor** bonded from the

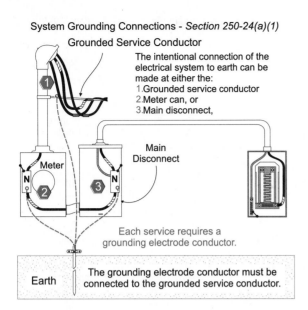

Fig. 11-14 System Grounding at the Services

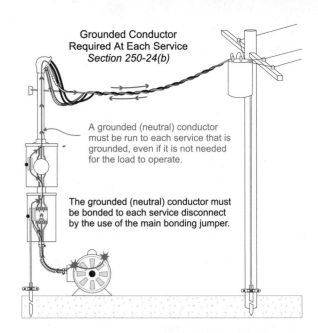

Fig. 11-15 A Grounded (Neutral) Conductor is Required at Each Service Disconnect to Provide a Low Impedance Path for Fault Current

service system grounded (neutral) conductor to a **grounding electrode** as listed in Sections 250-50 or 250-52.

(1) Grounded Conductor Termination. The connection of the service system grounded (neutral) conductor to the grounding electrode by the use of a **grounding electrode conductor** can be at any **accessible** location, from the load end of the service-drop or service-lateral, up to and including the service disconnecting means, Fig. 11-14.

> CAUTION: Some inspectors will only permit the grounding electrode conductor to originate from the meter socket enclosure, and other inspectors only permit the connection in the service disconnect.

(5) Neutral-to-Ground Connections. A neutral-to-ground connection shall not be made on the *load side* of the service disconnecting means except as permitted for separately derived systems [250-30(a)(1)], separate buildings or structures [250-32(b)(2)], or meter enclosures [250-142 Ex. 2].

A neutral-to-ground connection at any other location will cause neutral current to flow on the conductive metal parts of the electrical equipment.

This violates Section 250-6(a) and can create electric shock, fires as well as power quality problems. See Section 250-142 in this book for more information.

(b) Grounded (neutral) Conductor. Where an alternating current system is required to be grounded [250-20(b)], a grounded (neutral) conductor must be run to each service disconnect. The grounded (neutral) conductor must be bonded to the each service disconnect by the use of a **main bonding jumper**, Fig. 11-15.

> CAUTION: The **main bonding jumper** must be supplied by the manufacturer for equipment marked "service equipment" [230-66]. If you fail to install the main bonding jumper at the service, the equipment grounding system will not have a low impedance path for ground-fault current [250-2(b)].

(1) Minimum Size Grounded (neutral) Conductor. The grounded conductor must be routed with the phase conductor and it must must be sized not smaller than specified in Table 250-66, and it must have the capacity to carry unbalanced neutral current [220-22].

Service Neutral Size For A Dwelling Unit
Section 250-24(b)(1)

4/0 THHN Service Phase Conductors

100-ampere maximum unbalanced load.

Use No. 2 THHN

The neutral must be capable of carrying the 100 ampere unbalanced load but it cannot be smaller than the grounding electrode conductor size from Table 250-66.

Fig. 11-16 Service Neutral Sized to Carry Maximum Unbalanced Load

Question. What is the minimum size service grounded (neutral) conductor, if the service phase conductors are No. 4/0, Fig. 11-16.

Answer.   No. 2, Table 250-66

(2) Parallel Grounded Conductor. Where **service-entrance** conductors are installed in parallel, the size of the grounded (neutral) conductor in each raceway shall be based on the size of the largest phase conductor in the raceway. But in no case can the grounded (neutral) conductor be smaller than No. 1/0 [310-4].

(c) Grounding Electrode Conductor. A **grounding electrode conductor** shall be used to connect the equipment grounding conductor of the service, the service equipment, and the grounded (neutral) conductor to one of the electrodes as listed in Part C of this Article. The grounding electrode conductor must be sized to Table 250-66, based on the size of the service phase conductors.

250-30 Grounding Separately Derived Systems

Note. A separately derived system is a premises wiring system that derives its power from battery, solar photovoltaic system, a generator, transformer, or converter winding. Separately derived systems have

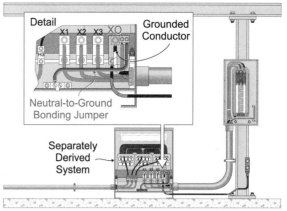

Grounding and Bonding of Separately Derived Systems
Section 250-30(a)(1)

Detail

X1 X2 X3 XO

H3

Grounded Conductor

Neutral-to-Ground Bonding Jumper

Separately Derived System

Secondary neutral-to-ground bonding jumper connects the metal case of the transformer to the grounded (neutral) conductor.

Fig. 11-17 Grounding and Bonding of Separately Derived Systems Must Be Done at One Location

no direct electrical connection to supply conductors of another system [Article 100 and 250-20(d)].

(a) Methods of Grounding System. The grounding of separately derived systems must comply with (1) through (3).

(1) Neutral-to-Ground Bonding Jumper. The metal case of the derived system (equipment grounding conductor) must be bonded to the system grounded (neutral) conductor, at the separately derived system or at the first system disconnect, Fig. 11-17.

The **bonding jumper** for this purpose is sized based on the phase conductor size, using Table 250-66.

Exception. Bonding at the separately derived system and at the first system disconnect is permitted if this does not permit parallel paths for neutral current to flow through metal raceways, building steel, piping, framing, etc.

(2) Grounding Electrode Conductor. A grounding electrode conductor must connect the system grounded (neutral) conductor to a suitable grounding electrode.

To ensure proper single-point grounding of separately derived systems, the termination of the grounding electrode conductor must be at the same

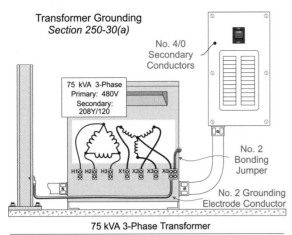

Transformer Grounding
Section 250-30(a)

No. 4/0 Secondary Conductors

75 kVA 3-Phase
Primary: 480V
Secondary: 208Y/120

No. 2 Bonding Jumper

No. 2 Grounding Electrode Conductor

75 kVA 3-Phase Transformer

Bonding Jumper: 250-30(a)(1), based on the largest secondary phase conductor size, Table 250-66, use No. 2 wire.

Grounding Electrode Conductor: 250-30(a)(2), based on secondary conductor size, Table 250-66, use No. 2 wire.

Fig. 11-18 Grounding Electrode Conductor Sized to Secondary Conductors, Use Table 250-66

point on the separately derived system where the neutral-to-ground connection is made as required in (1) above.

The grounding electrode conductor for this purpose is sized according to the size of the ungrounded secondary conductors, using Table 250-66, Fig. 11-18.

(3) Grounding Electrode (Nearest Available). The grounding electrode conductor must terminate to a grounding electrode located as close as practicable, and preferably in the same area of the neutral-to-ground termination of the derived system in (2) above.

The grounding electrode shall be the nearest one of the following:

(a) Effectively grounded metal member of the building structure.

(b) Effectively grounded metal water pipe, but only if it's within 5 feet from the point of entrance of the water pipe into the building.

(c) Where electrodes specified by (a) or (b) above are not available, then any of the following can be used:

• Grounded metal structure [250-50(b)]
• Concrete-encased electrode [250-50(c)]
• No. 2 ground ring [250-50(d)]
• Made electrode, ground rod [250-52]

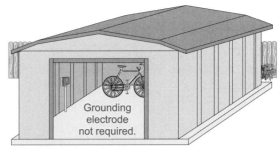

Grounding Electrode at Separate Building or Structure
Section 250-32(a) Exception

Grounding electrode not required.

A grounding electrode is not required at a separate building or structure that has only one branch circuit.

Fig. 11-19 Grounding Electrode at Separate Building

FPN: Interior metal water piping in the area served by a separately derived system must be bonded to the grounded (neutral) conductor at the separately derived system [250-104(a)(4)].

250-32 Buildings or Structures Supplied by a Feeder or Branch Circuit

(a) Grounding Electrode. Each building or structure supplied by a feeder or branch circuit shall have the building disconnecting means [225-31] grounded to a grounding electrode.

Exception: One Branch Circuit. A grounding electrode at separate buildings or structures is not required where only one branch circuit supplies the building or structure. However, the single branch circuit must include an equipment grounding conductor, Fig. 11-19.

(b) Grounded Systems. An electrical connection is required from the metal enclosure of the building disconnecting means to the grounding electrode in accordance with either (1) or (2).

(1) Equipment Grounding Conductor. An equipment grounding conductor shall be run with the supply conductor and it must terminate to the metal enclosure of the building or structure disconnecting means and to a grounding electrode.

Under this condition, a neutral-to-ground connection must not be made to the equipment

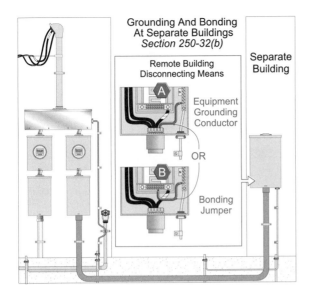

Fig. 11-20 Grounding and Bonding at Separate Buildings

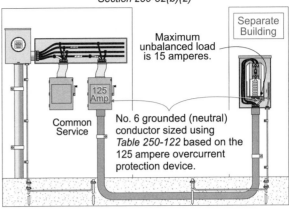

Fig. 11-22 Sizing Grounded Conductor at Separate Buildings When No Equipment Ground Is Provided

grounding conductor or to the grounding electrode conductor, Fig. 11-20.

(2) Neutral-to-Ground Connection. Where an equipment grounding conductor is not run with the supply conductors, the grounded (neutral) conductor must be **bonded** to the building or structure disconnecting means and to the grounding electrode(s), Fig. 11-21.

This method of grounding separate building or structure disconnecting means is only permitted

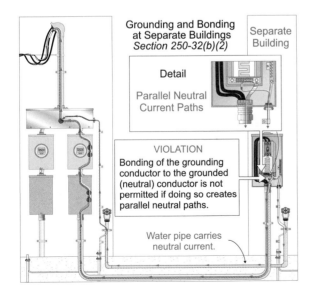

Fig. 11-21 Grounding at Separate Buildings Must Permit Parallel Neutral Current Paths

where a neutral-to-ground connection does not create parallel paths for neutral current, Fig. 11-22.

CAUTION: The 1996 *NEC* specified that when no equipment grounding conductor was run to a separate building or structure, the grounded (neutral) conductor was required to be sized not smaller than that specified in Table 250-122, based on the size of feeder overcurrent protection device. This requirement was accidentally omitted when Article 250 was rewritten, Fig. 11-22.

Note. Interior metal piping system must be bonded to a low impedance path in accordance with Section 250-104(c).

(e) Buildings or Structures Containing Livestock. Buildings or structures containing livestock must have a copper equipment grounding conductor to the remote building or structure disconnect enclosure. That portion of the equipment grounding conductor run underground to the separate building or structure must be insulated or covered copper, Fig. 11-23.

Note. Agricultural buildings have additional grounding and bonding requirements in Section 547-9(b), Fig. 11-24.

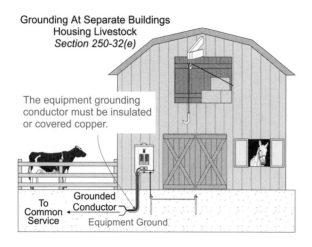

Fig. 11-23 Buildings Housing Livestock Must Have Insulated Equipment Grounding Conductor

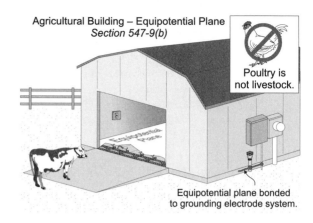

Fig. 11-24 Agricultural Building Equipotential Plane Bonded to Grounding Electrode System

(f) Grounding Conductor to Electrode. The grounding conductor from the disconnect to the grounding electrode must be sized based on the feeder or branch circuit supply overcurrent protection device in accordance with Table 250-122. In addition, the grounding conductor to the grounding electrode is not required to be larger than the ungrounded supply conductors, Fig. 11-25.

Note. The grounding electrode conductor for service equipment is sized to Table 250-66 [250-24(c), 250-50].

CAUTION: An exception to this Section in the 1996 NEC stated that the grounding conductor to a "made electrode" was not required to be larger than No. 6, and not larger than No. 4 to a "concrete-encased electrode." This exception was accidentally omitted during the 1999 Code rewrite of Article 250.

250-34 Generators-Portable and Vehicle-Mounted

(a) Portable Generators. Portable generators are used in a variety of conditions, such as construction sites, stand-by power, or emergency power. The frame of a portable generator frame can serve as the grounding electrode if both (1) and (2) apply.

(1) The generator supplies only equipment mounted on the generator or cord- and plug-con-

nected equipment to receptacles mounted on the generator.

(2) The noncurrent-carrying metal parts of generator equipment and generator receptacle equipment grounding terminals are bonded to the generator frame.

(b) Vehicle-Mounted Generators. Vehicle-mounted generators are often used for carnivals, military, television stations for sporting events, and so on. The frame of a vehicle-mounted generator can serve as the grounding electrode if all of the following conditions are met, Fig. 11-26:

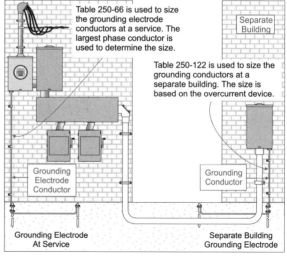

Fig. 11-25 Sizing Grounding Conductor to Electrode at Separate Building

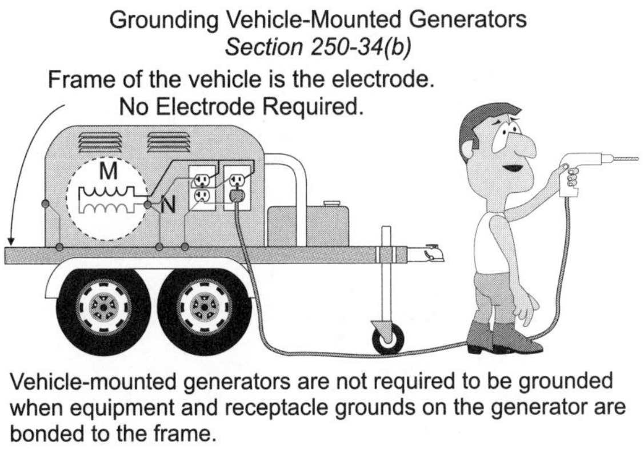

Grounding Vehicle-Mounted Generators
Section 250-34(b)
Frame of the vehicle is the electrode.
No Electrode Required.

Vehicle-mounted generators are not required to be grounded when equipment and receptacle grounds on the generator are bonded to the frame.

Fig. 11-26 Grounding Vehicle-Mounted Generators

(1) The vehicle-mounted generator is bonded to the vehicle frame.

(2) The generator supplies only equipment mounted on the generator or cord- and plug-connected equipment to receptacles mounted on the generator.

(3) The noncurrent-carrying metal parts of generator equipment and generator receptacle equipment grounding terminals are bonded to the generator frame.

PART C. GROUNDING ELECTRODE SYSTEM AND GROUNDING ELECTRODE CONDUCTOR

The intentional earth grounding of a system is to limit voltages due to lightning, line surges, or unintentional contact with higher voltage lines. In addition, systems are grounded to stabilize the voltage to ground during normal operation [250-2(a)].

Note. A grounding electrode system serves as a terminal to the earth for the purpose of grounding systems. It is not intended to carry fault-current to open an overcurrent protection device to clear the fault, Figure 0029.

250-50 Grounding Electrode System

At each building or structure, all electrodes listed in (a) through (d) of this Section that are

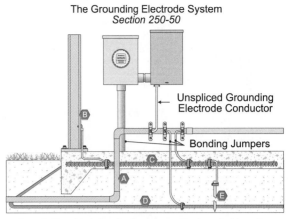

The Grounding Electrode System
Section 250-50

Unspliced Grounding Electrode Conductor

Bonding Jumpers

If available, the following items must be bonded together to form a Grounding Electrode System:
A: Metal Underground Water Pipe, 250-50(a).
B: Metal Frame of Building, 250-50(b).
C: Concrete-Encased Electrode, 250-50(c).
D: Ground Ring, 250-50(d).
E: Made Electrode, 250-52(c).

Fig. 11-27 Grounding Electrode System

available must be bonded together to form the grounding electrode system[250-58], Fig. 11-27.

> CAUTION: Section 250-50 specifies that all electrodes that are available on the premises must be bonded. Does the NEC require you to open up the concrete after it has been poured to get to the building steel? Careful, some inspectors require this.

The available electrodes must be bonded together by a bonding jumper sized in accordance with Table 250-66. The bonding jumper must be installed in accordance to the requirements of Section 250-64.

Attachment Fitting. The attachment fittings for the grounding electrodes can be installed underground or encased in concrete [250-52(c)(3), 250-68(a) Ex.] and it must be listed and identified for this use [250-70].

Grounding Electrode Conductor. The grounding electrode conductor can terminate to any convenient grounding electrode [250-64(c)] of the grounding electrode system, Fig. 11-27.

The grounding electrode conductor must be sized based or the largest grounding electrode con-

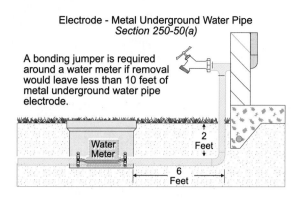

Fig. 11-28 Bonding Jumper for Metal Underground Water Pipe Electrode

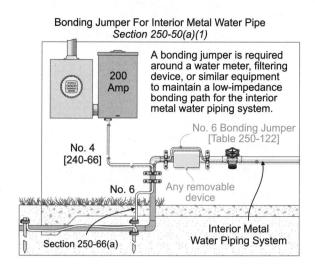

Fig. 11-29 Bonding Jumper for Interior Metal Water Pipe System

ductor required for any of the electrodes in accordance with Section 250-66.

Splicing of the grounding electrode conductor is permitted only by the use of exothermic welding or by irreversible compression-type connectors that are listed for the purpose [250-64(b)].

Interior Metal Pipe. Interior metal water pipe located more than 5 feet from the point of entrance to the building shall not serve as a conductor for the bonding of electrodes together.

Note. Interior metal water piping further than 5 feet from the point of entrance shall not be used as a conductor to interconnect electrodes because of the increasing use of nonmetallic pipe for repairs.

Exception: Industrial and Commercial Buildings. Interior metal water pipe located more than 5 feet from the point of entrance to the building can serve as a conductor for the bonding of electrodes, where conditions of maintenance and supervision ensure that only **qualified persons** will service the installation, and the entire length of the interior metal water pipe used for the bonding is **exposed**.

(a) Metal Underground Water Pipe Electrode. Underground metal piping used as a grounding electrode must be electrically continuous for a minimum of 10 feet. Bonding jumpers of sufficient length must be installed around any metal parts such as water meters or filters that may be removed that would reduce the grounding electrode to less than 10 feet, Fig. 11-28.

The bonding jumpers for (1) and (2) that follows must be sized according to Table 250-66 [250-50(a)].

> CAUTION: Don't confuse the grounding electrode conductor requirements which is used to divert lightning to the earth, with the interior metal water pipe bonding requirements [250-104(a)].

(1) Interior Piping Continuity. Where it is necessary to maintain the *electrical continuity* of the interior metal water pipe, bonding jumpers must be installed around insulated joints, filtering devices, and similar equipment [250-68(b)] in accordance with Section 250-104(a), Fig. 11-29.

Note. Section 250-104(a) specifies that the bonding jumper must be sized to the circuit overcurrent protection device size in accordance with Table 250-122.

> CAUTION: Don't confuse the bonding of the interior water pipe system, which is intended to carrying ground-fault-current [250-104(a)], with the requirements of using the underground metal water pipe as a grounding electrode [250-50(a)].

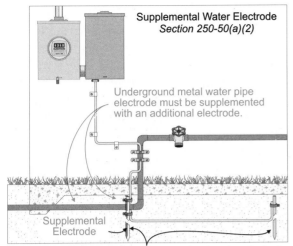

Fig. 11-30 Supplemental Electrode for
Underground Water Pipe Electrode

Note. Controversy about using the metal
underground water pipe as an electrode has
existed for many years. Plumbers feel that
neutral current flowing on the metal water
pipe system corrodes the metal (they're
right).

**(2) Supplemental Electrode Bonding
Jumper.** Because plastic water pipe is often used
to replace damaged underground metal water pipe,
the underground metal water pipe electrode must be
supplemented with an additional electrode. The
additional electrode can the metal frames of buildings that are **effectively grounded** [250-50(b)]
concrete-encased electrodes [250-50(c)] or a made
electrode such as a ground rod [250-52(c)].

Where the supplemental electrode is a single
made electrode and it does not have a resistance to
ground of 25 *ohms* or less, it must be bonded to an
additional made electrode [250-56], Fig. 11-30.

Note. For all practical purposes, a single
made electrode such as a ground rod generally will not provide a resistance of 25 ohm
or less.

The supplemental electrode for the water pipe
electrode can be bonded to:
• Grounded service conductor,
• Grounded service raceway or enclosure, or
• Any other grounding electrode.

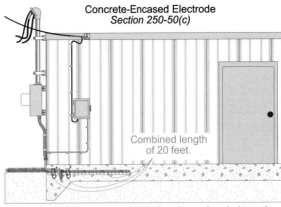

Steel reinforcing rods (rebar) ½ inch or larger bonded together
by tie wire that have at least a combined length of 20 feet can
serve as the grounding electrode.

Fig. 11-31 Concrete-Encased Electrode

The bonding conductor for the supplemental
electrode must be sized in accordance with Table
250-66, but it is not required to be larger than No. 6
copper to a made electrode.

(b) Metal Frame of Building. The effectively
grounded metal frame of a building can service as a
grounding electrode.

The grounding electrode conductor to this type
electrode is sized according to Section 250-66.

(c) Concrete-Encased Electrode. Bare, galvanized, or other electrically conductive coating,
reinforcing steel rods not smaller than ½ inch in
diameter can be used as an electrode if the total
length of the steel is not less than 20 feet. In addition the steel must be encased in at least 2 inches of
concrete, and it must be located near the bottom of a
foundation or footer that is in direct contact with the
earth. Reinforcing steel re-bar can be bonded
together by steel tie wires or other effective means
to create the 20 foot concrete-encased electrode.

The largest grounding electrode conductor
required for this type of electrode is No. 4 copper
[250-66(b)].

Question. Does the concrete foundation or
footer steel rebar grounding electrode have to be
one continuous length of 20 feet, Fig. 11-31?

Answer. No

Note. Concrete-encased re-bar meeting the
above requirements is also often called a

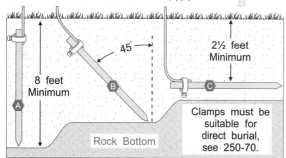

Made Electrode (Ground Rod) Installations
Section 250-52(c)(3)

45°

2½ feet
Minimum

8 feet
Minimum

B

C

A

Clamps must be
suitable for
direct burial,
see 250-70.

Rock Bottom

Top end of electrode must be flush with or below ground level
unless the aboveground end and the grounding electrode
conductor attachment are protected against physical damage.

Fig. 11-32 Made Electrode Installation
Requirements

"Ufer Ground," named after Herb Ufer, the person who determined its usefulness as a grounding electrode.

250-52 Made and Other Electrodes

If none of the electrodes specified in Section 250-50 are available, then local metal underground systems (not gas), ground rods, ground pipes, or plate electrodes can be used. Where practical, made electrodes must be installed below the permanent moisture level, and where more than one grounding electrode is installed, the grounding electrodes must not be installed any closer than 6 feet from each other [250-56].

(a) Underground Metal Gas Pipe System. Metal underground gas piping systems shall not be used as the grounding electrode. However, the metal gas piping system above the ground must be effectively grounded in a manner specified in Section 250-104(b).

(b) Other Underground Metal Systems or Structures. Made electrodes can be underground piping systems or structures such as underground tanks.

The grounding electrode conductor for this type of electrode shall be sized according to Section 250-66.

(c) Made Electrode.

(2) Ground Rod(s). Ground rod(s) made of galvanized iron or steel must be at least 5/8 inch in diameter. *Nonferrous* rods such as copper, brass, bronze, or stainless steel must be at least ½ inch in diameter, and they must be listed.

The grounding electrode conductor for this type of electrode is not required to be larger than No. 6 [250-66(a)].

> **Note.** The diameter of a ground rod has an insignificant effect on the resistance of the grounding electrode. However, larger diameter ground rods (to 1 inch) are often installed to compensate for the loss of metal due to corrosion or for mechanical strength needed for its installation.

(3) Ground Rod Installation. Ground rod electrodes must have a minimum of 8 feet of soil contact and an 8-foot rod must be driven all the way into the earth [250-68(a) Ex.]. When solid rock is encountered, the electrode shall be driven at a maximum angle of 45°, or if this is not possible, they may be buried at least 2½ feet deep, Fig. 11-32.

The top end of the ground rod can be above, flush with, or below the ground level. When the top end of the electrode is above the ground, the grounding electrode attachment fitting must not be subjected to physical damage [250-10]. When the grounding electrode attachment fitting is located underground, it must be listed for direct burial [250-70], Fig. 11-32.

250-54 Supplemental Grounding Electrodes

Additional electrodes can be installed to supplement the equipment grounding conductor. As a matter of fact, manufacturers of computer controlled machine tools often suggest an addition connection to the earth to reduce RF noise to improve the performance of the equipment. Additional electrodes for this purpose is permitted by the *NEC* and they are not required to be bonded to the building grounding electrode system.

Earth for Ground. However, grounding electrodes cannot serve as the sole equipment grounding conductor because the resistance of the earth is so high [250-2, 250-50, 250-56], Fig. 11-33.

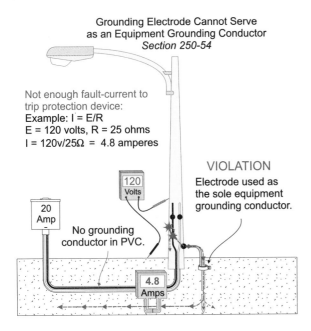

Fig. 11-33 The Grounding Electrode Cannot be Used as an Equipment Grounding Conductor

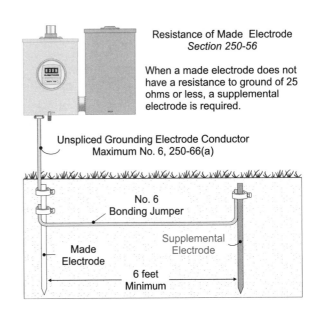

Fig. 11-34 Where Resistance of Made Electrode Exceeds 25 Ohms, Supplement Electrode Required

250-56 Resistance to Ground for Made Electrodes

The resistance of a made electrode is affected by the resistivity of the soil by the following factors:

- The soil type (topsoil vs. sandy loam),
- Moisture content,
- Salt content, and
- Temperature

When the resistance to ground for a single made electrode is over 25 ohm, one additional made electrode must be installed, but not closer than 6 feet from any other electrode. In addition, both electrodes must be bonded together [250-58], Fig. 11-34.

FPN: Spacing electrodes farther than 6 feet apart will lower the total resistance to ground. Ideally, electrodes should be spaced apart 2 times the length of the electrode. For example, two 8-foot ground rods should be spaced 16 feet apart for optimum performance.

Note. The resistance also can be decreased by driving longer ground rods.

250-58 Bonding Electrodes Together

When a building has more than one feeder or service as permitted in Sections 225-30 and 230-2(a), the same grounding electrode system shall be used to reduce the difference of potential between the different systems. This is often accomplished by using the effectively grounded metal frame of metal buildings [250-50(b)] or the concrete-encased structural reinforcing steel of the building [250-50(c)], Fig. 11-35.

Electrodes that are effectively bonded together are considered as one electrode.

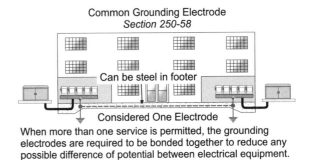

Fig. 11-35 Common Grounding Electrode Required for Multiple Services

250-60   Lightning Protection

Lightning protection air terminal conductors or their electrodes shall not be used as the grounding electrode for a building or structure as required in Section 250-52. However, the lightning protection system electrode must be bonded to the building or structure grounding electrode system [250-106].

250-62   Material

(a) Grounding Electrode Conductor Material. The grounding electrode conductor shall be copper, aluminum, or copper-clad aluminum and it can be solid or stranded, as well as insulated, covered, or bare.

> **Note.** The *NEC* does not have any color identification requirement for the grounding electrode conductor, but most electricians use the color green.

250-64   Installation

(a) Aluminum Grounding Electrode Conductor. Aluminum grounding electrode conductors cannot be in contact with earth, masonry, or other corrosive conditions. When aluminum conductors are used outdoors, the connection to the grounding electrode shall be made at least 18 inches above the earth.

(b) Protection. Grounding electrode conductors, No. 8 and smaller, must be installed in rigid metal conduit, intermediate metal conduit, nonmetallic conduit, electrical metallic tubing, or armored cable. Grounding electrode conductors, No. 6 and larger and not subject to physical damage, can be run exposed, if run properly along the surface of the building.

> **Note.** Metal raceways that contain the grounding electrode conductor must have both ends bonded [250-92(a)(3), 250-102(c)].

(c) Grounding Electrode Conductor Splice Permitted. The grounding electrode conductor can not be spliced except by the use of exothermic

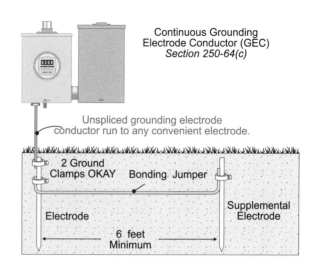

Fig. 11-36 Unspliced Grounding Electrode Conductor

welding or irreversible compression-type connectors that are listed for the purpose.

In addition, the grounding electrode conductor can terminate to any convenient grounding electrode on the grounding electrode system [250-50], Fig. 11-36.

(d) Grounding Electrode Conductor Taps. When a service has two or more disconnects as permitted in Section 230-40, a grounding electrode tap is permitted if:

(1) The tap conductor is connected from each service disconnect grounded (neutral) conductor to the grounding electrode conductor,

(2) The tap is sized to Section 250-66, based on the largest ungrounded (phase) conductor, and

(3) The tap is connected to the grounding electrode conductor in a way that there are no splices or joints in the grounding electrode conductor itself.

(e) Raceway for Grounding Electrode Conductor. Raceways used for the protection of the grounding electrode conductor shall be installed according to the *Code* requirements for that raceway system. Metal raceways that contain the grounding electrode conductor must have both ends bonded to the grounding electrode conductor [250-92(a)(3), 250-102(c)].

250-66 Grounding and Bonding Conductor Sizing (Line-Side)

The grounding electrode conductor or bonding jumper for the following applications must be sized based on the total area of the phase conductors, using Table 250-66:

Grounding Electrode Conductor
• Services, 250-50
• Separately derived systems 250-30(a)(2)

Bonding Jumper
• Grounding electrode, 250-50
• Service conductor 250-28, 250-79(d)
• Service raceways, 250-94(4) and 250-102(c)
• Water pipe, 250-104

Table 250-66
Grounding Conductor Requirements

Service Entrance Conductor or Equivalent Area for Parallel Conductors		Size of Conductor	
Copper	Aluminum	Copper	Aluminum
No. 2 or smaller	1/0 or smaller	8	6
No. 1-1/0	2/0-3/0	6	4
2/0-3/0	4/0-250 kcmil	4	2
4/0-350 kcmil	300 kcmil-500 kcmil	2	1/0
400 kcmil-600 kcmil	600 kcmil-900 kcmil	1/0	3/0
700 kcmil-1,100 kcmil	1,000 kcmil-1,750 kcmil	2/0	4/0
Over 1,100 kcmil	Over 1,750 kcmil	3/0	250 kcmil

Note. Grounding and bonding conductors on the load-side of service equipment are sized based on the overcurrent protection device size in accordance with Section 250-122.

Parallel Installations. If the conductors are run in parallel, the grounding or bonding jumper for the above applications is sized according to the total area of the parallel conductors.

Note. See Section 310-4 for additional rules for grounding conductors in parallel.

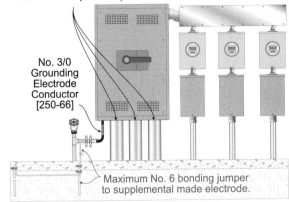

Connections to Made Electrode
Section 250-66(a)

600 kcmil per raceway x 3 raceways equals 1800 kcmil equivalent phase conductors

No. 3/0 Grounding Electrode Conductor [250-66]

Maximum No. 6 bonding jumper to supplemental made electrode.

Fig. 11-37 No. 6 Maximum Size Grounding Electrode Conductor to a Made Electrode

(a) Made Grounding Electrodes. The grounding electrode conductor to a made electrode is not required to be larger than No. 6 copper [250-50(a)(2)].

Question. What size grounding electrode conductor is required to a water pipe electrode for a service that is installed with three parallel sets of 600 kcmil conductor per phase? What size bonding jumper is required to a made electrodes, Fig. 11-37?

Answer. No. 3/0 and No. 6
No. 3/0 to the water pipe and a No. 6 to the supplemental made electrodes.

(b) Concrete-Encased Electrodes. The grounding electrode conductor to a concrete-encased electrode shall not be required to be larger than No. 4 copper [250-50(c)].

250-68 Grounding Electrode Connection

The attachment of the grounding electrode conductor to the grounding electrode shall be **accessible** and must assure an effective grounding connection.

Exception. Concrete-encased, or buried grounding electrode connections, are not required to be accessible.

Note. The termination fitting that is concrete-encased or buried must be listed for this purpose [250-50, 250-53(c)(3), 250-70].

250-70 Grounding Electrode Fitting

The grounding electrode conductor shall be connected to the grounding electrode fitting by exothermic welding, listed *lugs*, listed pressure con- nectors, listed clamps, or by other listed means. The ground clamps must be listed for the materials of the grounding electrode and the material of the grounding electrode conductor.

Direct Burial. Grounding termination fittings that are concrete-encased or buried in the earth must be listed for direct burial and marked "DB," see Sections 250-50, 250-52(c)(3)].

No more than one conductor can terminate on a single clamp, unless the clamp is approved for more than one conductor.

SUMMARY

Introduction to Grounding   ☐ Grounding electrical metal enclosures is to remove dangerous voltage that could exist during a ground-fault condition.   ☐ People die when voltage pushes elec- trons through their bodies, particularly through the heart.   ☐ Ground-faults can create a fire and electric shock.

Part A. General   ☐ System grounding is to limit voltage on metal parts from nearby lightning strikes, static charge, and voltage transients.   ☐ Equipment grounding removes dangerous voltage from ground-faults by neutral-to-ground connection at the power supply.   ☐ Metal water piping and structural steel members likely to become energized shall be bonded to the system grounded (neutral) conductor.   ☐ Grounding path must be mechanically and electrically continuous and have the capacity to withstand fault-currents without damage to itself.   ☐ The earth is a high impedance path for ground-fault current and cannot be used for equipment grounding.   ☐ Electrical systems and equipment must be grounded to prevent objectionable current from flowing through the grounding conductors or the grounding path.   ☐ Currents that cause electrical noise, electromag- netic interference, or data errors are not considered objectionable currents.   ☐ Equipment grounding conductors, grounding electrode conductors, and bonding jumpers must be connected by exothermic welding, listed pressure connectors, listed clamps, or other listed means. ☐ Grounding fittings shall be protected from physical damage.   ☐ The contact surface for ground- ing fittings must be clean so it can provide proper electrical continuity.

Part B. System Grounding   ☐ Grounded AC systems must have a grounded (neutral) conductor bonded to a grounding electrode.   ☐ Generally a neutral-to-ground connection shall not be made to any grounded circuit conductor on the load side of the service disconnecting means.   ☐ Equipment grounding conductors of a derived system must be bonded to the system grounded (neutral) con- ductor at the separately derived system and/or the first system disconnect, so there are no parallel paths for neutral current. ☐ Grounding electrode conductor must be connected to a grounding elec- trode in the same area of the derived system.   ☐ Building or structure supplied by a feeder shall have the building disconnecting means grounded to a grounding electrode.   ☐ The frame of a portable and vehicle-mounted generators are not required to be grounded and the case can serve as the grounding electrode.

Part C. Grounding Electrode System and Grounding Electrode Conductor
☐ The grounding electrode conductor can terminate to any convenient grounding electrode of the grounding electrode system.   ☐ Splicing of the grounding electrode conductor is permitted by exothermic welding or by irreversible compression-type connectors listed.   ☐ Interior metal water

pipe located more than 5 feet from the point of entrance to the building shall not serve as a conductor for the bonding of electrodes together.

REVIEW QUESTIONS

Introduction to Grounding

1. Why are electrical metal enclosures grounded?

2. How do ground-faults currents create a fire?

3. How do ground-fault currents create a condition for electric shock?

Part A. General

4. What is the purpose of system grounding?

5. What is the purpose of equipment grounding?

6. What must we do with metal water piping and structural steel members that are likely to become energized?

7. Grounding path must be mechanically and electrically continuous and must have _____ impedance path to fault-current.

8. Why can't the earth be used as the sole equipment grounding conductor?

9. Give an example of objectionable current flowing through a grounding conductor.

10. Are currents that cause electrical noise, electromagnetic interference, or data errors considered objectionable currents?

11. Equipment grounding conductors, grounding electrode conductors, and bonding jumpers must be connected by _____, listed pressure connectors, listed clamps, or other listed means.

12. Grounding fitting shall be protected from _____.

13. The contact surface for grounding fittings must be _____ so it can provide proper electrical continuity.

Part B. System Grounding

14. To what must the grounded (neutral) conductor of a AC system be bonded?

15. Where can neutral-to-ground connection be made on the load side of the service disconnecting means?

16. Equipment grounding conductors of a derived system must be bonded to the system grounded (neutral) conductor at _____ and/or _____, so there are no parallel paths for neutral current.

17. The grounding electrode conductor of a separately derived system must be connected from the grounded system conductor to the _____ in the same area of the derived system.

18. Buildings or other structures supplied by a feeder shall have the building or structure disconnecting means grounded to a _____.

19. The frame of portable and vehicle-mounted generators _____ required to be grounded and the case can serve as the _____.

Part C. Grounding System and Grounding Electrode Conductor

20. Where can the grounding electrode conductor from the service grounded (neutral) conductor terminate?

21. When can the grounding electrode conductor be spliced?

22. Interior metal water pipe located more than 5 feet from the point of entrance to the building shall not serve as _____.

23. In your own words explain the meaning of the following Article 100 definitions:

    Accessible
    Bonded
    Bonding jumper
    Effectively grounded
    Equipment
    Grounding conductor
    Grounding electrode
    Qualified persons
    Service

24. In your own words explain the meaning of the following Glossary terms:

    Delta-connected
    Electrically continuous
    Exothermic welding
    Fault-current
    Ground-fault
    Impedance
    Load side
    Low impedance
    Nonferrous
    Parallel
    Phase
    Wye-connected

Unit 12

Article 250
Grounding and Bonding
Part II (Sections 250-80 through 250-148)

OBJECTIVES

After studying this unit, the student should be able to understand:
- why service raceways and enclosures bond differently from nonservice equipment.
- the method of bonding service equipment.
- the method of bonding EMT containing service conductors.
- when a receptacle does not require an equipment bonding jumper.
- special bonding requirements for 480Y/277 volt systems.
- the purpose of bonding the interior metal water pipe system.
- when the grounding electrode conductor runs in a metal pipe, why both ends of the metal raceway must be bonded.
- the grounding electrode requirements for underground water pipes.
- the rules for the supplemental grounding electrode for the underground water pipe.
- the requirements of installing ground rods used as the grounding electrode.
- how and when to apply Table 250-66.
- the rules for attachment of the grounding electrode conductor to the grounding electrode.

Definitions and Glossary Terms

To better understand the *NEC* rules contained in this unit, review the following:

Definitions Article 100 - Unit 2
Accessible	Approved
Bonding	Feeder
Raceway	Service equipment

Glossary Terms
Bonding bushings	Bonding locknut
Bosses	Circular mil
Elbows	Electrical continuity
Enclosures	Flex

Hubs	Knockout
Load side	Parallel
Pigtail	Plaster ring
Studs	Yoke

PART D. ENCLOSURE GROUNDING

250-80 Service Enclosures

Metal raceways and enclosures containing service conductors shall be grounded.

Exception: Isolated Metal Elbows. Isolated metal *elbows* installed underground isolated from

153

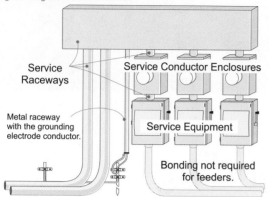

Service Bonding - *Section 250-92(a)*
Service bonding requirements apply to all service enclosures, service raceways, and the metal raceway containing the grounding electrode conductor.

Fig. 12-1 Service Equipment and Raceways Must Be Bonded

possible contact to persons by a minimum of 18 inches of cover are not required to be grounded.

Note. See Section 300-5 for the installation requirement of underground raceways.

250-86 Other Enclosures

All metal raceways and enclosures containing electrical conductors shall be grounded.

Exception No. 2: Cables in Raceways. Short sections of metal raceways are not required to be grounded if used for the support or physical protection of cables.

Exception No. 4: Isolated Metal Elbows. Isolated metal elbows installed underground isolated from possible contact to persons by a minimum 18 inches of cover to any part of the elbow are not required to be grounded.

PART E. BONDING

250-90 General

In order to maintain an effective grounding path, conductive metal parts of electrical equipment must be bonded together. In addition, bonding is required if the grounding path is not continuous or if the grounding path does not have sufficient capacity to conduct safety the available fault-current [250-96(a)].

Note. Special bonding rules apply to service equipment and raceways [250-92], enclosures and raceways containing circuits over 250 volts [250-97], raceways in hazardous (classified) locations [250-100], and for swimming pool, spa, and hot tub equipment [680-22].

250-92 Bonding Service and Communications Systems

(a) Service Enclosure and Raceways.
A fault in service equipment, enclosures and raceways containing service conductors can cause tremendous destruction if not cleared quickly. Because of this danger, the *NEC* requires the following parts to be bonded according to the requirements of 250-94, Fig. 12-1:

(1) Metal raceways, cable trays, or cables containing **service conductors**,

(2) *Enclosures* containing service conductors such as disconnects, meter enclosures, and gutters or wireways,

(3) The metal **raceway** that contains the grounding electrode conductor.

WARNING:
If metal conduit for the grounding electrode conductor is not properly bonded, the grounding electrode conductor effectiveness is reduced. Lightning current flowing through a conductor creates a magnetic field around the conductor that expands and collapses with the rising and falling current. When an expanding and collapsing magnetic field passes through the surrounding metal raceway, it creates an induced voltage and current in the raceway. A single conductor carrying lightning current in a metal raceway causes the raceway to act as an inductor, which can severely limit, or choke, the current flow through the grounding electrode conductor during a lightning strike.

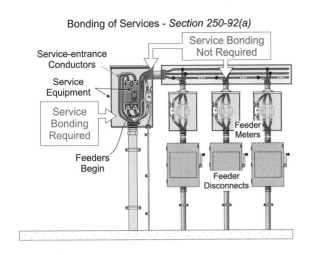

Fig. 12-2 Service Bonding Doesn't Apply to Feeder Enclosures or Raceways

Note. The bonding requirements of Section 250-92 only apply to **service equipment** and service raceways, not to **feeder** equipment or feeder raceways, Fig. 12-2.

(b) Communications Systems. To prevent electric shock and fires from lightning, all communications systems that enter a building must be bonded to the building grounding electrode system.

At each service, at separate building disconnects, and at each mobile home, an accessible and external means must be provided for the **bonding** of communications systems, such as CATV, telephone, satellite dish, etc. to the any of the following:

(1) A metal service raceway,

(2) A grounding electrode conductor,

(3) **Approved** external connection.

> **WARNING:**
> Failure to bond communications systems to the building or structure grounding electrode system can result in an electric shock, destruction of electrical components, as well as fires from lightning.

FPN No. 2: To reduce the differences of potential between different systems from lightning, all communications systems that enter a building must be properly bonded

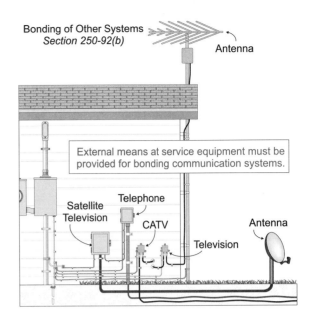

Fig. 12-3 Communications Systems Must Be Properly Bonded

according to the following requirements, Fig. 12-3:

System	Section
• CATV	820-40(d)
• Telephone circuits	800-40(d)
• Antennas/satellite dishes	810-21(j)

250-94 Bonding Service Enclosures and Raceways

Service equipment, enclosures, and raceways containing service conductors must be bonded by any of the following methods [250-92], Fig. 12-4:

(1) Bonded to the Grounded (Neutral) Service Conductor. Service enclosures and service raceways must be bonded to the grounded (neutral) service conductor by the use of a *bonding bushing* with a bonding jumper sized in accordance with Table 250-66 [250-24(b), 250-28, 250-102(c)].

(2) Threaded Metal Conduit Fittings. Raceways threaded wrenchtight into enclosures, couplings, *hubs*, *bosses*, conduit bodies, or other threaded entries, are considered bonded to the threaded fitting or enclosure.

(3) Raceway Fittings. Set screw and compression couplings and connectors made wrenchtight are considered bonded to the raceway.

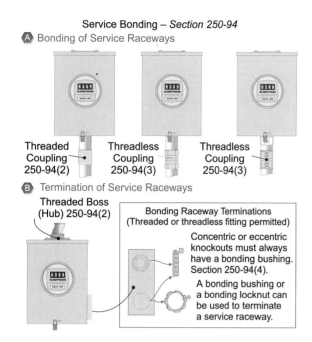

Fig. 12-4 Methods of Service Bonding

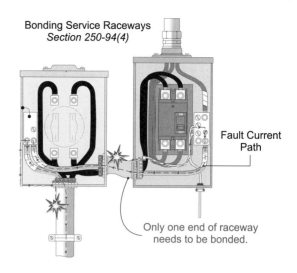

Fig. 12-5 Only One End of a Service Raceway Is Required to Be Bonded

However, the locknut on the connector is not considered suitable to bond the connector to the enclosure. Therefore all threadless connectors on service raceways must be bonded in accordance with (4) below.

(4) Bonding Bushings and Locknuts.

Bonding Bushings. When metal raceways or connectors are installed through a ringed *knockout*, the raceway or raceway connector must be bonded with a bonding jumper. The bonding jumper must terminate to the grounded (neutral) service conductor, or to the service enclosure. The bonding jumper must be sized according to Table 250-66 based on the size of the ungrounded (phase) service conductors in the raceway or cable [250-102(c)].

Bonding Locknuts. When metal conduits or connectors are installed in enclosures without ringed knockouts, the raceway may be bonded with either a bonding jumper or a *bonding locknut*. A bonding locknut has a special screw that drives through the metal enclosure to bond the locknut to the enclosure.

Note. Only one end of a service raceway is required to be bonded, Fig. 12-5.

250-96 Bonding Other Enclosures

(a) **General.** Metal parts that serve as the grounding conductor path such as raceways, equipment, and enclosures must be effectively bonded to assure *electrical continuity* [300-10]. The grounding path must be sized to safely handle the available fault-currents of the circuit [250-122].

Nonconductive coatings such as paint, enamel, tarnish, etc., on contact points and surfaces that are part of the grounding path must be removed.

CAUTION: For the overcurrent protection device to operate properly (clear the fault), the grounding system must provide a low impedance path that can carry the ground-fault current. *Reducing washers* (donuts) are not capable of or listed to carry fault-current; therefore, bonding jumpers are required around this high impedance grounding path, Fig. 12-6.

Note. Fittings such as locknuts automatically remove the paint when they are installed [250-12].

(b) **Isolated Ground Circuits.** Where required for the reduction of electrical noise (electromagnetic interference) on the grounding circuit, equipment supplied by a branch circuit can be isolated from a metal raceway by a listed nonmetallic

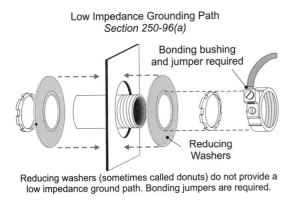

Fig. 12-6 Reducing Washers Do Not Provide Low Impedance Grounding Path

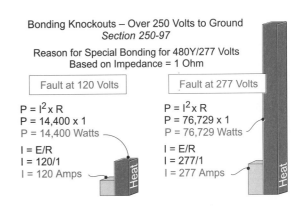

Fig. 12-8 Why Bonding Required Over 250 Volts to Ground Circuits

raceway fitting located at the point of attachment of the raceway to the equipment enclosure. The raceway must be supplemented by an internal insulated equipment grounding conductor installed in accordance with Section 250-146(d) to ground the equipment enclosure.

FPN: Under this condition, the metal raceway must still be properly grounded.

250-97 Bonding Circuits Over 250 Volts to Ground

Metal raceways or connectors containing 277 or 480 volt circuits through a ringed knockout must be bonded with a bonding jumper in accordance with Section 250-94(a)(4). The bonding jumper for this application must be sized in accordance with the circuit overcurrent protection device setting using table 250-122, Fig. 12-7.

Note. Special bonding is required because a 277 volt ground-fault generates 5 times as much heating as a 120 volt ground-fault (I^2R or E/R), Fig. 12-8.

Exception. Raceways or cables can terminate with standard locknuts and fittings where ringed knockouts are not encountered, or where the box is listed for this purpose, Fig. 12-9.

250-100 Hazardous (Classified) Locations Bonding

Because of the special dangers of electrical installations in hazardous (classified) locations, all metal raceways must be bonded either by a bonding jumper or a bonding type locknut.

This special bonding requirement extends from the hazardous (classified) location to the service or separately derived system. Locknuts alone are not

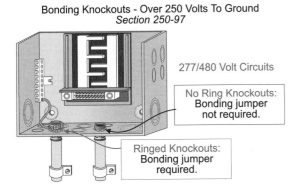

Fig. 12-7 Bonding Jumper Requirements Over 250 Volts to Ground Circuits

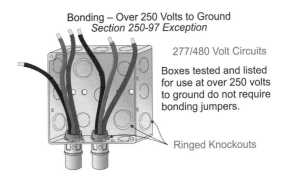

Fig. 12-9 Bonding Jumper Not Required Over 250 Volts to Ground Circuits

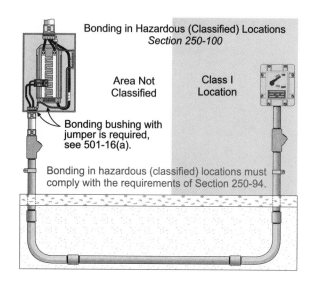

Fig. 12-10 Bonding of Raceways in Hazardous (Classified) Locations

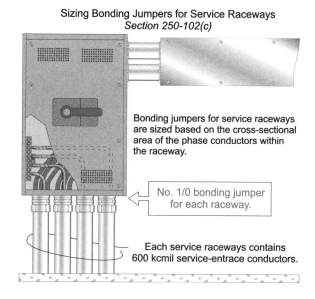

Fig. 12-11 Bonding Jumpers for Service Raceways Sized According to Table 250-66

acceptable, even if ringed knockouts are not used, Fig. 12-10.

> **Note.** See Sections 501-16 for Class I (gas and vapors) installation, Section 502-16 for Class II (dust) installation, and Section 503-16 for Class III (fiber) installation.

250-102   Equipment Bonding Jumpers

(a) **Material.**   Equipment bonding jumpers shall be of copper or other corrosive-resistant material.

(b) **Bonding Jumpers Attachment.**   The attachment of bonding jumper shall be by exothermic welding, listed pressure connectors, listed clamps, or other listed means and must be **accessible** [250-8]. The attachment fitting must be listed for the grounding material (copper or aluminum); see Section 250-70 for the proper connection to the grounding electrode.

(c) **Supply Side Size.**   Bonding jumpers for service raceways shall be sized based on the cross-sectional area of the phase conductors within the raceway according using Table 250-66.

Question. What size service bonding jumper is required for each raceway containing 600 kcmil conductors, Fig. 12-11?

Answer.   No. 1/0

Grounding Electrode Raceway.   A metal raceway containing a grounding electrode conductor must have both ends bonded [250-92(a)(3), 250-64(d)]. The bonding jumper for the grounding electrode raceway must not be smaller than the enclosed grounding electrode conductor as selected from Table 250-66, Fig. 12-12.

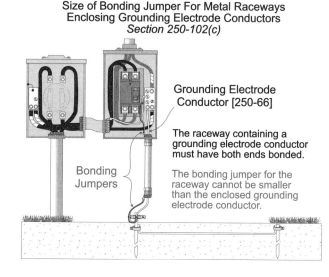

Fig. 12-12 Bonding Required for Metal Raceway Enclosing the Grounding Electrode Conductor

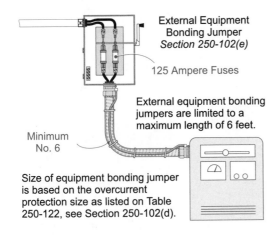

Fig. 12-13 External Equipment Bonding Jumper

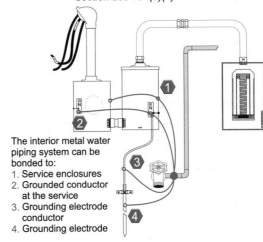

The interior metal water piping system can be bonded to:
1. Service enclosures
2. Grounded conductor at the service
3. Grounding electrode conductor
4. Grounding electrode

Fig. 12-14 Bonding Location of Interior Metal Water Piping System

(d) Load Side Equipment Bonding Jumper Size. Where an **equipment bonding jumper** is required to bond a single raceway or to bond two or more raceways, the bonding jumper must be sized to the largest overcurrent protection device for any circuit contained in the raceways using Table 250-122.

(e) Equipment Bonding Jumper Installation. The equipment bonding jumper can be installed either inside or outside the raceway. But when installed outside the raceway, the bonding jumper shall be no more than 6 feet in length and it must be routed with the raceway, Fig. 12-13.

When the equipment bonding jumper is installed inside the raceway there is no maximum length.

250-104 **Bonding of Piping and Building Steel**

(a) Interior Metal Water Piping System. The interior metal water piping system must be bonded so that it provides a low impedance path to help in the clearing of ground-fault current in accordance with (1) through (4) below.

(1) General. The interior metal piping system must be bonded to any one of the following locations, Fig. 12-14:

• Service enclosure,
• Grounded (neutral) service conductor,
• Grounding electrode conductor*,
• Grounding electrode*.

*WARNING:
When bonding the interior metal water piping system to the grounding-electrode conductor or to the grounding electrode, the grounding-electrode conductor must not be smaller than the interior bonding jumper.

Note. Hot and cold water pipes that are not electrically connected require a bonding jumper between the pipes.

Bonding Jumper Size. The bonding jumper for the interior metal water piping system must be sized according to the size of the service conductors using Table 250-66, and it must be installed is accordance with Section 250-64.

(2) Multiple Occupancy Building. When the

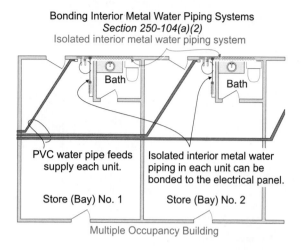

Bonding Interior Metal Water Piping Systems
Section 250-104(a)(2)
Isolated interior metal water piping system

Bath

Bath

PVC water pipe feeds supply each unit.

Isolated interior metal water piping in each unit can be bonded to the electrical panel.

Store (Bay) No. 1

Store (Bay) No. 2

Multiple Occupancy Building

Fig. 12-15 Bonding Interior Metal Water Piping Systems for Multiple Occupancy Buildings

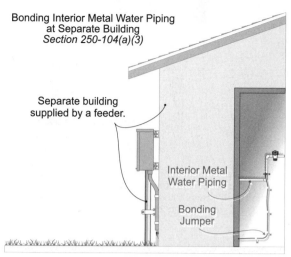

Bonding Interior Metal Water Piping at Separate Building
Section 250-104(a)(3)

Separate building supplied by a feeder.

Interior Metal Water Piping

Bonding Jumper

Where a separate building is supplied by a feeder, the interior metal piping system must be bonded to the building disconnecting means, the grounding electrode conductor, or the grounding electrode.

Fig. 12-16 Bonding Interior Metal Water Piping at Separate Building

interior metal water piping system is metallically isolated from each occupancy, the isolated interior metal water piping system for each unit may be bonded to the equipment grounding conductor of the occupancy panelboard, or it must be bonded in accordance with Section 250-104(a).

The bonding jumper to the equipment grounding conductor must be sized to the overcurrent protection rating of the panelboard feeder, using Table 250-122, Fig. 12-15.

(3) Separate Building or Structure. The interior metal water piping system of a separate building or structure that is supplied by a feeder or branch circuit must be bonded to any of the following locations, Fig. 12-16:

• Building disconnect enclosure
• Supply equipment grounding conductor
• Grounding electrode system

Bonding Jumper Size. The metal water pipe bonding jumper for the separate building or structure must be sized to the feeder or branch circuit supply overcurrent protection rating, using Table 250-122.

(4) Separately Derived Systems. To prevent a hazardous voltage difference between the metal water piping systems and metal electrical equipment from ground-faults or voltage transients, metallic water pipe systems in the area served by the separately derived system must be bonded to the grounded (neutral) conductor of the separately

derived system, Fig. 12-17.

This bonding connection must be made at the same point on the separately derived system where the grounding electrode conductor is connected to the system grounded (neutral) conductor.

The bonding jumper for this purpose must be sized to the secondary conductors in accordance with Table 250-66.

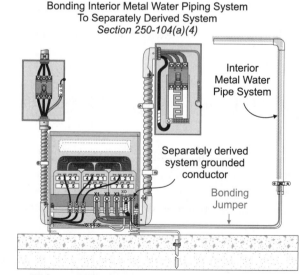

Bonding Interior Metal Water Piping System To Separately Derived System
Section 250-104(a)(4)

Interior Metal Water Pipe System

Separately derived system grounded conductor

Bonding Jumper

The grounded conductor of a separately derived system must be bonded to the nearest available point of the interior metal water piping system in the area served by the separately derived system.

Fig. 12-17 Bonding Interior Metal Water Piping System to Separately Derived System

(b) ˣGas Piping Systems. Metal gas piping system upstream from the equipment shutoff valve must be bonded to the grounding electrode system.

Note. This is a new requirement in the 1999 *NEC* (extracted from NFPA 54 – *National Fuel Gas Code*) and is very vague. It does not specify where the connection can be made, the method of the connection or the method of sizing the bonding jumper. It should be bonded in the same manner as the interior metal water piping system as contained in Section 250-104(a).

(c) Other Metal Piping Systems. Other interior metal piping systems such as compressed air or sprinklers that "may become energized," must be bonded in the same manner as the interior metal water piping system as listed in 250-104(a).

> CAUTION: The phrase "that may become energized" is subject to interpretation by the inspector. All piping systems, air ducts, metal structural members, or any other metal surface in a building could become energized if the conditions are right (or, perhaps, if the conditions are wrong). I know of a case where the gas piping for a restaurant that contained no electrical appliances was not bonded to the electrical system. A rat ate the insulation of nonmetallic-sheathed cable (naturally, the hot side), and the exposed conductor energized the gas metal piping system, killing a kitchen worker.

The bonding conductor for this purpose shall be sized to the rating of the largest overcurrent device of the circuit that may energize the piping system, using Table 250-122.

The equipment grounding conductor for the circuit that may energize the piping can serve as the bonding means.

FPN: Bonding of all piping and metal air ducts within the building provides an additional degree of safety.

(d) Structural Steel. Exposed building steel

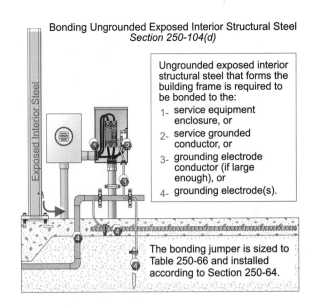

Fig. 12-18 Bonding of Ungrounded Exposed Interior Structural Steel

can become and remain energized during a ground-fault condition. The high impedance path of the building steel can result in low levels of fault-current which prevents the clearing of the fault by circuit protective device. To protect against this hazard, ungrounded exposed interior structural steel of building frames "which may become energized," must be bonded to any one of the following locations, Fig. 12-18:

- Service enclosure,
- Grounded (neutral) service conductor,
- Grounding electrode conductor*,
- Grounding electrode*.
  * Where the grounding electrode conductor is sized to Table 250-66.

The bonding jumper used for this purpose must be accessible, sized in accordance with Table 250-66, and it must be installed in accordance with the requirements of Section 250-64.

Note. This requirement is intended to apply to exposed interior structural steel building frames, not interior metal *studs* or isolated metal bar joists.

250-106 Lightning Protection System

The lightning protection system grounding electrode shall be bonded to the building or struc-

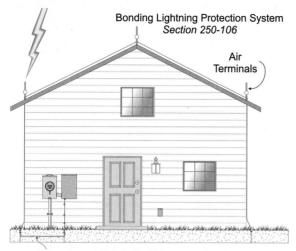

Lightning protection ground terminals must be bonded to the building or structure grounding electrode system.

Fig. 12-19 Bonding of Lightning Protection System to Building Grounding Electrode System

ture grounding electrode system, Fig. 12-19.

FPN No. 2: Metal raceways, enclosures, frames, and other noncurrent-carrying metal parts of electric equipment installed on a building equipped with a lightning protection system may require bonding or spacing from lightning protection conductors in accordance with NFPA-780 *Installation of Lightning Protection Systems*.

PART F. EQUIPMENT GROUNDING

250-110 When Grounding Required

Exposed metal parts of fixed equipment "likely to become energized" must be grounded under any of the following conditions:

(1) Located within 8 feet vertically or 5 feet horizontally of ground or grounded objects if subject to contact by persons.

(2) Installed in wet or damp locations if not isolated.

(3) The equipment is in electrical contact with metal.

(4) Where in a hazardous (classified) location.

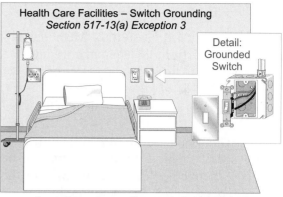

An equipment grounding conductor is required for a switch in patient care areas.

Fig. 12-20 Insulated Grounding Conductor Required for Patient Care Area Switches

(5) The wiring methods provide a ground path. Grounding is also required for:

• Metal mounting *yokes* for switches, 380-9(b),
• Switch metal faceplates, 380-12,
• Switch health care, 517-13(a), Fig. 12-20,
• Receptacles, 250-146,
• Receptacle metal faceplates, 410-56(d),
• Receptacle health care, 517-13(a),
    Fig. 12-21.

250-118 Equipment Grounding Conductors

Equipment grounding conductors must be one

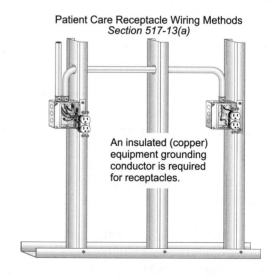

An insulated (copper) equipment grounding conductor is required for receptacles.

Fig. 12-21 Insulated Grounding Conductor Required for Patient Care Area Receptacles

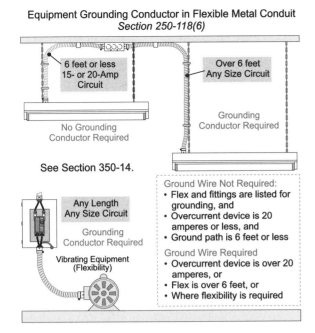

Fig. 12-22 Use of Flexible Metal Conduit as an Equipment Grounding Conductor

or more of the following:

(1) Conductor*
(2) Rigid Metal Conduit
(3) Intermediate Metal Conduit
(4) Electrical Metallic Tubing
(6) Flexible Metal Conduit [350-14]**
(8) Liquidtight Flexible Metal Conduit ***
(9) Armored Cable
(10) Mineral Insulated Cable
(11) Metal Clad Cable
(12) Cable Trays
(13) Cablebus
(14) Other metal raceways

* The equipment grounding conductor can be solid, stranded, bare, covered, or insulated. When covered or insulated, the covering or insulation along its entire length must be green [250-119].

** Flexible metal conduit can serve as the equipment grounding conductor, if the ground return path of the *flex* does not exceed 6 feet, the circuit conductors are protected by overcurrent protection devices rated 20 ampere or less, and the fittings are listed for grounding, Fig. 12-22.

Note. All listed flex fittings sized from 3/8 inch through ¾ inch are listed for grounding.

*** Liquidtight flexible metal conduit can serve as the equipment grounding conductor, if the ground return path of the liquidtight does not exceed 6 feet for 3/8 inch and ½ inch liquidtight contains conductors protected by a protection device rated 20 ampere or less, or ¾ inch through 1¼ inch liquidtight contains conductors protected by a 60 ampere or less protection device, and the liquidtight termination fittings are listed for grounding.

CAUTION: Flexible metal and liquidtight flexible metal conduit cannot serve as an equipment grounding path within hazardous (classified) locations. See Section 501-16(b) for Class I locations (gas or vapor), 502-16(b) for Class II locations (dust), and 503-16(b) for Class III locations (fiber).

250-119 Identification of Grounding Conductor

The equipment grounding conductor can be bare, covered, or insulated. Individually covered or insulated equipment grounding conductors shall have a continuous outer finish that is either green or green with one or more yellow stripes.

Note. The following *Code* Sections require an insulated equipment grounding conductor:

• Health care receptacles/switches, 517-13
• AC cable in places of assembly, 518-4(a)
• Pool wet-niche lights, 680-20(b)(1)
• Pool equipment, 680-25

(a) Conductors Larger than No. 6. Equipment grounding conductors larger than No. 6 are not required to have a green insulation or covering, but they must be identified at every point where the conductor is accessible. The identification can be by:

(1) Removing the insulation so that the conduc-

tors are bare.

(2) Coloring the exposed insulation with the color green.

(3) Installing green phase tape or green adhesive labels on the exposed accessible insulation.

250-120 Equipment Grounding Conductors Installation

(a) Wiring methods that serve as the equipment grounding conductor must be installed so as to be electrically and *mechanically continuous* [300-12]. Fittings shall be for the specific wiring method [300-15(a)], and the fittings shall be made tight using suitable tools.

250-122 Sizing Grounding and Bonding Conductors

(a) General Requirements. Table 250-122 is used to size the equipment grounding conductor [250-102(d)], Fig. 12-23.

In addition, Table 250-122 is used to size the interior metal water pipe bonding conductor for multiple occupancy buildings [250-104(a)(2)].

However, the equipment grounding and bonding conductor are not required to be larger than the circuit conductors.

WARNING:
Grounding conductors must be capable of safely conducting ground-fault current likely to be imposed on them [250-2(d)]. A continuous low impedance ground path is necessary for the operation of the overcurrent protection device to clear the ground-fault. If the grounding conductor is not sized to withstand the ground-fault currents, the conductor may melt or vaporize before the protective device responds. The conductors listed in Table 250-122 are minimum sizes; larger conductors may be necessary to conduct ground-fault currents safely. See Section 110-10 and the Note at the bottom of Table 250-122.

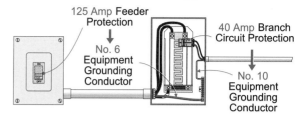

Sizing Equipment Grounding Conductors
Section 250-122(a)

Equipment Grounding Conductors are sized based on the overcurrent protection device using Table 250-122.

Fig. 12-23 Equipment Grounding Conductors Sized According to Table 250-122

(b) Voltage Drop. When ungrounded conductors are increased in size to compensate for voltage drop, the equipment grounding conductor must be increased in equal proportion [210-19(a) FPN No. 4, 215-2(d) FPN No. 2, 310-15(a) FPN No. 1].

Question. If phase conductors are increased in size from a No. 8 to a No. 6 because of voltage drop, what size equipment grounding conductor is required? The overcurrent protection device is rated 40 ampere.

Answer.   No. 8

No. 6 has a *circular mil* area 59 percent greater than the No. 8 (26,240/16,510) [Chapter 9 Table 8].

The equipment grounding conductor for 40 ampere circuit is No. 10 [Table 250-122] which has a circular mil area of 10,380, which must be by 59 percent, 10,380 CM × 1.59 = 16,500, CM = No. 8.

(c) Multiple Circuits. When multiple circuits are installed in the raceway, a single equipment grounding conductor can be used. The single equipment grounding conductor must be sized to the largest overcurrent device protecting the conductors in the raceway, based on Table 250-122.

Question. What size equipment grounding conductor is required for a nonmetallic raceway that contains the following circuits, Fig. 12-24?

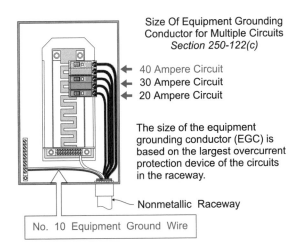

Fig. 12-24 Equipment Grounding Conductor for Multiple Circuits, Table 250-122

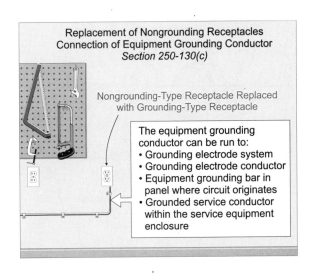

Fig. 12-26 Connection of Equipment Grounding Conductor, Where No Ground Exists in Outlet Box

Answer. No. 10

(f) **Parallel Runs.** When the circuit conductors are run in *parallel* in accordance with Section 310-14, an equipment grounding conductor must be run in each raceway or cable.

(1) **Raceways.** The equipment grounding conductor in each raceway must be sized to the circuit overcurrent protection device, using Table 250-122.

Question. What size equipment grounding conductor is required for a 1,000 ampere feeder

installed with 4 raceways, Fig. 12-25?

Answer. No. 2/0

Based on the 1,000 ampere setting of the circuit overcurrent protection device in accordance with Table 250-122.

PART G.   METHODS OF EQUIPMENT GROUNDING

250-130 Equipment Grounding Conductor Connections

Separately Derived Systems. Equipment grounding conductor connections at a separately derived systems shall be bonded to the system grounded (neutral) service conductor [250-142(a)] and to the system grounding electrode conductor [250-30].

(a) **Service Equipment.** Equipment grounding at service equipment shall be made by bonding the equipment grounding conductor to the grounded (neutral) service conductor [250-142(a)] and to the grounding electrode conductor [250-24(a)].

(c) **Replacing Nongrounding Type Receptacles.** When replacing a nongrounding-type receptacle with a grounding-type receptacle at an outlet box that does not contain an equipment grounding conductor, the grounding contacts of a replacement receptacle must be connected to any of

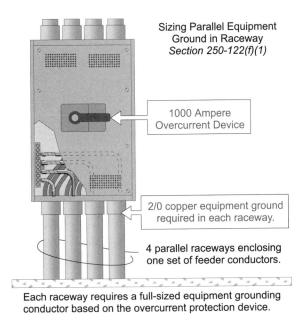

Fig. 12-25 Equipment Grounding Conductor for Parallel Circuits, Table 250-122

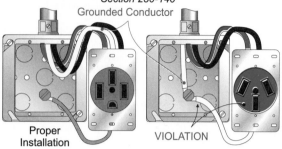

Neutral Not Permitted As Grounding Conductor
Section 250-140
Grounded Conductor

Proper Installation   VIOLATION

For new installations, the grounded conductor (neutral) cannot be used for grounding. This means that 4-wire cords and receptacles are required.

Fig. 12-27 The Grounded (Neutral) Conductor Not Permitted to Serve as a Grounding Conductor

the following locations, Fig. 12-26:

(1) Grounding electrode system [250-50]

(2) Grounding electrode conductor

(3) Panelboard equipment grounding terminal

(4) Grounded service conductor

FPN: Section 210-7(d) permits nongrounding-type receptacles to be replaced with a grounding-type receptacles without a ground connection, if the receptacle is GFCI protected.

250-134 Fixed Equipment Grounding

Metal parts of electrical equipment, raceways, and enclosures must be grounded according to (a) or (b).

(a) Equipment Grounding Conductor Types. A bare or insulated equipment grounding conductor of the type listed in Section 250-118 that has the capacity to carry the fault-current in accordance with Section 250-2(d).

(b) Equipment Grounding Circuit Conductor. Equipment grounding conductors must be installed within the same raceway, cable tray, trench, cable, or cord with the circuit conductors.

Note. The grouping of all circuit conductors prevents inductive heating of metal

parts and reduces the impedance of the ground-fault path [300-3(b), 300-5(i), 300-20(a)].

250-136 Equipment Considered Effectively Grounded

(a) Equipment Secured to Grounded Metal Supports.   Equipment shall be considered grounded when mounted to a metal rack or frame that is grounded according to Section 250-134. However, the grounded structural metal frame of a building cannot be used as an equipment grounding conductor.

250-140 Grounding of Ranges, Clothes Dryers, and Their Outlet Boxes

The *Code* no longer permits the grounded (neutral) conductor to be used to ground the frame or junction box of electric ranges, wall-mounted ovens, counter-mounted cooking units, or clothes dryers. However, the use of the grounded conductor in existing installations can continue to be used ground the frame or junction box of electric ranges, wall-mounted ovens, counter-mounted cooking units, or clothes dryers [250-142(b) Ex.1], Fig. 12-27.

> CAUTION: Ranges, dryers and ovens have their metal case bonded to the grounded (neutral) conductor at the factory. This bonding jumper must be removed when these appliances are installed in new construction.

250-142 Neutral-to-Ground Connections

(a) When Permitted. Neutral-to-ground connections are required or permitted at only the following locations [250-24(a)(5)]:

Service Disconnect Equipment. Because electric utilities are not required to install a safety ground wire, a neutral-to-ground connection must be made at the service disconnecting means so that the grounded (neutral) conductor can serve to carry

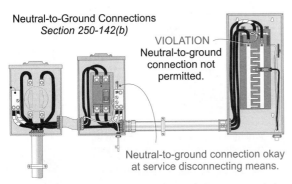

Neutral-to-Ground Connections
Section 250-142(b)

VIOLATION
Neutral-to-ground
connection not
permitted.

Neutral-to-ground connection okay
at service disconnecting means.

A grounding connection shall not be made to any grounded
conductor on the load side of the service disconnecting means.

Fig. 12-28   Neutral-to-Ground Connections Not
Permitted at the Load Size of Service Equipment

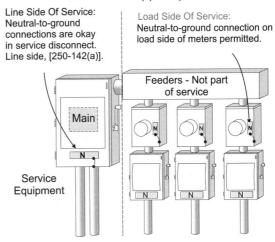

Neutral-to-Ground Connections - Load Side Of Service
Section 250-142(b) Exception 2

Line Side Of Service:
Neutral-to-ground
connections are okay
in service disconnect.
Line side, [250-142(a)].

Load Side Of Service:
Neutral-to-ground connection on
load side of meters permitted.

Feeders - Not part
of service

Main

N

Service
Equipment

Fig. 12-29   Neutral-to-Ground Connection at Meter
Enclosures on the Load Side of Service

> **DANGER:**
> The accepted practice of omitting the
> safety ground for service conductors and
> using the neutral-to-ground connection at
> the electric service is very dangerous
> because if the grounded (neutral) service
> conductor is open, we lose our safety
> ground. This can cause all grounded metal
> parts to become energized resulting in
> electric shock as well as fires.

fault-current in accordance with Section 250-24(b).

Separately Derived Systems. A neutral-to-ground connection is required at separately derived systems, such as transformers or generators in accordance with Section 250-30(a).

Separate Building or Structure Disconnect. The *NEC* permits, but does not require, neutral-to-ground connections at separate building and structures disconnecting means that have power supplied from another building. See Section 250-32(b)(2) for details.

Note. This is a dangerous practice and should not be done.

(b) Not Permitted. Neutral-to-ground connections are not permitted on the *load side* at any location except as listed in (a) above for service equipment, separately derived systems, and separate

building disconnects, Fig. 12-28.

Exception No. 1: Existing Installations. Existing neutral-to-ground connections for ranges, dryers, and ovens are permitted [250-140].

Exception No. 2: Meter Enclosures. Grounded (neutral) conductor can be used to ground meter enclosures on the load side of the service disconnect if the meter enclosures are located near the service disconnect and no ground-fault protection is installed at the service, Fig. 12-29.

The size of the grounded (neutral) conductor for this purpose must not be less than the maximum unbalanced load as calculated in Section 220-22 and not smaller than the size specified in Table 250-122 for equipment grounding conductors.

THE DANGER OF NEUTRAL-TO-GROUND CONNECTIONS

Improper neutral-to-ground connections can result in stray neutral currents traveling on metal parts of electrical equipment and building structures. This can result in electric shock, fires, and power quality issues results from elevated ground voltage and excessive electromagnetic *fields*.

Electric Shock. Electric shock can occur if the neutral conductor's path is open and a person gets in

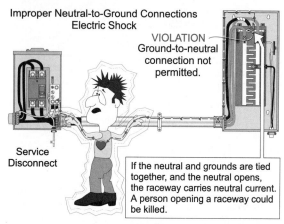

VIOLATION
Ground-to-neutral connection not permitted.

Improper Neutral-to-Ground Connections Electric Shock

Service Disconnect

If the neutral and grounds are tied together, and the neutral opens, the raceway carries neutral current. A person opening a raceway could be killed.

A grounding connection shall not be made to any grounded conductor on the load side of the service disconnecting means.

Fig. 12-30 Improper Neutral-to-Ground Connections - Electric Shock

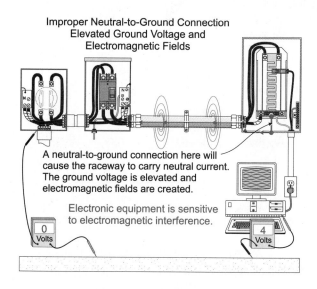

Improper Neutral-to-Ground Connection Elevated Ground Voltage and Electromagnetic Fields

A neutral-to-ground connection here will cause the raceway to carry neutral current. The ground voltage is elevated and electromagnetic fields are created.

Electronic equipment is sensitive to electromagnetic interference.

0 Volts

4 Volts

Fig. 12-32 Improper Neutral-to-Ground Connection - Elevated Ground Voltage and Electromagnetic Fields

series with the safety ground path [250-24(a)(5)], Fig. 12-30.

Fires. Metal raceways and equipment can cause *arcing* at loose connections, which has resulted in fires. Arcing is particularly dangerous in areas containing hazardous classified gases, vapors, dust, or other areas containing combustible materials, Fig. 12-31.

Elevated Ground Voltage. An improper neutral-to-ground connection can cause the safety ground wire to carry neutral current. When this occurs, the ground voltage is elevated from zero depending on the magnitude of the current and the resistance of the conductors, Fig. 12-32.

Electromagnetic Fields. Neutral current on the metal parts of electrical equipment and buildings' can create electromagnetic fields that can impact some electronic devices, particularly video monitors. In addition, there's always the concern on the health issues of electromagnetic fields, Fig. 12-32.

Note. Office buildings have large quantities of single-phase **nonlinear loads** such as personal computers and laser printers. These loads produce harmonic currents that add on the neutral conductor (instead of canceling) which can cause the neutral conductor to carrying as much as 200 percent of its maximum expected load. This results is significantly higher elevated ground voltage, particularly when high amperage loads, such as laser printers or copiers cycle on and off.

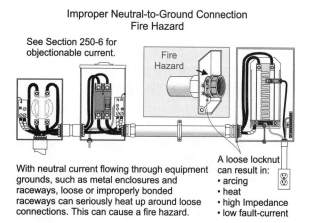

Improper Neutral-to-Ground Connection Fire Hazard

See Section 250-6 for objectionable current.

Fire Hazard

With neutral current flowing through equipment grounds, such as metal enclosures and raceways, loose or improperly bonded raceways can seriously heat up around loose connections. This can cause a fire hazard.

A loose locknut can result in:
• arcing
• heat
• high Impedance
• low fault-current

Fig. 12-31 Improper Neutral-to-Ground Connection - Fire Hazard

250-146 Bonding – Receptacles

Grounding type receptacles must have their grounding terminals connected to an equipment grounding conductor [210-7(b)] by a bonding jumper, Fig. 12-33.

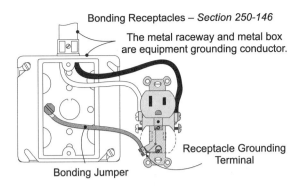

Fig. 12-33 Receptacle Equipment Bonding Jumper

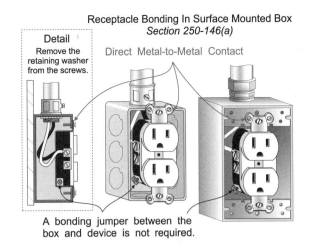

Fig. 12-34 Receptacle Bonded to Surface Box

(a) Surface-Mounted Box. Where an outlet box is mounted on or at the surface, a bonding jumper is not required for the receptacles where there is direct metal-to-metal contact between the receptacle mounting yoke and the metal box, Fig. 12-34.

Bonding jumpers must be installed on receptacles that are supported to metal covers [410-56(f)(3)], Fig. 12-35.

(b) Self-Grounding Receptacles. Contact devices that are designed and listed for the purpose can be used to establish the grounding circuit between the device yoke and flush or surface boxes, Fig. 12-36.

> **WARNING:**
> Outlet boxes are sometimes set back more than ¼ inch from the finish surface (violation of Section 370-20), and the electrician replaces the self-grounding receptacle mounting screws with longer screws to reach the box, which is a violation of Section 110-3(b).

(d) Isolated Ground Receptacles. Isolated ground receptacles are not required to have their grounding terminal grounded to the metal box [250-148 Ex.]. The grounding terminal of the receptacle can be grounded to an insulated equipment grounding conductor. This insulated grounding conductor can pass through panelboards [384-20 Ex.], and terminate directly to the equipment grounding terminal at the building service equipment, or separately derived system, Fig. 12-37.

For additional grounding requirements of isolated ground receptacles, see Sections 250-6(d), 250-148 Ex., and 410-56(c).

FPN: Metal raceways and enclosures containing an insulated equipment grounding conductor must still be grounded [250-86].

250-148 Equipment Grounding Conductors in Boxes

Where one or more equipment grounding conductors enter a box, all grounding conductors shall be spliced together in the box or to the metal box. Conductors must be spliced or terminated to the box with devices suitable for the use [110-14(b)].

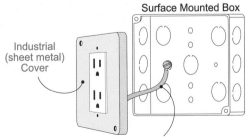

Fig. 12-35 Equipment Bonding Jumper Required for Cover-Mounted Receptacles

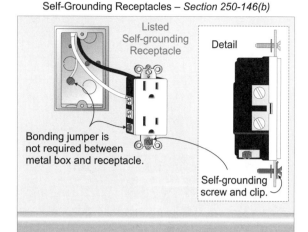

Fig. 12-36 Equipment Bonding Jumper Not Required for Self-Grounding Receptacles

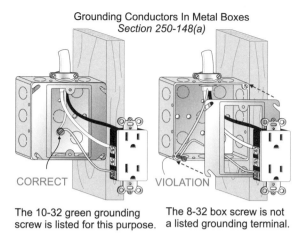

Fig. 12-38 Grounding Conductors in Metal Boxes Must Be Properly Terminated

Grounding Conductor Continuity. The removal of a device, such as a receptacle, shall not interfere with the grounding continuity of the circuit. This means that the equipment grounding conductors must be spliced, and a *pigtail* brought out for the receptacle grounding terminal.

Exception: Isolated Ground Receptacles. The equipment grounding conductor for isolated ground receptacles [250-146(d)] is not required to be connected to any equipment grounding conductor or to the metal box.

(a) Metal Boxes. When equipment grounding conductors are installed in metal boxes, an electrical connection is required between the equipment grounding conductors and the metal box by a screw or listed grounding connection devices that is used for no other purpose, Fig. 12-38.

Note. This means that grounding conductor for nonmetallic-sheathed cable cannot terminate to the cable connector or clamp, nor can a grounding conductor terminate to a 8-32 screw that is used to secure a *plaster ring* to the metal box, Fig. 12-38.

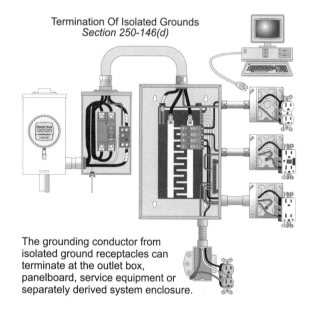

Fig. 12-37 Termination of Equipment Grounding Conductor - Isolated Ground Receptacles

SUMMARY

Part D. Enclosure Grounding ☐ Short sections of metal raceways used for the support or physical protection of cables are not required to be grounded.

Part E. Bonding ☐ External means must be provided for the bonding of CATV, telephone, and antennas for satellite, radio, and television equipment. ☐ Only one end of a service raceway must be bonded. ☐ Metal service raceways or connectors installed through a ringed knockout must be bonded with a bonding jumper. ☐ Metal service conduits or connectors installed in enclosures without ringed knockouts may be bonded with either a bonding jumper or a bonding type locknut. ☐ Metal raceways or connectors containing 480Y/277 volt circuits through a ringed knockout must be bonded with a bonding jumper. ☐ A metal raceway containing the grounding electrode conductor must be bonded at both ends. ☐ The equipment bonding jumper can be installed outside the raceway for a maximum of 6 feet. ☐ The interior metal water piping system must be bonded to provide a low impedance path to help in the clearing of the fault-current. ☐ Above ground gas piping system upstream from the equipment shutoff valve shall be bonded to the grounding electrode system. ☐ Interior metal piping systems, such as compressed air, that may become energized, shall be bonded. ☐ The lightning protection system grounding electrode shall be bonded to the building grounding electrode system.

Part F. Equipment Grounding ☐ The equipment grounding conductor can be solid, stranded, bare, covered, or insulated. ☐ When ungrounded conductors are increased in size to compensate for voltage drop, the equipment grounding conductor must be proportionately increased in size. ☐ When multiple circuits are installed in a raceway, a single equipment grounding conductor can be used. ☐ When the circuit conductors are run in parallel, the equipment grounding conductor must also be run in parallel in each raceway.

Part G. Methods of Equipment Grounding ☐ A neutral-to-ground connection on the load side of service equipment can result in electric shock or fires. ☐ Improper neutral-to-ground connection can result in elevated ground voltage. ☐ Receptacles must have their grounding terminals grounded. ☐ All equipment grounding conductors that enter a box shall be spliced together in the box or to the metal box. ☐ Equipment grounding conductors in metal boxes must have an electrical connection to the box.

REVIEW QUESTIONS

Part D. Enclosure Grounding

1. When are short sections of metal raceways not required to be grounded?

Part E. Bonding

2. Why must we bond CATV, telephone, and antennas for satellite, radio, and television equipment to the building grounding electrode system?

3. How many ends of a service raceway must be bonded?

4. When metal service raceways or connectors are installed through a ringed knockout, how must they be bonded?

5.   Metal service conduits or connectors installed in enclosures without ringed knockouts can be bonded with _____ or _____.

6.   Metal raceways or connectors containing 480Y/277 volt circuits through a ringed knockout must be bonded with a _____.

7.   A metal raceway containing the _____ must be bonded at both ends.

8.   What are the rules that permit an equipment bonding jumper to be installed outside the raceway?

9.   Why must we bond the interior metal piping system to the grounded (neutral) conductor at service equipment?

10. When must gas piping systems be bonded to the grounding electrode system?

11. Interior metal piping systems, such as compressed air that may become _____, shall be bonded to _____.

12. The lightning protection system grounding electrode _____ bonded to the building grounding electrode system. (shall or shall not)

Part F. Equipment Grounding

13. The equipment grounding conductor can be solid, stranded, bare, covered, or insulated.

14. When ungrounded conductors are increased in size to compensate for _____, the equipment grounding conductor must be proportionately increased in size.

15. What are the equipment ground conductor rules when installing multiple circuits in a raceway?

16. What are the equipment ground conductor installation rules when the circuit conductors are run in parallel?

Part G. Methods of Equipment Grounding

17. Explain how an improper neutral-to-ground connection can result in electric shock or fire.

18. Explain how an improper neutral-to-ground connection can result in elevated ground voltage.

19. When is an equipment bonding jumper not required for grounding type receptacles?

20. What must we do with equipment grounding conductors that enter a box?

21. What must we do with equipment grounding conductors that enter metal boxes?

22. In your own words explain the meaning of the following Article 100 definitions:

> Accessible
> Approved
> Bonding
> Equipment bonding jumper
> Feeder
> Raceway
> Service conductors
> Service equipment

23. In your own words explain the meaning of the following Glossary terms:

> Bonding bushings
> Bonding locknut
> Bosses
> Circular mil
> Elbows
> Electrical continuity
> Enclosures
> Flex
> Hubs
> Knockout
> Load side
> Parallel
> Pigtail
> Plaster ring
> Studs
> Yoke

CHAPTER 3
WIRING METHODS AND MATERIALS

Scope of Chapter 3 Articles

Chapter 2 was a bit of an uphill climb. There were many technical requirements and some complicated ideas involving theory. Many rules had a kind of abstract or elusive quality to them. Chapter 3, however, involves mostly conductors, cables, boxes, raceways, and fittings, and the installation requirements and restrictions involved with these different types of wiring methods. The actual type of wiring method used depends on several factors: *Code* (safety) requirements, environment, need, and cost. The materials most frequently used are single conductors, cables, raceways, and enclosures.

ARTICLES 300 AND 310

Conductors. Single conductors are generally installed in raceways by the electrician, or assembled into cables from the manufacturer. Some single conductors are listed for installation directly in the earth without protection of a raceway.

ARTICLES 328 THROUGH 339

Cables. Cables generally have two or more conductors, the most common types being nonmetallic-sheathed cable (NM), armored cable (AC), metal-clad cable (MC), service-entrance cable (SE), and underground feeder and branch circuit cable (UF).

ARTICLES 345 THROUGH 362

Raceways. The most common types of raceways are electrical metallic tubing (EMT), rigid nonmetallic conduit (PVC), rigid metallic conduit (rigid), and flexible conduits (flex and liquidtight).

Note. Electrical nonmetallic tubing is covered in Article 331.

ARTICLES 370 THROUGH 384

Enclosures. Cables and raceways enter into and exit from boxes or enclosures. These can be outlet boxes, junction boxes, cabinets (panels), cutout boxes (disconnects), switches, equipment enclosures, and so forth.

Unit 13

Article 300
Wiring Methods

OBJECTIVES

After studying this unit, the student should be able to understand:
- whether Article 300 applies to fire alarm, CATV, telephone, or low-voltage control wiring.
- that generally single conductors must be installed in a raceway or cable.
- which raceways or cables require special protection when run through or parallel to wood framing members.
- which raceways or cables require special protection when run through metal framing members.
- the *NEC* requirements for supporting computer, telephone, and control cables.
- when raceways must be sealed and what type of seal is required.
- why all conductors of a circuit should be in the same raceway.
- the restrictions on installing components inside electrical raceways.
- the general requirements of installing raceways and cables to enclosures.
- how steel raceways and enclosures are heated by alternating current circuits.

Definitions and Glossary Terms

To better understand the *NEC* rules contained in this unit, review the following:

Definitions Article 100 - Unit 2

Cabinets	Conduit body
Identified	Nonlinear load
Plenum	Service-lateral

Glossary Terms

Backfill	Boss
Bushing	Corrosion
Eddy currents	Ferrous
Fire resistance	Harmonic current
Hertz	Hub

Inductive choke	Low impedance
Pigtail	Studs

PART A. GENERAL REQUIREMENTS

300-1 Scope

(a) Wiring Installations. Article 300 contains the general requirements for all wiring methods listed in the *NEC*.

Note. *NEC* Chapter 7 and Chapter 8 wiring and cables are not required to comply with the wiring method requirements contained in Article 300, unless a Section in those

Articles refers specifically to a Section in Article 300:

System	Section
CATV	90-3
Class 1, 2, and 3 Circuits	725-3
Communications (telephone)	90-3
Fiber Optics	770-3
Fire Alarm Circuits	760-3
Radio and Television Equipment	90-3

CAUTION: It is very important to understand that the requirements of Article 300 do not apply to control, signal, data, fiber optic, fire alarm, CATV, communications, or radio and television wiring. The wiring methods for these systems are covered in Chapter 7 or Chapter 8 of the *Code*.

(b) Integral Parts of Equipment. The requirements of Article 300 do not apply to the integral parts of electric equipment, such as the internal wiring of motors, controllers, motor control centers, or factory-assembled control equipment [90-6].

300-2 Voltage and Temperature Limitations

(a) Voltage Limitations. The wiring methods listed in Chapter 3 (Articles 318 through 384) are for 600 volt systems, nominal, or less, including 600Y/347, 480Y/277, 208Y/120, and 120/240.

(b) Temperature Limitations. Conductors must not be installed in any way that causes the conductor's temperature insulation rating to be exceeded. See Sections 310-10 and 310-15 in this book for details and examples.

300-3 Conductors

(a) Conductors. Conductors are permitted only as part of a wiring method in Chapter 3 and must be installed in a raceway, cable, or enclosure. However, single underground service-entrance conductors (Type USE) can be directly buried in the earth. See Table 310-13 in the *NEC* for details.

Grouping Conductors Of The Same Circuit
Section 300-3(b)
Ground wire of circuit must be inside the raceway.

VIOLATION

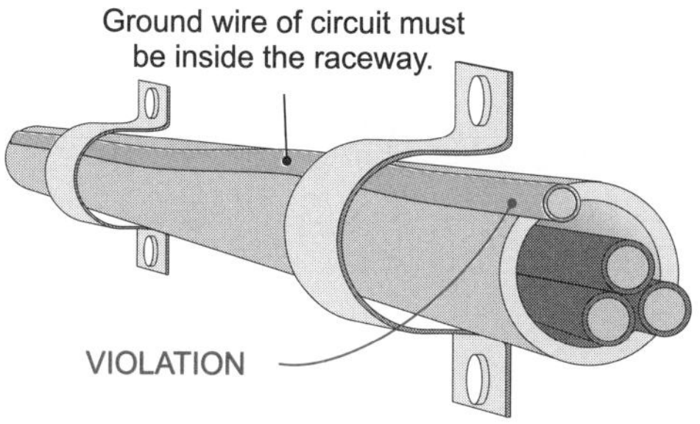

All conductors of a circuit must be grouped together in the same raceway, cable, trench, cord, or enclosure. This includes the ground and neutral conductors.

Fig. 13-1 All Conductors of the Circuit Must Be Grouped Together

(b) Conductors Together. All conductors of a circuit must be grouped together in the same raceway, cable, trench, cord, or cable tray. This includes ungrounded, grounded, and equipment grounding conductors of the same circuit [250-119 and 300-5(i)] in accordance with (1) through (3), Fig. 13-1.

This requirement is particularly important in order to minimize *induction* heating of metallic raceways and enclosures, and to maintain a low impedance grounding path. See Sections 300-5(i), and 300-20 for more details.

(1) Parallel Installations. Parallel conductors installed in accordance with Section 310-4 must have all conductors of the circuit installed in each raceway.

(2) Grounding Conductors. Grounding conductors installed in the same raceway with all of the circuit conductors; except as permitted for grounding type receptacles [250-130] and for equipment bonding jumpers on the outside of raceways shall be permitted [250-102(e)].

(3) Isolated Phase Installations. "Phase A" conductors can be installed in a separate nonmetallic raceway from "Phase B," "Phase C," or "Neutral" conductors [300-5(i) Ex. 2].

(c) Conductors of Different Systems.

(1) 600 Volt or Less. Electrical conductors of different systems can occupy the same raceway, cable, or equipment enclosure, if all conductors have an insulation rating not less than the maximum circuit voltage. Low-voltage or limited-energy conductors must be installed as follows:

Class 1 Conductors. Class 1 circuit conductors cannot occupy the same cable, enclosure, or raceway with other low-voltage or limited-energy circuits. But Class 1 control circuits and power conductors associated with the same equipment can occupy the same raceway [725-26(b)].

Fiber Optical Cable. Nonconductive optical fiber cable can occupy the same raceway, cable tray, or enclosure with power, or any other low-voltage or limited-energy cables that operate at 600 volt or less [770-52(b)].

Nonpower Limited Fire Alarm Conductors. Nonpower limited fire alarm conductors can be installed in the cable, enclosure, or raceway with power conductors used solely for supplying power to the equipment to which fire alarm conductors are connected [760-26 and 760-28(c)].

Other Low-Voltage and Limited-Energy Conductors. All other low-voltage or limited-energy wiring must be separated from power, Class 1, non-power limited fire alarm conductors so that they are not accidentally energized by the higher voltage conductors. The following *Code* Sections prohibit the mixing of low-voltage and limited-energy conductors with power, Class 1, or nonpower-limited fire alarm circuit conductors in the enclosure:

System	Section
CATV	820-10(f)(1)
Class 2 and 3 Control	725-54(a)(1)
Fire Alarm	760-54(a)(1)
Intrinsically Safe Systems	504-30(a)(2)
Instrument Tray Cable	727-5
Network Powered Systems	830-58(c)
Radio/Television	810-18(c)
Sound Systems	640-9(c)
Telecommunications	800-52(a)(1)

Exceptions to the above *Code* Sections permit power conductors to terminate on to listed low-voltage and limited-energy equipment, if the power conductors maintain a minimum of 0.25 inch separation from the low-voltage and limited-energy conductors.

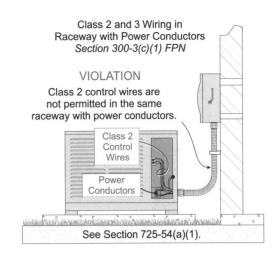

Fig. 13-2 Low-Voltage Cables Not Permitted in Raceways or Enclosures with Power Conductors

Question. Can Class 2 thermostat control cables be installed in the same raceway with the air-conditioning power conductor [725-54(a)(1)], Fig. 13-2?

Answer. No

300-4 Protection against Physical Damage

Conductors and equipment must be protected when subject to physical damage [110-27(b)].

(a) Cables and Raceways through Wood Members. The following cables and raceways installed through wood members must comply with items (1) and (2) of this Section:

Armored Cable – 333
Electrical Nonmetallic Conduit – 331
Flexible Metal Conduit – 350
Liquidtight Flexible Conduit – 351
Metal-clad Cable – 334
Nonmetallic-sheathed Cable – 336
Service-entrance Cable – 338

(1) Drilling Holes in Wood Members. When drilling holes through wood framing members, the holes must be done so the edge of the bored hole is at least 1¼ inch from the nearest edge of the wood member. If the 1¼ inch distance cannot be maintained, a 1/16 inch thick steel plate of sufficient

Raceways and Cables in Wooden Notches
Section 300-4(a)(2)

Steel Plate

If notching is permitted, a steel plate
is required for:
• Armored Cable, 333
• Electrical Nonmetallic Tubing, 331
• Flexible Metal Conduit, 350
• Liquidtight Flexible Conduit, 351
• Metal-clad Cable, 334
• Nonmetallic-sheathed Cable, 336
• Service-entrance Cable, 338

Fig. 13-3 Some Raceways and Cables in Wooden
Notches Must Be Protected from Nails or Screws

length and width must protect the wiring method
from screws and nails.

(2) Notching Wood Members. Notching of
wood framing members is permitted only if there is
no weakening of the building structure. Cables and
raceways laid in these wood notches must have a
1/16 inch thick steel plate of sufficient length and
width to protect the wiring method from screws and
nails, Fig. 13-3.

> CAUTION: When drilling or notching wood
> members, be sure to check with the building
> inspector; you can significantly damage and
> weaken the structure if not done properly.

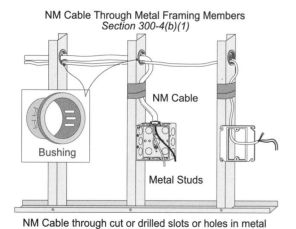

NM Cable Through Metal Framing Members
Section 300-4(b)(1)

NM Cable

Bushing

Metal Studs

NM Cable through cut or drilled slots or holes in metal
framing members must be protected by bushings or
grommets securely fastened before cable installation.

Fig. 13-4 NM Cable through Metal Framing
Members Must Be Protected

(b) Metal Framing Members.

**(1) Nonmetallic-Sheathed Cable through
Metal Framing Members.** When nonmetallic-
sheathed cables pass through cut or drilled slots or
holes in metal framing members, the cable needs to
be protected by bushings or grommets securely fastened in the opening prior to installation of the
cable. This requirement does not apply to low-voltage or limited energy cables [90-3, 725-3], Fig.
13-4.

Note: Separation is not required between
low-voltage or limited-energy cables, and
raceways or cables containing power conductors, Fig. 13-4.

System	Section
CATV	820-52(a)(2)
Control/Signaling	725-54(a)(3)
Fire Alarm	760-54(a)(3)
Intrinsically Safe	504-30(a)(2)
Network Broadband	830-58(a)(2)
Radio and Television	810-18(b)
Telecommunications	800-52(a)(2)

**(2) Electrical Nonmetallic Tubing or NM
Cable Protection from Screws.** ENT or NM
cable installed through metal framing members
must be protected by a 1/16 inch thick steel plate, if
they are likely to be penetrated by nails or screws.

(c) Behind Suspended Ceilings. All wiring
installed behind panels that allow access must be
supported. This means that cables and raceways
cannot lie on the suspended ceiling and they must
be supported as required for the wiring method
involved.

Similar rules apply to low-voltage and limited-
energy cables in the following *Code* Sections.

System	Section
CATV	820-5
Control and Signaling	725-5
Fiber Optical Cable	770-7
Fire Alarm	760-5
Network Broadband	830-6
Sound Systems	640-5
Telecommunications Systems	800-5

**(d) Cables and Raceways Parallel to Fram-
ing Members.** Protection must be provided for

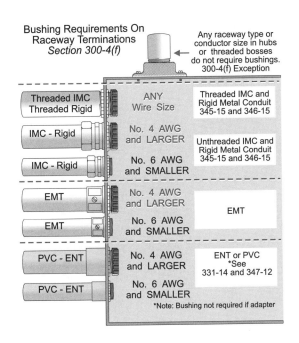

ANY Wire Size — Threaded IMC and Rigid Metal Conduit 345-15 and 346-15

No. 4 AWG and LARGER / No. 6 AWG and SMALLER — Unthreaded IMC and Rigid Metal Conduit 345-15 and 346-15

No. 4 AWG and LARGER / No. 6 AWG and SMALLER — EMT

No. 4 AWG and LARGER / No. 6 AWG and SMALLER — ENT or PVC *See 331-14 and 347-12

*Note: Bushing not required if adapter

Fig. 13-5 Insulating Bushing Required for Some Raceway Terminations

cables or raceways run parallel to framing members if they are likely to be penetrated by nails or screws. The protection can be by installing the wiring method so that they are not less than 1¼ inch from the nearest edge of the framing member, or they must be protected by a 1/16 inch thick steel plate.

Note. When running nonmetallic-sheathed cable (Type NM) on the interior of masonry walls, keep the cable 1¼ inch from the furring strips (framing member).

(e) Cables and Raceways Installed in Shallow Groove. Cables and raceway-type wiring methods installed in a groove must be protected by a 1/16 inch thick steel plate, sleeve, or by 1¼ inch free space.

Sometimes NM cable is installed by laying the cable in a shallow groove cut into the styrofoam type insulation building block structure and then covered with wallboard. Anyone could drive a nail or screw through the wallboard or other type finish into the NM cable, unless it is properly protected.

Exception: This rule does not apply when the cable is installed in rigid metal conduit, intermediate metal conduit, rigid nonmetallic conduit or electrical nonmetallic conduit.

(f) Insulating Bushings. Conductors No. 4 and larger entering an enclosure must be protected from abrasion during and after its installation by a fitting that provides a smooth, rounded insulating surface, such as an insulating *bushing*, Fig. 13-5.

Note. Not all PVC male adapters provide the smooth rounded insulating surface required by this Section. So check with your inspector on his or her interpretation on the suitability of PVC without an insulating bushing.

Exception. Insulating bushings are not required where conductors enter threaded *hubs* or *bosses* of a cabinet, box, or enclosure that provides a smooth, rounded, or flared entry for the conductors.

300-5 Underground Installations

(a) Minimum Burial Depths. When cables and raceways are run underground, they must have a minimum earth cover according to Table 300-5. The cover requirement is dependent on the wiring method and the installation location.

Note. Cover is defined as the shortest distance from the top surface of a wiring method to the surface of finish grade, Fig. 13-6.

(c) Cables under Buildings. Cables run under a building must be in a raceway, and the raceway must extend past the outside walls of the building.

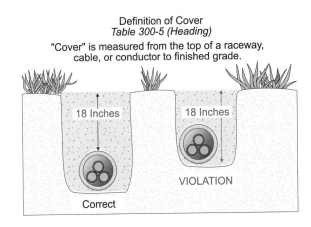

Fig. 13-6 Definition of Cover

Table 300-5   Minimum Cover Requirements: 0-600 Volts

| Type of Wiring Method or Circuit | | | | | |
Column 1	Column 2	Column 3	Column 4	Column 5	Column 6
Under The Following Locations:	(UF-USE)	(Rigid-IMC)	Nonmetallic Raceways	Residential 5-20A, GFCI Branch Circuit	Landscape Lighting Max 30V
1. Locations not listed below	24	6	18	12	6
2. TRENCH not less than 2 inches of concrete	18	6	2	6	6
3. BUILDING	Raceway only	No Depth	No Depth	Raceway only	Raceway only
4. SLAB not less than 4 inches of concrete	18	4	4	6" Direct Burial 4" Raceway only	6" Direct Burial 4" Raceway only
5. STREET, Driveways, Parking Lots	24	24	24	24	24
6. One-Two-Family DRIVEWAYS (not garage floors)	18	18	18	12	18
7. AIRPORT RUNWAYS	18	18	18	18	18
8. SOLID ROCK, Covered not less than 2 inches of concrete	2" Raceway only	2	2	2" Raceway only	2" Raceway only

9. Raceway approved for burial only when concrete enclosed, require at least 2 inches of concrete.

Depths less than shown in the Table are permitted for terminations, splices, and where access is required.

When the wiring methods listed in Columns 3 and 4 are combined with a conduit type in columns 5 and 6, the lesser depth is permitted.

(d) Protecting Underground Conductors and Cables. Where direct-buried conductors or cables emerge from underground, they must be protected by enclosures or raceways. This protection must extend below grade to a depth specified by Table 300-5, but the depth is not required to be greater than 18 inches below grade. The protection above grade must extend to at least 8 feet.

Service-Laterals. **Service-laterals** not encased in concrete and buried 18 inches or more, must have their location identified by a warning ribbon placed in the trench at least 12 inches above the underground installation.

Note. Commercially available warning ribbon are often red with black letters stating: CAUTION: BURIED ELECTRIC LINE.

Physical Protection. Where enclosures containing direct-buried conductors or cables are subject to physical damage, they must be installed in rigid metal conduit, intermediate metal conduit, or Schedule 80 PVC.

(e) Splices and Taps Underground. Direct-buried conductors or cables can be spliced or tapped without the use of a splice box [300-15(g)]. The splice or tap must be made in accordance with the requirements of Section 110-14(b).

(f) Backfill. Backfill used for underground wiring must not damage the wiring method, prevent compacting of the fill, or contribute to the corrosion of the raceway. If underground raceways or cables require protection from physical damage, it must be provided using granular material, running boards, sleeves, or other approved means.

Large rocks, chunks of concrete, steel rods, mesh, and other sharp-edged objects shall not be used as *backfill* because they can damage the underground conductors, cables, or raceways.

(g) Raceway Seals. Where moisture could enter a raceway and contact energized live parts, seals or plugs are required at one or both ends of the raceway. This is a common problem for equipment located downhill or in underground equipment rooms. See Section 230-8 for service raceway seals, and Section 300-7(a) for different temperature area seals.

FPN: Hazardous explosive gases or vapors require the sealing of underground conduits

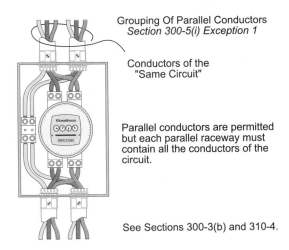

Grouping Of Parallel Conductors
Section 300-5(i) Exception 1

Conductors of the
"Same Circuit"

Parallel conductors are permitted
but each parallel raceway must
contain all the conductors of the
circuit.

See Sections 300-3(b) and 310-4.

Fig. 13-7 Parallel Conductors Permitted in Different Raceways

or raceways entering the building in accordance with Sections 501-5 and 501-11.

Note. It is not the intent of this FPN that sealoffs of the types required in Hazardous (Classified) locations be installed in unclassified locations, except where required in Chapter 5.

(h) Bushing. Raceways that terminate underground must have a bushing or fitting at the end of the raceway to protect emerging cables or conductors.

(i) Conductors Grouped Together. All conductors of a circuit must be installed in the same raceway or placed close together in the same trench. This includes the ungrounded, grounded, and the grounding conductors.

The purpose of grouping conductors is to ensure a *low impedance* grounding path, and to reduce inductive heating of metal parts due to *hysteresis*.

Note. Other *Code* Sections that require the conductors to be grouped are 250-102(f), 300-3(b), 300-20(a), and 318-8(d).

Exception No. 1: Parallel Conductor Installations. Conductors are permitted to be parallel in accordance with the requirements of Section 310-4, Fig. 13-7.

Note. When there are more than 3 current-carrying conductors in a raceway, the conductor ampacity as listed in Table 310-16 must be reduced. See Sections 310-10 and 310-15 of this book for details.

Exception No. 2: Isolated Phase Installations. Each phase conductor and each neutral can be installed in nonmetallic raceways in close proximity where conductors are paralleled as permitted in Section 310-4 and the conditions of Section 300-20 are met. This type of installation makes it easier to terminate large conductors, Fig. 13-8.

> CAUTION: When installing isolated phase installations there is no canceling of the magnetic fields. I know of a case where computer monitors would not display properly, but when the monitors were moved to other locations in the building the monitors worked fine. It was determined that the magnetic field was caused by an isolated phase installation of the feeder conductors in the concrete slab. Be aware that some equipment can be very sensitive to electromagnetic fields.

(j) Ground Movement. Direct buried conductors, cables, or raceways subject to movement by settlement or frost must be arranged so as to prevent damage to conductors or equipment connected to the wiring.

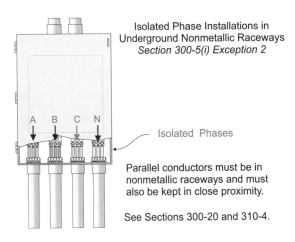

Isolated Phase Installations in
Underground Nonmetallic Raceways
Section 300-5(i) Exception 2

Isolated Phases

Parallel conductors must be in
nonmetallic raceways and must
also be kept in close proximity.

See Sections 300-20 and 310-4.

Fig. 13-8 Isolated Phase Installations in Nonmetallic Raceways

FPN: "S" loops in underground direct burial to raceway transitions, expansion joints in pipe risers, and flexible connections to equipment are methods of providing extra cable to accommodate ground movement.

300-6 Protection against Corrosion

Metal electrical equipment must be suitable for the environment in order to prevent *corrosion* of the equipment.

(a) What Requires Protection. Metal fittings, raceways, cables, and enclosures must be protected from corrosion inside and out. This includes supports, straps, and screws. The protective coating must be an approved corrosion-resistant material, such as zinc (galvanized), cadmium coated, or enamel.

Exception: Raceway Threads. An identified electrically conductive compound (such as zinc chromate paste) can be used to coat threads to reduce the resistance of the joints.

(b) Concrete or Direct Contact with the Earth. Equipment in concrete, in direct contact with the earth, or other areas of severe corrosive influences shall be of a material **identified** for the condition. If the equipment is not identified for the condition, then it must be provided with corrosion protection approved for the particular condition.

(c) Indoor Wet Locations. Exposed surfaces of metallic equipment, raceways, and cables mounted in indoor wet locations must have a ¼ inch of air space between the equipment and the wet indoor mounting surface.

Exception: Nonmetallic Parts. The ¼ inch air space does not apply to nonmetallic equipment, raceways, and cables.

Note. For outdoor installations of enclosures, see Section 373-2(a) for the requirement of a ¼ inch air space.

300-7 Raceways Exposed to Different Temperatures

(a) Sealing. Circulation of warm and cold air causes condensation of moisture resulting in water; this must be prevented in raceways and enclosures. When raceways are placed across widely different temperature zones (walk-in coolers and freezers), a seal must be installed to prevent the warm air from mixing with the cold air. The use of putty or any other method that will prevent the air mixture is permitted. This seal can be installed at a box or **conduit body,** Fig. 13-9.

Question. Is a sealoff fitting, like that required in hazardous classified locations, necessary?

Answer. No
Section 300-7(a) requires a compound seal, not a sealoff fitting. If the intent of 300-7 was to require sealoff fittings, a reference would have been made to Section 501-5.

(b) Expansion Fittings. Expansion fittings or flexible raceways are required to compensate for thermal expansion and contraction of raceways or building structures, Fig. 13-10.
In addition, expansion fittings might be required by the plans, specifications, or by the building code. See Section 347-9 for rigid nonmetallic conduit expansion fitting requirement.

Note. If expansion fittings are used for metal conduit, bonding jumpers are required to maintain the equipment grounding path [250-98].

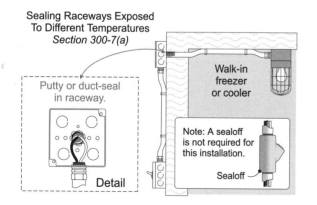

Fig. 13-9 Raceways Exposed to Different Temperatures Must Be Sealed

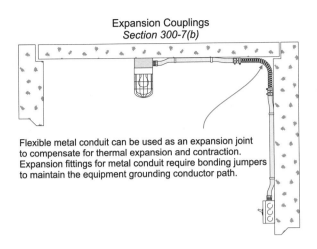

Expansion Couplings
Section 300-7(b)

Flexible metal conduit can be used as an expansion joint
to compensate for thermal expansion and contraction.
Expansion fittings for metal conduit require bonding jumpers
to maintain the equipment grounding conductor path.

Fig. 13-10 Expansion Couplings

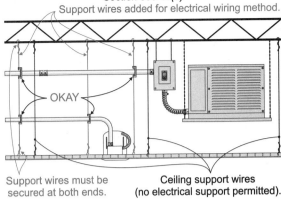

Wiring Supported To Suspended Ceilings
Section 300-11(a)
Support wires added for electrical wiring method.

OKAY

Support wires must be
secured at both ends.

Ceiling support wires
(no electrical support permitted).

Note: In *fire-rated assembies*, support wires must be identified
so they are distinguishable from the ceiling support wires.

Fig. 13-11 Wiring Not Permitted to Be Supported
to Suspended Ceilings

300-8 Not Permitted in Raceways

Raceways are designed for the exclusive use of electrical conductors and cables, and must not contain any pipe, tube, or equal for steam, water, air, gas, drainage, etc.

300-10 Electrical Continuity

All metal raceways, armored cable, boxes, fittings, **cabinets,** and metal enclosures for conductors must be metallically joined to form a continuous low impedance grounding path. The path must have the capacity to carry the fault-current to assist in opening the overcurrent protection device, thereby removing dangerous voltages [250-2(d)].

To maintain an effective grounding path, fittings must be made up tight [250-120], and must be mechanically secured to boxes and other enclosures [300-12]. Electrical continuity must be maintained by the use of proper bonding in accordance with Part E of Article 250.

Exception: Cables in Raceway. Short sections of metal enclosures used for the support or protection of cables are not required to be electrically continuous, nor are they required to be grounded [250-86 Ex. 2, and 300-12 Ex.].

Question. If NM cable is installed within electrical metallic tubing, will this cause an *inductive choke* if there is a short-circuit or ground-fault?

Answer. No

When the cables are installed inside a metal raceway, current travels in both directions within the metal enclosure, thus canceling the magnetic inductive fields.

Reduction of Electric Noise. Section 250-96(b) specifies that metal raceways that supply circuits to electronic equipment can terminate to a nonmetallic fitting on the electronic equipment. In addition an insulated equipment grounding conductor must be installed within the metal raceway to ground the equipment.

300-11 Securing and Supporting

(a) Secured in Place. Raceways, cable assemblies, boxes, cabinets, and fittings shall be securely fastened in place.

(1) Fire-Rated Assembly. Electrical wiring located within the cavity of a fire-rated floor/ceiling or roof/ceiling assembly shall not be secured to, or supported by, the ceiling grid support wires. Independent support wires secured at both ends installed in addition to the ceiling grid support wires can be used for the support of electrical wiring. The independent support wires shall be distinguishable from the suspended ceiling framing support wires by color, tagging, or other effective means, Fig. 13-11.

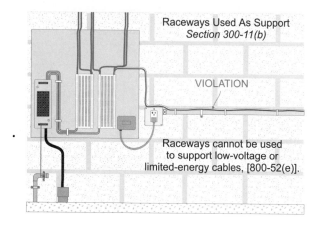

Fig. 13-12 Raceways Not Permitted to Be Used for Support

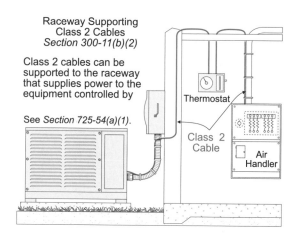

Fig. 13-13 Raceway Permitted to Supporting Class 2 Cable

Exception. Fire-rated ceiling systems tested to support wiring and equipment can support wiring.

(2) Non Fire-Rated Assembly. Independent support wires secured at both ends installed in addition to the suspended ceiling grid support wires can be used to support electrical wiring. Support wires within non fire-rated assemblies are not required to be distinguishable from the suspended ceiling framing support wires, Fig. 13-11.

Exception. Branch circuit wiring and associated equipment can be supported to the ceiling system according to the ceiling system manufacturer's instructions.

Note. Outlet boxes can be supported to the suspended ceiling support wires or framing members [370-23(d)], and lighting fixtures can be supported to the suspended ceiling where securely fastened to the ceiling framing member [410-16(c)].

(b) Raceways Used as Support. Raceways cannot be used for the support of other raceways, cables, or other equipment such as plumbing pipes, ducts, etc. In addition, low-voltage and limited-energy cables cannot be strapped, taped, or attached to electrical raceways, not because of this rule but because of the following *Code* Sections, Fig. 13-12:

System	Sections
CATV	820-6
Control/Signaling	725-7, 725-54(d)
Fire Alarm	760-8, 760-54(d)

Optical Fiber Cables	770-8
Network Broadband	830-7, 830-58(d)
Radio and Television	810-12
Telecommunications	800-6, 800-52(e)

(2) Control Conductors. Class 2 and 3 cables can be supported to the raceway that supplies power to the equipment controlled by the Class 2 or 3 cables. Since Class 2 and Class 3 cables cannot be installed in the same raceway with the power conductors [725-54(a)(1)], the next best thing is to attach them to the raceway, Fig. 13-13.

(3) Boxes Supported by Conduits. Raceways can support threaded boxes and conduit bodies under specific conditions [370-23(e) and (f)] and some lighting fixtures [410-16(f)].

300-12 Mechanical Continuity

Raceways and cable sheaths must be mechanically continuous between all points of an electrical system. See Sections 300-10 and 370-17(b) and (c) for more details, Fig. 13-14.

Exception: Cables in Raceway. Raceways used for the support or protection of cables are not required to be mechanically continuous or grounded. See Sections 250-86 Exception No. 2, 300-10 Exception, and 300-15(c) for more details.

Note. *Mechanical continuity* is not required for underground raceways used for the protection of cables [300-5(h)], or under-

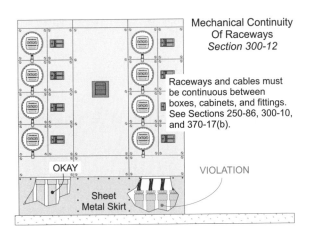

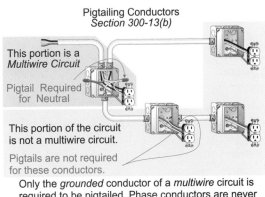

Fig. 13-14 Mechanical Continuity of Raceways
Must be Maintained

Fig. 13-16 Pigtailing Required for Grounded
(Neutral) Conductor--Multiwire Circuit

ground raceways stubbed within switch-boards [300-16(b) and 384-10].

300-13 Splices and Pigtails

(a) Splices. Conductors in raceways must be continuous between all points of the system and splices are not permitted in the raceway, Fig. 13-15.

A box is required at each conductor splice connection point, outlet, switch, junction, or pull point in accordance with Section 300-15, except for the following wiring methods:

Wiring Method	Section
Boxes	300-5(a)
Cabinets	373-8
Conduit Bodies	370-16(c)
Lighting Fixtures	410-31
Surface Raceways	352-7
Wireways	362-7

(b) Pigtail Neutrals. The removal of a device must not cause the grounded (neutral) conductor of

a multiwire branch circuit to open. The grounded (neutral) conductors must be spliced together and a separate lead wire (*pigtail*) must be taken to the device. For additional multiwire branch circuit requirements see Section 210-4, Fig. 13-16.

Question. If two wires enter an outlet box and terminate on a receptacle, can we terminate two more wires on the same device to supply another receptacle, Fig. 13-16?

Answer.   Yes
The grounded (neutral) conductors of two-wire circuits are not required to be pigtailed, and the ungrounded (hot) conductors are never required to be pigtailed. The opening of the hot, or grounded (neutral) conductor of a two-wire circuit during the replacement of a device, does not cause a safety hazard; therefore, pigtailing of these conductors is not required.

CAUTION: Don't remove the grounded (neutral) conductor from the grounded terminal bar in a panelboard if the phase conductors are energized. The grounded (neutral) conductor could be part of a multiwire branch circuit, resulting in the destruction to electrical equipment due to overvoltage.

Danger of Multiwire Branch Circuits

The reason the grounded (neutral) conductor of a multiwire branch circuit must be pigtailed

Fig. 13-15 No Splices in Raceways

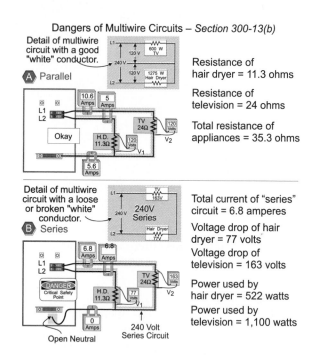

Fig. 13-17 Dangers of Multiwire Branch Circuit

is that the opening of the grounded (neutral) conductor can cause overvoltage at electrical equipment, Fig. 13-17.

Step 1. Find the resistance of these appliances:
Hair Dryer Resistance.
$R = E^2/P$
$R = 120 \text{ volt}^2/1,275 \text{ watts}$
$R = 11.3 \text{ ohm}$

Television Resistance.
$R = E^2/P$
$R = 120 \text{ volt}^2/600 \text{ watts}$
$R = 24 \text{ ohm}$

Step 2. Current of the circuit using the formula.
$I = E/R$
$I = 240 \text{ volt}/35.3 \text{ ohm}$
$(11.3 \text{ ohm} + 24 \text{ ohm})$
$I = 6.8 \text{ ampere}$

Step 3. Operating voltage for each appliance.
Hair Dryer.
$E = I \times R$
$E = 6.8 \text{ ampere} \times 11.3 \text{ ohm}$
$E = 76.84 \text{ volt}$

Television.
$E = I \times R$
$E = 6.8 \text{ ampere} \times 24 \text{ ohm}$
$E = 163.2 \text{ volt}$

Step 4. Power consumed by each appliance.
Hair Dryer.
$P = E^2/R$
$P = 76.8 \text{ volt}^2/11.3 \text{ ohm}$
$P = 522 \text{ watts}$

Television.
$P = E^2/R$
$P = 163.2 \text{ volt}^2/24 \text{ ohm}$
$P = 1,110 \text{ watts}$

The television (rated 600 watts at 120 volt) will consume 1,110 watts at 163 volt. You can kiss this TV good-bye.

> **WARNING:**
> Be sure when you terminate the line conductors of multiwire branch circuits that each ungrounded (phase) conductor is terminated to a different phase in the panelboard. If the ungrounded conductors are not terminated to different phases, then the grounded (neutral) conductor could be overloaded and the insulation destroyed. Now you know why white neutral conductors turn black at the neutral bar, Fig. 13-18.

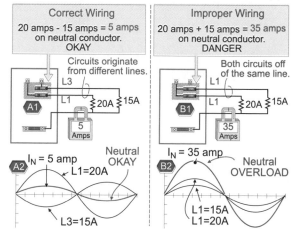

Fig. 13-18 Neutral Current on Multiwire Circuits

Length Of Free Conductor - *Section 300-14*

A minimum of 3 inches outside of opening required.

6 Inches from point of entry.

4 in.

2 in.

At least 6 inches of free conductor is required at each outlet measured from the point in the box where the conductors emerge from the raceway or cable sheath.

Fig. 13-19 Minimum 6 Inches of Free Conductor

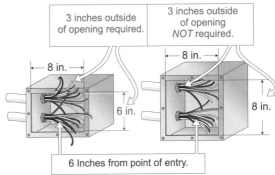

Length Of Free Conductor - *Section 300-14*

3 inches outside of opening required.

3 inches outside of opening *NOT* required.

8 in.

8 in.

6 in.

8 in.

6 Inches from point of entry.

6 inches of free conductor is required at each outlet. Enclosures with openings less than 8 inches require 3 inches of conductors outside the opening.

Fig. 13-20 Length of Free Conductor Dependent on the Size of the Enclosure Opening

300-14 Length of Free Conductors

At least 6 inches of free conductor, measured from the point in the box where the conductors enter the enclosure, shall be left at each outlet, junction, and switch point for splices or the connection of lighting fixtures or devices. Where the opening to an outlet, junction, or switch point is less than 8 inches in any dimension, each conductor shall be long enough to extend at least 3 inches outside the opening of the enclosure, Fig. 13-19.

Openings less than 8 inches. Outlet boxes that have openings less than 8 inches shall have at least 3 inches of free conductor outside the opening, and no less than 6 inches of free conductor measured from the point in the box where the conductors enter the enclosure.

Example. A box that is sized 6 inches high × 8 inches wide × 4 inches deep would require about 7 inches of free conductor (4 inches in the box and 3 inches outside the box), Fig. 13-20.

Note. There's no limit on the number of *extension rings* on an outlet box, however there must be at least 3 inches of free conductors outside the opening of the outlet box.

Opening 8 inches or more. Outlet boxes with opening 8 inches or more shall have at least 6 inches of free conductor, measured from the point in the box where the conductors enter the enclosure.

Example A. 8 inches × 8 inches × 4 inches deep junction/splice box would only require 6 inches of

free conductor measured from the point in the box where the conductors enter the enclosure. The 3 inch outside the box rule does not apply to this enclosure, Fig. 13-20.

Exception: Conductors that Pass Through. Conductors that pass through an outlet, junction, or switch point, and are not used for splices or termination, do not require 6 inches of free conductor outside the box.

300-15 Boxes or Conduit Bodies Required

(a) Required. A box or conduit body shall be installed at each conductor splice connection point, outlet, switch, junction, or pull point, except as permitted in the following:

Wiring Method	Section
Cabinet or Cutout Boxes	373-8
Conduit Bodies	370-16(c)
Lighting Fixtures	410-31
Surface Raceways	352-7
Wireways	362-7

Question. Can a splice be made in a conduit body?

Answer. Yes

Yes, if the conduit body is marked with its cubic inch capacity [370-16(c)].

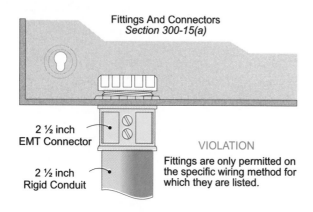

Fig. 13-21 Fittings and Connectors Must Be for the Specific Wiring Method

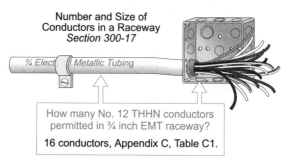

Fig. 13-22 Number and Size of Conductors in a Raceway Can Be Determined by Using Appendix C

Question. Can a conductor be spliced in a panelboard cabinet?

Answer.   Yes

Yes, if the conductors, splices, and taps do not exceed 75 percent of the area of the cabinet wiring space [373-8].

Fittings and Connectors. Fittings shall be used only with the specific wiring methods for which they are designed and listed. Electricians often violate this *Code* Section by using NM fittings with armored cable, PVC fittings with liquidtight, or EMT fittings with rigid metal conduit, Fig. 13-21.

> **Note.** PVC couplings and connectors can be used with Electrical Nonmetallic Tubing, but be careful to use the proper glue [331-8].

> CAUTION: Conduits 2½ inches and larger all have the same outer diameter. This means the fittings for rigid, EMT, PVC, and IMC are mechanically interchangeable. EMT fittings are permitted only on EMT, rigid fittings on rigid, and so on, unless the fitting is listed for other raceways (some are).

(c) Raceways for Support or Protection. When a raceway is used for the support or protection of cables, a fitting such as a coupling or connector is required where cables enter or exit the raceway. See Sections 250-86 Ex. 2, 300-5(c), 300-10 Ex., 300-12 Ex. for more details.

(g) Underground Splices. Underground splices without a box is permitted if the installation are installed in accordance with the requirements of Sections 110-14(b) and 300-5(e).

300-17 Number and Size of Conductors in Raceway

The number and size of conductors in a raceway must permit the installation or removal of the conductors without damaging the conductor insulation. The maximum percentage of conductor fill listed in Table 1 of Chapter 9 is based on conditions where the length of the conductor and number of raceway bends are within reasonable limits [Table 1 of Chapter 9, FPN].

When all of the conductors are the same size and insulation, the number of conductors permitted in a raceway can be determined simply by looking at the tables in Appendix C.

Question. How many No. 12 THHN conductors can be installed in a ¾ inch electrical metallic tubing, Fig. 13-22?

Answer.   16 conductors
Appendix C, Table C1

300-18 Inserting Conductors in Raceways

(a) Complete Runs. Before installing conductors in a raceway, the raceway must be mechanically completed between the pulling points

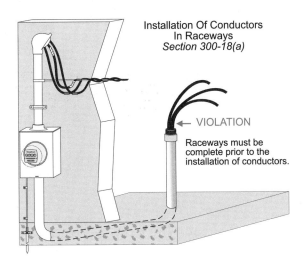

Installation Of Conductors
In Raceways
Section 300-18(a)

← VIOLATION

Raceways must be
complete prior to the
installation of conductors.

Fig. 13-23 Raceway Must Be Complete Before Conductors Can Be Installed

[300-10, 300-12]. The purpose of this rule is to protect the conductor insulation from abrasion in the event the conductors were pulled before the raceway is completed, Fig. 13-23.

However, conductors can be pulled through flexible conduit before the termination fittings are installed. In addition, prewired assemblies such as listed manufactured prewired electrical nonmetallic tubing [331-3] and listed manufactured prewired liquidtight flexible nonmetallic assemblies [351-23(a)] are also permitted.

300-19 Supporting Conductors in Vertical Raceways

(a) Spacing Intervals. Cable support shall be provided at the top of the vertical raceway or as close to the top as practical. An additional intermediate support(s) shall be provided as necessary to support conductors, according to Table 300-19(a).

The weight of long vertical runs of conductors can cause the conductors to actually drop out of the raceway. There are many cases where conductors in a vertical raceway were released from the pulling basket (at the top), and the conductors fell down the raceway, injuring the electrician.

Question. How many vertical supports are required for a 75-foot vertical rise for four 250 kcmil cu conductors?

Answer. Two
Two supports are required, one at the top, and a second one not more than 60 feet from the top or bottom of the raceway.

300-20 Induced Currents in Metal Parts

(a) Conductors Grouped Together. All circuit conductors, including the phase, grounded (neutral) and equipment grounding conductor. This reduces electromagnetic fields that could be a health problem, and it's intended to reduce the inductive and hysteresis heating of the surrounding metal raceway.

Induction Heating. When alternating current flows through a conductor, a pulsating or varying magnetic field is created around the conductor. This magnetic field is constantly expanding and contracting with the frequency of the alternating current. In the United States, the frequency is 60 cycles per second, assuming we are not supplying **nonlinear loads.** Since it is the nature of alternating current to reverse polarity (120 times per second), the magnetic field surrounding the conductor reverses its direction as well. This expanding and contracting magnetic field will induce *eddy currents* in the metal raceway, causing the metal to heat.

Table 300-19(a) Maximum Distance Between Supports

Size of Wire	Aluminum	Copper
8 AWG through 18 AWG	100 feet	100 feet
6 AWG through 1/0	200 feet	100 feet
2/0 through 4/0	180 feet	80 feet
250 through 350 kcmil	135 feet	60 feet
400 through 500 kcmil	120 feet	50 feet
600 through 750 kcmil	95 feet	40 feet
Over 750 kcmil	85 feet	35 feet

> WARNING:
> There has been much discussion of the effects of electromagnetic fields on humans. According to the Institute of Electrical and Electronic Engineers (IEEE), there is insufficient information to define safe and unsafe field levels. In general, there is not enough relevant scientific data to establish that common exposure to power-frequency fields (i.e. 60-400 Hz) should be considered a health hazard.

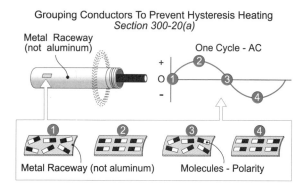

Grouping Conductors To Prevent Hysteresis Heating
Section 300-20(a)

Fig. 13-24 Grouping of Conductors Helps Prevent Hysteresis Heating of Ferrous Metal Raceways

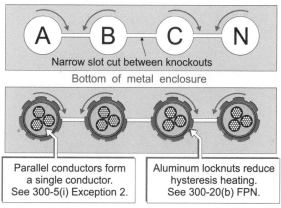

Reduction Of Inductive Heating Of Metal
Section 300-20(b)

Slots between knockouts allow "air" (high reluctance material) to reduce the electromagnetic field flux which reduces induction heating (hysteresis).

Narrow slot cut between knockouts

Bottom of metal enclosure

Parallel conductors form a single conductor. See 300-5(i) Exception 2.

Aluminum locknuts reduce hysteresis heating. See 300-20(b) FPN.

Fig. 13-25 How to Reduce Inductive Heating of Metal Parts

Hysteresis Heating. Hysteresis affects *ferrous* metals that have magnetic properties, such as steel and iron, but not aluminum. Simply put, the molecules of steel and iron align to the polarity of the magnetic field. When the magnetic field reverses, the molecules reverse their polarity as well. This back and forth alignment of the molecules in steel or iron can generate tremendous heat due to friction, Fig. 13-24.

When conductors are grouped together, the magnetic fields of the different conductors tend to cancel each other. Since the magnetic field is reduced, lower induced currents and hysteresis heating will be produced in the surrounding metal enclosure.

> **Note.** Aluminum conduit, locknuts, and enclosures can be used to reduce inductive heating [300-20(b) FPN].

Low Impedance Path. The grouping of all circuit conductors lowers the impedance of the ground-fault or short-circuit path which permits the proper operation of the overcurrent protection device to clear the fault, thereby removing dangerous voltages [250-2(b)]. Other *Code* Sections that require conductors to be grouped include Sections 250-102(f), 300-3(b), 300-5(i), and 300-20(b).

> **Note.** Nonlinear loads produce *harmonic currents* on the grounded (neutral) conductor which can result in tremendous induction heating of the metal raceway.

(b) Conditions for Single Conductors.
When all conductors of a single phase is installed in rigid nonmetallic conduit and it passes through a *ferrous* (steel or iron) metal wall, the inductive heating can be minimized by grouping all conductors of the circuit within the same metallic opening. This can be accomplished by cutting slots in the metal between the individual holes through which the individual phase conductors pass or by passing all of the circuit conductors through one opening that is sufficiently large for all of the circuit conductors [300-5(i) Ex. 2], Fig. 13-25.

> FPN: Aluminum is not a magnetic metal and is not affected by inductive heating; therefore, grouping of conductors, or special treatment in passing conductors through aluminum wall sections of enclosures is not required.

300-21 **Spread of Fire or Products of Combustion**

Electrical circuits and equipment in hollow spaces, vertical shafts, ventilation, or air handling ducts, must be installed in a way that the possible spread of fire, or products of combustion, will not be substantially increased. This means that openings made in fire-rated walls, floors, and ceilings for electrical equipment must have an approved fire-stop. The fire-stop must maintain the *fire resistance* of that structure, and the installation of fire-stop materials must be done according to specific instructions listed with the material. Fire-stop

Fire-Rated Walls, Ceilings, and Floors - *Section 300-21*

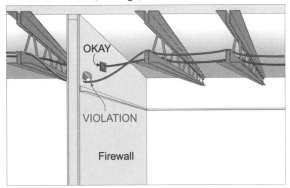

Openings around electrical penetrations through fire-resistant rated walls, partitions, floors, or ceilings are required to be firestopped using approved methods to maintain the fire resistance rating.

Fig. 13-26 Fire-Rating of Walls, Ceilings, and Floors Must Be Maintained

material is approved for specific types of wiring methods (rigid, PVC, ENT, NM cable, etc.), and types of construction structures, such as drywall, concrete, etc.

FPN: According to Underwriter Laboratory Fire Resistance Directory, outlet boxes must have a horizontal separation of not less than 24 inches. To overcome these limitations, fire resistant "Puddy Pads" are available.

The fire-stop requirements of Section 300-21 applies to low-voltage and limited-energy cables because the following *Code* Sections required this compliance, Fig. 13-26.

System	Section
CATV	820-3(b)
Control and Signaling	725-3(a)
Fiber Optical Cable	770-2(a)
Fire Alarm	760-3(a)
Network Broadband	830-3(a), and 830-58(b)
Sound Systems	640-3(a)
Telecommunications	800-52(b)

FPN: Directories published by qualified testing laboratories contain many listing installation restrictions necessary to maintain the fire resistive rating of assemblies where penetrations or openings are made.

Ducts Used For Dust And Loose Stock
Section 300-22(a)

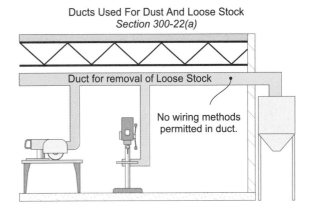

Fig. 13-27 No Wiring Is Permitted in Ducts Used for Dust, Loose Stock, or Vapor

Question. Can a ¾ inch EMT raceway be run through a 1 inch hole in a fire-rated assembly without a fire seal?

Answer. No
You need to have an opening that is large enough so that sufficient fire-sealant material can be installed. Check with the product installation instructions.

300-22 Ducts, Plenums, and Air Handling Spaces

This Section limits the installation of electric wiring and equipment in ducts, plenums, and other areas used for environmental air.

(a) Ducts Used for the Removal of Dust, Loose Stock, or Vapor. Manufactured ducts that transport dust, loose stock, flammable vapors, or are used for the ventilation of commercial cooking equipment, shall not have any wiring method installed within the duct, Fig. 13-27.

Question. Can rigid conduit be run inside a restaurant cooking hood to wire lighting fixtures?

Answer. No
Exposed electrical wiring is not permitted within commercial cooking hoods, and this includes rigid metal conduit [410-4(c)].

(b) Ducts and Plenums. This subsection permits some wiring methods (see the *NEC* for the

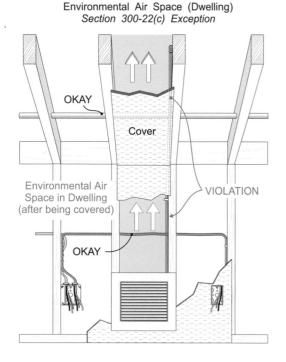

Environmental Air Space (Dwelling)
Section 300-22(c) Exception

OKAY

Cover

Environmental Air
Space in Dwelling
(after being covered)

VIOLATION

OKAY

Fig. 13-28 Spaces Between Studs Used for
Environmental Air Permitted to Have Wiring--
Dwelling

specific wiring methods) to be installed within ducts
and plenums specifically fabricated to transport
environmental air.

This is very rare, however some commercial
and industrial buildings have large supply or return
air **plenums** with ducts which require the installation of damper motors, control equipment, and
lighting fixtures.

> CAUTION: Section 300-22(b) does not apply
> to the dropped ceiling space above a sus-
> pended ceiling used to move environmental
> air. See the Fine Print Note to Section 300-
> 22(c).

**(c) Space Above Drop Ceilings Used for
Environmental Air.** The space above a dropped
ceiling used for environmental air must comply
with (1) and (2) of this subsection. Areas of
buildings whose main purpose is not air handling
are not considered "other space used for environmental air."

Question. An electrical equipment room contains some air handling equipment and the door to

the equipment room is louvered (vented) for the
return air. Can nonmetallic raceways or cables be
used in this room?

Answer. Yes
Since the main purpose of the room is electrical
equipment, nonmetallic raceways and cables are
permitted.

Exception: Spaces Between Studs. The restrictions of Section do not apply to the space between
framing members (*studs* or joists) used as an air
handling path where the wiring methods pass per-
pendicularly through the space, Fig. 13-28.

(1) Wiring Methods Permitted. The most
common wiring methods permitted in the dropped
ceiling area used for environmental air are: electrical metallic tubing, rigid metal conduit,
intermediate metal conduit, armored cable of any
length, and flexible metal conduit of any length.
Totally enclosed nonventilated insulated busway
having no provisions for plug-in connections, and
wiring methods consisting of Type MI cable, and
Type MC cable without an overall nonmetallic covering, are also permitted in dropped ceiling areas.

Exception: Liquidtight. Liquidtight flexible
conduit, in single lengths not over 6 feet, can be
installed in the dropped ceiling area used for environmental air.

Wiring Methods Not Permitted. Because the
area above the dropped ceiling is used to transport
environmental air, nonmetallic conduits and cables
are not permitted in these areas because they create
deadly toxic fumes when they burn.

Low-voltage and limited-energy cables
installed in drop ceiling space used to move environmental air must be plenum rated, Fig. 13-29.

System	Sections
CATV	820-51(a)
Control and Signaling	725-61(a)
Fire Alarm	760-61(a)
Network Broadband	830-54(b), and 830-55(b)
Optical Fiber Cables	770-53(a)
Sound System Cables	640-9(c), and 725-61(a)
Telecommunications	800-53(a)

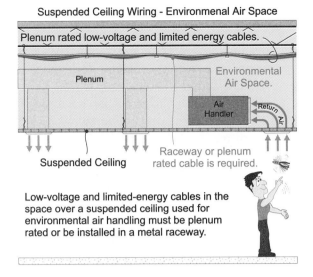

Low-voltage and limited-energy cables in the space over a suspended ceiling used for environmental air handling must be plenum rated or be installed in a metal raceway.

Fig. 13-29 Wiring Restrictions Apply to Area Above Suspended Ceilings Used for Environmental Air

(2) Electrical Equipment. Listed electrical equipment having adequate fire-resistance and low smoke-producing characteristics, such as metal boxes and lighting fixtures, can be installed in the space used for environmental air movement.

(d) Data Processing Systems (Information Technology Equipment). The area beneath raised floors for data processing systems is not considered a duct, a plenum, or other space used for environmental air and is not required to comply with the requirements of Section 300-22. Wiring methods beneath computer raised floors must comply with the requirements of Section 645-5(d).

The general requirement of Section 645-5(d) is that cables installed within raised floors of a computer room shall be listed for data processing rooms by having a marking of "Type DP." Type DP cable is constructed to have adequate fire-resistance characteristics suitable for use under raised floors of a computer room.

Low-voltage and limited-energy cables in the area beneath raised floors for Information Technology Equipment (data processing rooms) are not required to be plenum rated if the cables are marked as follows [645-5(d)(5)(c)], Fig. 13-30:

System	Cable Type
CATV	CATV
Control and Signaling	CL2, CL3, PLTC
Fiber Optic	OFC and OFN

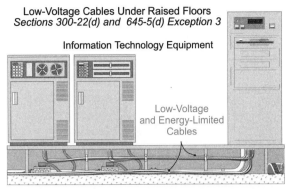

Low-voltage and limited-energy cables in the area beneath raised floors for computers are not required to be plenum rated. Note: Support is not required, see Section 645-5(b).

Fig. 13-30 Low-Voltage Cables under Raised Computer Floors Not Required to Be Plenum Rated

System	
Fire Alarm	NPLF and FPL
Sound System	CL2, CL3, PLTC
Telecommunication	CM and MP

> CAUTION: To be classified as an information technology equipment room, the room must comply with all of the requirements of Section 645-2.

300-23 Panels Designed to Allow Access

Wiring methods and equipment installed behind suspended ceiling panels must be arranged to permit the removal of the panels to give access to equipment. Access to equipment must not be prohibited by an accumulation of cables that prevent the removal of suspended ceiling panels, Fig. 13-31.

Cables must be located so that the suspended ceiling panels can be moved to provide access to electrical equipment.

System	Section
CATV	820-5
Class 2 and 3	725-5
Fiber Optical Cable	770-7
Fire Alarm	760-5
Network Broadband	830-6
Sound Systems	640-5
Telecommunications	800-5

SUMMARY

☐ Article 300 covers wiring methods for all wiring installations except control, signal, data, low voltage, security, fire alarm, fiber optic, communications and telephone cables as listed in the Exceptions.   ☐ The wiring methods in Chapter 3 are for nominal voltages 600 volt or less. ☐ Single conductors can be used only as part of wiring methods in Chapter 3 and cannot be run exposed.   ☐ All conductors of a circuit must be grouped together in the same raceway, cable, trench, cord, or cable tray.   ☐ Conductors that supply different circuits of 600 volt or less can occupy the same raceway, cable, or equipment enclosure.   ☐ Conductors must be protected when subject to physical damage.   ☐ Holes through a wood framing member must be bored so that any edge of the hole is at least 1¼ inch from the nearest edge of the wood member.   ☐ Cables and some raceways, when laid in notches, require a steel plate at least 1/16 inch thick over the notch to protect the wiring method from screws and nails.   ☐ When NM cable is installed through holes in metal framing members, bushing or grommets must be inserted before installing the conductors. ☐ Where screws may penetrate NM cable in metal framing members, steel plates, steel sleeves, or steel clips that are at least 1/16 inch thick must be installed to protect the cable.

☐ Use Table 300-5 for the minimum cover requirements for underground cables and raceways. ☐ Underground cables run under a building must be in a raceway.   ☐ Where direct-buried conductors and cables emerge from underground, they must be protected by enclosures or raceways. The protection must be run to at least 8 feet above grade.   ☐ Backfill for trenches used for underground wiring must not damage any part of the wiring method or contribute to corrosion. ☐ Where the possibility exists that moisture could enter a raceway and contact energized live parts, seals or plugs must be installed at one or both ends of the raceway.   ☐ A raceway that ends underground must have a bushing or fitting with a bushed opening at the end of the raceway to protect the emerging underground cable and conductors.   ☐ Metallic electrical equipment must be suitable for its environment to prevent corrosion.   ☐ Equipment installed in concrete, direct burial, or other areas of severe corrosive influences must be of a material identified for the condition, or the equipment must be provided with corrosion protection approved for the condition.   ☐ Equipment, including raceways and cable, in indoor wet locations must be mounted so there is ¼ inch of air space between the equipment and the mounting surface.   ☐ Expansion joints must be used on raceways where it is necessary to compensate for thermal expansion and contraction of a raceway or structure.

☐ Raceways or cable trays cannot contain nonelectrical components.   ☐ All metal raceways, armored cable, boxes, fittings, cabinets, and other metal enclosures for conductors must be metallically joined together to form a continuous electric path.   ☐ Raceways, cable assemblies, boxes, cabinets, and fittings must be securely fastened in place.   ☐ Equipment shall not be supported by a suspended ceiling or its support wires.   ☐ Raceways cannot support other raceways, cables, or nonelectric equipment.   ☐ All raceways and cables must be continuous between all points of an electrical system.   ☐ Conductors in raceways must be continuous between all points of the system, and conductor splices and taps are not permitted within a raceway.   ☐ Pigtailing of the grounded (neutral) conductors is required for multiwire branch circuits but not for 2-wire circuits.   ☐ A minimum of 6 inches of free conductor is required at all electrical outlets for making up splices and connection to devices.   ☐ Fittings and connectors must be used only with the specific wiring methods for which they are designed.   ☐ Raceway runs must be mechanically completed between pulling points before installing the conductors.   ☐ Vertical runs of conductors must be supported according to Table 300-19(a).   ☐ Alternating current conductors installed in metal enclosures must be grouped together to avoid heating the surrounding metal by induction.   ☐ Electrical circuits and equipment must be installed in such a manner that the possible spread of fire or products of combustion will not be substantially increased.   ☐ No wiring of any type can be installed in ducts used to transport flammable dust or vapors. ☐ The dropped ceiling space used for environmental air is not a plenum. ☐ Electrical wiring in air handling areas beneath raised floors for data processing systems (information technology systems) must comply with Article 645-5(d). ☐ The area under a raised floor for data processing is not considered a plenum.

REVIEW QUESTIONS

1. Do the requirements of Article 300 apply to fire alarm, CATV, telephone, or low voltage control wiring?

2. Generally conductors must be installed in a raceway or cable. Give a few examples of when a conductor can be run exposed.

3. Which raceways or cables require special protection when run through wood framing members? What are the special requirements?

4. Which raceways or cables require special protection when run through metal framing members? What are the special requirements?

5. Can a raceway be used to support fire alarm, computer, telephone, and control cables?

6. When are raceways required to be sealed? What type of seal is required?

7. Why should all conductors of a circuit (hots, neutral, and ground) be installed within the same raceway?

8. What are the restrictions on installing building systems, such as water or gas piping, within electrical raceways?

9. What are the *NEC* requirements for the termination of raceways and cables to enclosures?

10. What are the *Code* rules for securing wiring within suspended ceiling?

11. A 120/240 volt multiwire branch circuit has a cheap 600 watt television on one circuit, and an expensive 200 watt computer on the other circuit. If the neutral is opened, what will happen to each appliance?

12. Explain how and why steel raceways and enclosures are heated by alternating current circuits.

13. What Code Section applies to the wiring installed within dropped ceiling areas used for environmental air?

14. In your own words explain the meaning of the following Article 100 definitions:

Cabinets
Conduit body
Identified
Nonlinear load
Plenum
Service-lateral

15. In your own words explain the meaning of the following Glossary terms:

Corrosion
Eddy currents
Ferrous
Fire resistance
Harmonic current
Hertz
Inductive choke
Low impedance
Pigtail

Unit 14

Article 305
Temporary Wiring

OBJECTIVES

After studying this unit, the student should be able to understand:
- all rules of the NEC that apply to temporary wiring.
- when temporary wiring is permitted.
- which temporary wiring rules apply only to construction sites.
- the rules that apply specifically to receptacles on construction sites.
- the rules that apply specifically to temporary lighting on construction sites.

Definitions and Glossary Terms

To better understand the *NEC* rules contained in this unit, review the following:

Definitions Article 100 - Unit 2
Branch circuit
Feeder
Ground-fault circuit interrupter
Qualified persons

Glossary Terms
Authority having jurisdiction
Lamp
Splice

305-1 Scope

The requirements of Article 305 apply to temporary power and lighting for construction, remodeling, maintenance, repair, demolitions, and decorative lighting. This Article also applies when temporary wiring is necessary during emergencies or for tests and experiments.

Note. Temporary wiring for trade shows must comply with the requirements of Article 518 and temporary wiring for carnivals, circuses, fairs and similar events is covered in Article 525.

305-2 Other Articles

All *NEC* rules apply to temporary wiring unless specifically modified in this Article.

Note. This Article contains only a handful of rules that modify the other rules of the *Code*.

CAUTION: Temporary wiring does not mean that you can rig up anything you want. Job safety is enforced by general contractors, insurance companies, and the Occupational Safety and Health Administration (OSHA). Temporary wiring is used by all trades and it must be kept in a safe working condition. This means that it must be installed to comply with all applicable *NEC* rules.

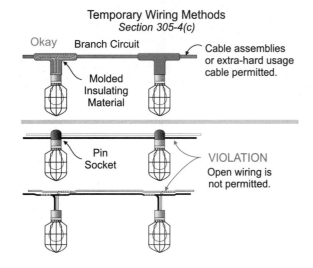

Fig. 14-1 Open Wiring Not Permitted for Temporary Lighting

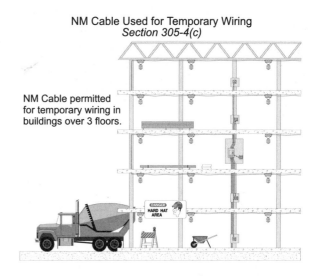

Fig. 14-2 NM Cable Can Be Used for Branch Circuit or Feeder Temporary Wiring

305-3 Time Constraints

(a) Construction Period. Temporary electrical power and lighting installations are permitted during the period of construction, remodeling, maintenance, repair, or demolition of buildings, structures, equipment, or similar activities.

(b) Decorative Lighting and Carnivals. Temporary electrical power and lighting installations are permitted for up to 90 days for decorative lighting or other similar purposes.

(c) Emergencies and Tests. Temporary electrical power and lighting installations are permitted during emergencies or for tests and experiments.

(d) Removal. Temporary wiring must be removed immediately upon the completion of the purpose for which it was installed.

305-4 General

(b) Feeder Circuits. Because of the concern for worker safety, open feeder conductors are not permitted for temporary wiring. However, cable assemblies and hard usage and extra hard usage cords are permitted.

Exception. Individual open **feeder** conductors can be used for temporary electrical power and lighting for emergencies and for tests, experiments, and developmental work, where the individual open feeder conductors are accessible only to **qualified persons.**

(c) Branch Circuits. Open **branch circuit** conductors, such as "festoon lighting" can not be used for temporary wiring [225-6(b)]. However, cable assemblies and hard usage and extra hard usage cords are permitted, Fig. 14-1.

Nonmetallic-sheathed cable can be used for temporary wiring in building of any height and the height limitations contained in Section 336-5(a)(1) does not apply, Fig. 14-2.

Christmas Lighting Not to Exceed 90 Days. Individual open Christmas and other holiday day lighting conductors can be used for a period not to exceed 90 days. This is only applied where the circuit voltage to ground does not exceed 150 volt, the conductors are not subjected to physical damage, and the conductors are supported on insulators at not more than 10 foot intervals.

(d) Receptacles. All receptacles shall be of the grounding type and must have the equipment grounding terminal of the receptacle grounded to an equipment grounding conductor [250-118]

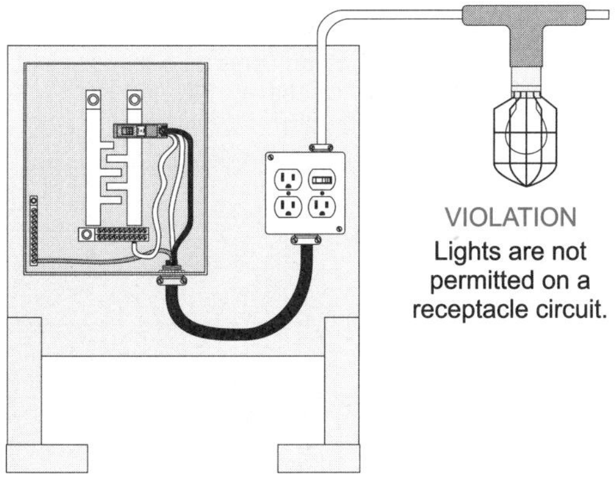

Temporary Lights Not Permitted On
Temporary Receptacle Circuits
Section 305-4(d)

VIOLATION
Lights are not
permitted on a
receptacle circuit.

Fig. 14-3 Temporary Lighting and Receptacles Are Not Permitted to Be on the Same Circuit

installed in accordance with the requirements of Section 250-146.

Temporary lighting and receptacles are not permitted to be on the same circuit, Fig. 14-3.

Note. You don't want the lights to go out if the GFCI circuit protection for the receptacles trips.

(e) Disconnecting Means. All ungrounded circuit conductors must have a disconnect, which can be a switch, a circuit breaker, or a plug connector.

Multiwire Branch Circuits. All ungrounded circuit conductors of multiwire branch circuits must have a disconnect that opens all of the ungrounded conductors simultaneously. This disconnect must be at the panelboard where the branch circuit originates. Where single-pole circuit breakers are used, an approved handle tie can be used to secure the trip handles together.

For additional requirements for multiwire branch circuits, see Sections 210-4 and 300-13(b).

(f) Lamp Protection. *Lamps* (bulbs) shall be protected from accidental contact by a suitable fixture or by the use of a lampholder with a guard.

(g) Splices. At construction sites boxes are not required for *splices* of cords, or nonmetallic cables [300-5(e), 300-15(g)]. However, all conductors

must be spliced or joined together by devices suitable for the purpose [110-14(b), 400-9], and the splices shall be covered with an insulated covering.

A metallic box or conduit body must maintain the equipment grounding conductor for metallic raceways and cables [300-15(a)].

(h) Protection from Accidental Damage. Cables and flexible cords shall be protected from accidental damage and from sharp corners and projections. Protection must also be provided to avoid damage to cables and flexible cords when passing through doorways or other pinch points.

(i) Cable Terminations. Cables entering panelboard cabinets, outlet boxes and other enclosures shall be secured to the enclosure with fittings designed for the purpose [400-10].

Note. The intent is to prevent stress on cables and cable termination.

(j) Cable Support. Cable assemblies, as well as flexible cords and cables, shall be supported at intervals that ensure protection from physical damage. Support shall be in the form of staples, cable ties, straps, or other securing fittings designed to not damage the cable or cord assembly.

The actual cable support requirements are determined by the *authority having jurisdiction (AHJ)* based on the job site conditions and any special requirements of the temporary installation [305-2(b)].

Note. Cables and cords must be protected from accidental damage, and sharp corners and projections must be avoided. Where cable or cords pass through doorways or other pinch points, protection shall be provided to avoid damage to the cable or cord [305-4(h)].

305-6 Ground-Fault Protection for Personnel

Ground-fault protection shall be provided for temporary power during construction, remodeling, maintenance, repair, or demolition of buildings, structures, equipment, or similar activities. Ground-fault protection must be accomplished by either (a) or (b).

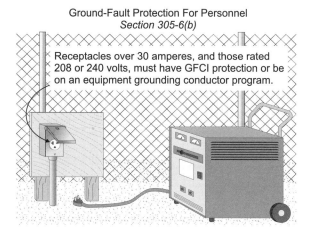

GFCI Protection for Personnel - *Section 305-6(a)*

GFCI Devices Permitted

All 125 volt 1-phase 15, 20, and 30 ampere receptacle outlets used by personnel for temporary power must have ground-fault circuit-interrupter protection for personnel. GFCI protection can be by a circuit breaker, receptacle, cord sets, or other devices such as GFCI adapters incorporating listed GFCI protection.

Fig. 14-4 GFCI Protection Required for 15, 20, and 30 ampere, 125 Volt Receptacles

Ground-Fault Protection For Personnel
Section 305-6(b)

Receptacles over 30 amperes, and those rated 208 or 240 volts, must have GFCI protection or be on an equipment grounding conductor program.

Fig. 14-5 GFCI or AEGCP Required for All Other Receptacles

(a) Ground-Fault Protection for 15, 20, and 30 Ampere. All 125 volt, 15, 20, and 30 ampere receptacle outlets used by personnel for temporary power shall have **ground-fault circuit-interrupter** protection for personnel. GFCI protection can be by the use of a circuit breaker, receptacle, cord sets, or other devices such as GFCI adapters incorporating listed GFCI protection, Fig. 14-4.

Note. This rule also applies to receptacles that are part of the permanent wiring of the building used for temporary power.

Exception No. 1: Generator Not Greater Than 5 kW. Ground-fault protection is not required for receptacles on a portable or vehicle-mounted generators rated not more than 5 kW [250-34].

Exception No. 2: Industrial Establishments. GFCI protection for personnel is not required in industrial establishments where only qualified personnel are involved in maintenance and an assured equipment grounding conductor program is utilized.

(b) Ground-Fault Protection Other Receptacles. Receptacles rated other than 125 volt, 15, 20, or 30 ampere utilized to supply temporary power to equipment used by personnel during construction, remodeling, maintenance, repair, or demolition of buildings, structures, equipment, or similar activities must have GFCI protection for personnel or be protected by an assured equipment grounding program, Fig. 14-5.

The Assured Equipment Grounding Conductors Program (AEGCP) requires grounding tests of cord sets, receptacles, and cord- and plug-connected equipment. The tests shall be performed before the first use, when there's evidence of damage, before equipment is returned to service following repairs, and at intervals not exceeding 3 months. The results of these tests shall be recorded and made available to the *authority having jurisdiction* (AHJ).

SUMMARY

☐ The requirements of Article 305 apply to temporary wiring for power and lighting. ☐ All of the rules in the *NEC* apply to temporary wiring, except as specifically modified in this Article. ☐ Temporary electrical power and lighting installations are permitted for up to 90 days for decorative lighting and similar purposes. ☐ Temporary wiring must be removed immediately upon completion of the purpose for which it was installed. ☐ All branch circuits must originate in an approved power outlet or panelboard. ☐ All receptacles shall be of the grounding type and must be connected to the circuit equipment grounding conductor. ☐ Receptacles cannot be connected to the same circuit that supplies temporary lighting. ☐ Lamps for lighting must be protected from accidental contact by a lampholder with a guard or a suitable fixture. ☐ Ground-fault circuit-interrupter (GFCI) protection for personnel must be provided for all 125 volt, 15, 20, and 30 ampere receptacle outlets.

REVIEW QUESTIONS

1. All rules of the *NEC* apply to temporary wiring, except as modified in this Article. Which rules are more restrictive and which are less restrictive than the other *Code* rules?

2. When is temporary wiring permitted?

3. Which temporary wiring rules apply only to construction sites?

4. Can NM cable be laid on the ground or floor of construction sites?

5. On construction sites, what are some of the rules that apply specifically to receptacles?

6. On construction sites, what are some of the rules that apply specifically to temporary lighting?

7. In your own words explain the meaning of the following Article 100 definitions:

   Branch circuit
   Feeder
   Ground-fault circuit interrupter
   Qualified persons

8. In your own words explain the meaning of the following Glossary terms:

   Authority having jurisdiction
   Lamp
   Splice

Unit 15

Article 310

Conductors for General Wiring

OBJECTIVES

After studying this unit, the student should be able to understand:
- when uninsulated conductors can be used.
- when the equipment grounding conductor must be insulated.
- when solid conductors larger than No. 10 must be installed in a raceway.
- what size conductor is required for parallel raceways.
- what the smallest size conductor for receptacles and lighting for commercial and industrial buildings is.

Definitions and Glossary Terms

To better understand the *NEC* rules contained in this unit, review the following:

Definitions Article 100 - Unit 2
Ampacity
Equipment grounding conductor
Grounded (neutral) conductor
Nonlinear load

Glossary Terms

Balanced system	Circular mils
Harmonic current	Nipples
Single-phase	Wye-connected

310-1 **SCOPE**

This Article contains the general requirements for conductors, such as insulation markings, ampacity ratings, and their use. Article 310 does not apply to conductors that are part of cable assemblies, flexible cords, fixture wires, or conductors that are an integral part of equipment [90-6, 300-1(b)].

CAUTION: The equipment grounding conductor must be insulated for patient care area receptacles and switches [517-13(b)], wet niche pool lights and pool equipment [680-25(b)(2)], 680-25], Fig. 15-1.

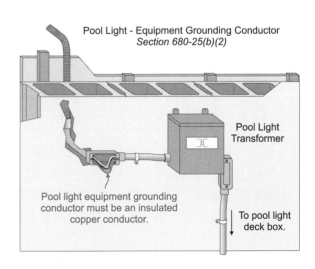

Fig. 15-1 Insulated Equipment Grounding Conductor Required for Pool Light

Fig. 15-2 Conductors No. 8 and Larger Must Be Stranded When Installed in a Raceway

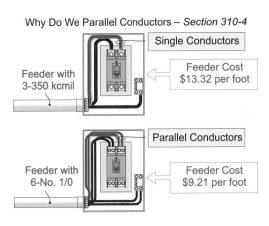

Fig. 15-3 Parallel Circuits Are Cost Effective

310-2 CONDUCTORS

(a) Insulation. All single conductors must be insulated and must be installed in an approved wiring method.

Exception. Uninsulated conductors are permitted for the equipment grounding [250-119] and the grounding electrode conductor [250-64].

310-3 STRANDED CONDUCTORS

Conductors No. 8 and larger must be of the stranded type (not solid) when installed in raceways, Fig. 15-2.

Note. Solid conductors are often used for the grounding electrode conductor [250-62] and for pool bonding conductors [680-22(b)].

310-4 CONDUCTORS IN PARALLEL

Ungrounded (hot) and grounded (neutral) conductors No. 1/0 and larger can be connected in parallel. Parallel means to join conductors together at both ends electrically, thereby forming a single electrical conductor.

Why Do We Parallel? Paralleling of conductors offers significant cost savings because it permits smaller conductors. A conductor's ability to carry current is a function of its resistance to the flow of electrons and its ability to dissipate heat. The larger the conductor, the lower the resistance, but larger conductors do not dissipate heat as easily

as smaller conductors (due to the ratio of mass to surface area).

The *circular mils* required to carry one ampere increases significantly with increased conductor size. Often it is more efficient and cost effective to parallel conductors rather than use larger conductors. For example a 300 ampere feeder can be installed in, Fig. 15-3:

• One 2½ inch raceway containing three 350 kcmil THHN conductors, rated 310 amperes. Cost per foot of $13.23.

• One 2 inch raceway containing six No. 1/0 conductors, each rated 150 amperes. Cost per foot of $9.21.

Exception No. 2: Control Circuits. Conductors smaller than No. 1/0 can be connected in parallel for control circuits, provided that the parallel conductors are all contained within the same raceway or cable, each parallel conductor has an ampacity sufficient to carry the entire load, and the circuit overcurrent protection device rating does not exceed the ampacity of any individual parallel conductor.

Note. Paralleling of low voltage control conductors can reduce the effects of voltage drop for long control runs. When an electric *coil* is energizing, the initial current can be very high (5 to 10 times the operating current) causing significant voltage drop in the control circuit. The reduced voltage at the coil (because of voltage drop) can cause the coil contacts to chatter (open and close like a buzzer), or not to close at all.

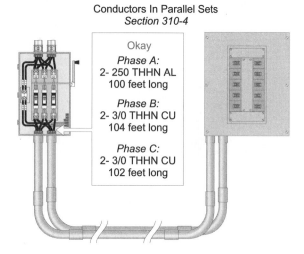

Conductors In Parallel Sets
Section 310-4

Okay

Phase A:
2- 250 THHN AL
100 feet long

Phase B:
2- 3/0 THHN CU
104 feet long

Phase C:
2- 3/0 THHN CU
102 feet long

Fig. 15-4 Conductors in Parallel Sets

Exception No. 4: Engineering Supervision. Under engineering supervision, existing neutral conductors larger than No. 2, but smaller than No. 1/0, can be paralleled.

FPN: This exception is intended to alleviate overheating of the neutral conductor in existing installations due to harmonic current from single-phase nonlinear loads.

Note. Many types of *single-phase* **nonlinear loads,** such as personal computers and laser printers, produce reflective harmonic currents that add on the neutral conductor of a 4-wire, 3-phase, *wye-connected* system. These additive triplen *harmonic currents* can result in the neutral conductor carrying as much as 2 times the maximum unbalanced load [220-22].

Parallel Rules. When conductors are run in parallel, the currents must be evenly distributed between the individual parallel conductors. This is accomplished by ensuring that each of the conductors within a parallel set has the same (impedance) length, material, mils area, insulation material, and terminate in the same method.

In addition all parallel raceways must have the same physical characteristics, such as metallic or nonmetallic. If one parallel conductor is run in steel conduit and another in nonmetallic conduit or aluminum, the conductors in the steel raceway will have higher alternating current resistance (because of inductive reactance). This will result in an unbal-

anced distribution of the currents between the parallel conductors.

Paralleling of conductors is done by sets; one phase, or neutral, is not required to be paralleled the same as those of another phase or neutral. For example a 400 ampere feeder that has a neutral load of 240 ampere can be in parallel as follows, Fig. 15-4:

  Phase A, 2 – 250 THHN Aluminum, 100 feet
  Phase B, 2 – No. 3/0 THHN Copper, 104 feet
  Phase C, 2 – No. 3/0 THHN Copper, 102 feet
  Neutral, 2 – No. 1/0 THHN *AL*, 103 feet
  Ground, 2 – No. 3 Copper, 101 feet

Parallel Equipment Grounding Conductors. When equipment grounding conductors are installed in parallel, each raceway must have a full-sized equipment grounding conductor installed according to Section 250-122. The **equipment grounding conductor** must comply with the parallel requirements in this Section, such as same length, material, circular mils, insulation, and termination, but they can be sized smaller than No. 1/0, Fig. 15-5.

┌─────────────────────────────────────┐
│   WARNING: │
│ When an equipment grounding conductor│
│ is required, the equipment grounding con-│
│ ductor must be run in each raceway of a│
│ parallel circuit [250-122]. Each raceway's│
│ grounding conductor is sized to the circuit│
│ overcurrent protection device size. │
└─────────────────────────────────────┘

Parallel Equipment Grounding Conductors
Sections 250-122 and 310-4
A Full-sized equipment grounding
conductor is required in each raceway.

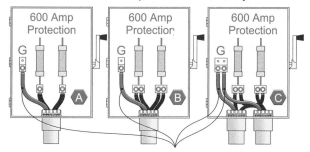

Table 250-122: Based on the 600 ampere protection, the equipment grounding conductor in each raceway is required to be a No. 1 CU (or 2/0 AL). The 1/0 minimum for parallel conductors does not apply.

Fig. 15-5 Full Size Equipment Grounding Conductor Required in Each Parallel Raceway

Ampacity Adjustment. When more than three current-carrying conductors are run in a raceway longer than 24 inches, the **ampacity** derating factors of Table 310-15(b)(2) must be applied, see Section 310-10.

310-5 MINIMUM SIZE CONDUCTORS

The smallest size conductor permitted for branch circuits for residential, commercial, and industrial locations is No. 14 copper [Table 310-5].

Note. There is a misconception that No. 14 copper is a residential conductor. This is not the case for the *NEC*, but it might be a local *Code* rule.

Exception: Smaller Than No. 14. Conductors smaller than No. 14 shall be permitted for Class 1 circuits [402-11 Ex. and 725-27], fixture wire [402-5 and 410-24], flexible cords [400-12], motor control circuits [430-72], and nonpower limited fire alarm circuits [760-27].

310-9 CORROSIVE CONDITIONS

Conductor insulation must be suitable for any substance that may have a detrimental effect on the conductor's insulation, such as oil, grease, vapor, gases, fumes, liquids, or other substances [110-11].

310-10 INSULATION TEMPERATURE LIMITATION

Conductors must not be used or installed in any way that the conductor insulation could be destroyed by excessive heat. What this means is that if 60°C, 75°C, and 90°C conductors are in the same raceways, the ampere load on all of the conductors must be limited to the lowest conductor temperature rating, Fig. 15-6.

This prevents the heat of the higher rated conductors (if they carry more current) from destroying the insulation of the lower rated conductors. Please review Table 310-16 in the *NEC* now.

FPN: The temperature rating of a conductor is the maximum operating temperature

Table 310-13 Conductor Properties

Type Letter	Insulation	Maximum Operating Temperature	Applications Provisions	Sizes Available	Outer Covering
THHN	Flame-retardant heat-resistant thermoplastic	90°C	Dry and damp locations	14-1,000	Nylon jacket or equivalent
THHW	Flame-retardant, moisture- & heat-resistant thermoplastic	75°C	Wet locations	14-1,000	None
		90°C	Dry and damp locations		
THW	Flame-retardant, moisture- & heat-resistant thermoplastic	75°C	Dry, damp, and wet locations	14-2,000	None
		90°C	By electrical discharge lighting equipment operating at 1000 v or less. Sizes 14-8 only [Section 410-31].		
THWN	Flame-retardant, moisture- & heat-resistant thermoplastic	75°C	Dry, damp, and wet locations	14-1,000	Nylon jacket or equivalent
TW	Flame-retardant, moisture-resistant thermoplastic	60°C	Dry, damp, and wet locations	14-2,000	None
XHHW	Flame-retardant cross-linked synthetic polymer	90°C	Dry and damp locations	14-2,000	None
		75°C	Wet locations		

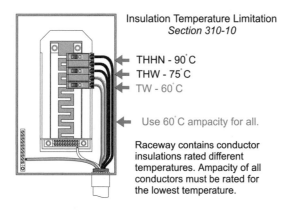

Insulation Temperature Limitation
Section 310-10

← THHN - 90°C
← THW - 75°C
← TW - 60°C

← Use 60°C ampacity for all.

Raceway contains conductor insulations rated different temperatures. Ampacity of all conductors must be rated for the lowest temperature.

Fig. 15-6 Ampacity Based on the Lowest Insulation Temperature Rating

The following list explains the lettering on the conductor insulation, Fig. 15-7.

-2	Suitable for 90°C in wet locations
F	Fixture wires (solid or 7 strand) [Table 402-3]
FF	Fixture Flexible (19 strands) [Table 402-3]
No H	60°C Insulation rating
H	75°C Insulation rating
HH	90°C Insulation
N	Nylon outer cover
T	Thermoplastic insulation
W	Wet or Damp

that the conductor insulation can withstand over a prolonged period of time without serious damage to the conductor insulation. The conductor ampacity listed in Table 310-16 must be adjusted for *ambient temperature* and conductor bundling.

310-12 CONDUCTOR IDENTIFICATION

(a) Grounded (neutral) Conductor. The grounded (neutral) conductor must be identified white or gray in accordance with Section 200-6.

(b) Equipment Grounding Conductor. The identification of the equipment grounding conductor must be green, green with a yellow stripe, or bare in accordance with Section 250-119.

(c) Ungrounded Conductors. Ungrounded (hot) conductors can be identified using any color, stripe, or mark, except white, natural gray, or green.

Note. The high-leg conductor of a delta-connected system is required to be identified with the color orange [215-8, 230-56].

310-13 CONDUCTOR CONSTRUCTION

Table 310-13 of the *NEC* contains information on insulation, operating temperature, applications, sizes, and outer covers. See Table 310-13 in this book.

CAUTION: The *NEC* does not require color coding of ungrounded conductors except for the feeder high-leg conductor when a grounded (neutral) conductor is present [215-8]. Although the *NEC* does not contain color code requirements for phase conductors, color coding is sometimes required by Section 210-4(d), local codes, blueprints, or specifications. Many existing buildings have an existing color code system with which you should continue. Electricians often use the following color code system:

120/240, 1-phase: Black, Red, White

120/240, 3-phase: Black, Orange, Blue, White

208Y/120: Black, Red, Blue, White

480Y/277: Brown, Orange, Yellow, Grey

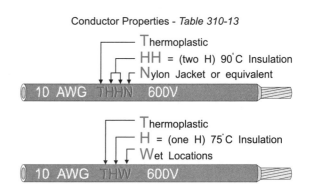

Conductor Properties - *Table 310-13*

Thermoplastic
HH = (two H) 90°C Insulation
Nylon Jacket or equivalent
10 AWG THHN 600V

Thermoplastic
H = (one H) 75°C Insulation
Wet Locations
10 AWG THW 600V

Fig. 15-7 Conductor Properties

Conductor Ampacity - Multiple Derating Factors
Section 310-15(a)(2)
Ambient temperature correction factors
from the bottom of Table 310-16.

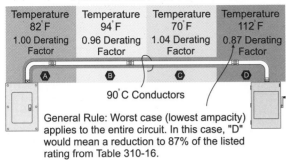

| Temperature 82°F | Temperature 94°F | Temperature 70°F | Temperature 112°F |
| 1.00 Derating Factor | 0.96 Derating Factor | 1.04 Derating Factor | 0.87 Derating Factor |

90°C Conductors

General Rule: Worst case (lowest ampacity) applies to the entire circuit. In this case, "D" would mean a reduction to 87% of the listed rating from Table 310-16.

Fig. 15-8 Conductor Ampacity Based on the Lowest Calculated Ampacity

Exception For Conductor Ampacity Derating For Bundling
Section 310-15(a)(2) Exception

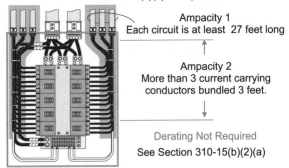

Ampacity 1
Each circuit is at least 27 feet long

Ampacity 2
More than 3 current carrying conductors bundled 3 feet.

Derating Not Required
See Section 310-15(b)(2)(a)

Where two ampacities apply to different portions of the same circuit, the higher ampacity (1 above) can be used if:
• ampacity 2 conductors are not more than 10 feet or,
• 10% of the circuit length figured at the higher ampacity

Fig. 15-9 Conductor Ampacity Based on the Highest Calculated Ampacity

310-15 CONDUCTOR AMPACITY

(a) General Requirements

(1) Tables or Engineering Supervision. There are two ways to determine conductor ampacity:

• Tables 310-16 [310-15(b)]
• Engineering formula

For all practical purposes, use the ampacities listed in Table 310-16.

FPN: The ampacities listed in Table 310-16 are based on temperature alone and don't take voltage drop into consideration. Voltage drop considerations are for efficiency of operation and not safety; therefore, sizing conductors for voltage drop is not a *Code* requirement; see Sections 215-2(d) FPN No. 2 for more details.

Ampacity Definition. The ampacity of a conductor is the current in ampere that a conductor can carry continuously without exceeding its temperature rating [Article 100].

(2) Use of Lower Ampacity. When two or more ampacities apply to a single conductor length, the lower ampacity value shall be used for the entire circuit.

For example this situation would occur when a conductor runs through different ambient tempera-

atures. The conductor ampacity will be different for the different ambient temperature areas; therefore, the lowest conductor ampacity must be used for the entire circuit, Fig. 15-8.

Exception: Use of Higher Ampacity. When two ampacities apply to a single conductor length, the higher ampacity can be used for the entire circuit if:

• The length of the conductor with the reduced ampacity does not exceed 10 feet, and
• No more than 10 percent of the length of the circuit conductor not reduced, Fig. 15-9.

(b) Table Ampacity. The allowable ampacities of a conductor is listed in Table 310-16. These ampacities are based on the condition where no more than 3 current-carrying conductors are bundled together at an ambient temperature of 86°F, Fig. 15-10.

Ambient Temperature Multiplier. The temperature surrounding the conductor may vary along the conductor length from time to time. When the ambient temperature differs from 86°F, the conductor's allowable ampacity as listed in Table 310-16 must be adjusted by the use of the correction factors in Table 310-16. The following values represent the ambient temperature correction factors for 90°C conductors such as THHN or XHHW.

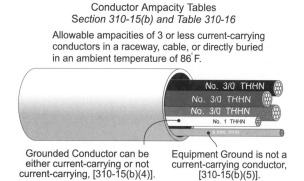

Conductor Ampacity Tables
Section 310-15(b) and Table 310-16

Allowable ampacities of 3 or less current-carrying conductors in a raceway, cable, or directly buried in an ambient temperature of 86°F.

Grounded Conductor can be either current-carrying or not current-carrying, [310-15(b)(4)].

Equipment Ground is not a current-carrying conductor, [310-15(b)(5)].

Fig. 15-10 Conductor Ampacity Based on No More Than 3 Conductor at Ambient Temperature of 86°F

Temperature	Multiplier
70-77°F	1.04
78-86°F	1.00
87-95°F	0.96
96-104°F	0.91
105-113°F	0.87
114-122°F	0.82
123-131°F	0.76
132-140°F	0.71

For example a No. 1 THHN conductor, rated 150 amperes at 90°C has an ampacity of less than 150 amperes when the ambient temperature is above 86°F, and more than 150 amperes when the ambient temperature is below 86°F, Fig. 15-11.

Note. When adjusting conductor ampacity for ambient temperature, use the ampacity of the conductor based on the con-

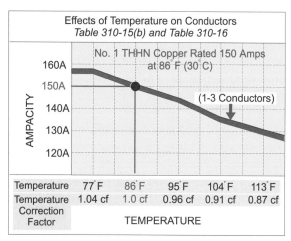

Effects of Temperature on Conductors
Table 310-15(b) and Table 310-16

No. 1 THHN Copper Rated 150 Amps at 86°F (30°C)

(1-3 Conductors)

AMPACITY: 160A, 150A, 140A, 130A, 120A

Temperature	77°F	86°F	95°F	104°F	113°F
Temperature Correction Factor	1.04 cf	1.0 cf	0.96 cf	0.91 cf	0.87 cf

TEMPERATURE

Fig. 15-11 Ampacity Changes with Different Ambient Temperatures

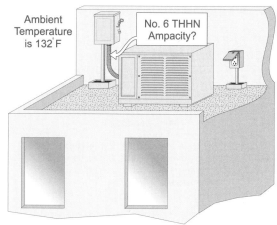

Ambient Temperature is 132°F

No. 6 THHN Ampacity?

75 Table Amps x 0.71 Correction = 53 Ampacity

Fig. 15-12 Ampacity Adjustment for Elevated Ambient Temperature

ductor temperature rating. For example the ampacity of No. 6 THHN is 75 amperes at the 90°C column [110-14(c)].

Question. What is the ampacity of three No. 6 THHN conductors installed in an ambient temperature of 132°F, Fig. 15-12?

Answer.   53 ampere
75 ampere × 0.71 = 53 ampere

(2)(a) Conductor Bundle Multiplier. When conductors are bundled together, the ability of the conductors to dissipate heat is reduced.

When 4 or more current-carrying conductors are bundled together, the conductor allowable ampacity listed in Table 310-16 must be reduced according to Table 310-15(b)(2) derating factors, Fig. 15-13.

Current-Carrying	Multiplier
1-3 Conductors	100%
4-6 Conductors	80%
7-9 Conductors	70%
10-20 Conductors	50%

Note. When adjusting conductor ampacity for ambient temperature, use the ampacity of the conductor based on the conductor temperature rating. For example the ampacity of No. 10 THHN is 40 amperes at the 90°C column [110-14(c)].

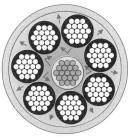

Ampacity Correction For Bundled Conductors
Table 310-15(b)(2)(a)

Ampacity at 100%
No Derating for 3 Conductors

Derated Ampacity at 70%
for 8 Conductors

Conductors have surface
space for heat dissipation.

Conductors in the
bundle have heat held
in by other conductors.

Fig. 15-13 Ampacity Decreases When More Than
Three Current-Carrying Conductors are Bundled

Question. What is the ampacity of four current-carrying No. 10 THHN conductors that
supplies a nonlinear load, Fig. 15-14?

Answer.   32 ampere
40 ampere × 0.8 = 32 ampere

Exception No. 3: Nipple. Ampacity factors do
not apply to conductors installed in raceway *nipples*
in lengths not exceeding 24 inches, Fig. 15-15.

Temperature and Bundling Adjustments

If there are more than three current-carrying
conductors, and the ambient temperature is not
86°F, the allowable ampacity listed in Table 310-16
must be adjusted for both conditions.

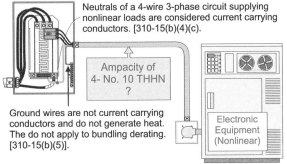

Conductor Ampacity - Conductor Bunching
Table 310-15(b)(2)(1) and Section 310-15(b)(4)(c)

Neutrals of a 4-wire 3-phase circuit supplying
nonlinear loads are considered current carrying
conductors. [310-15(b)(4)(c).

Ampacity of
4- No. 10 THHN
?

Ground wires are not current carrying
conductors and do not generate heat.
The do not apply to bundling derating.
[310-15(b)(5)].

Electronic
Equipment
(Nonlinear)

Table 310-16, No. 10 THHN rated 40 amperes
Table 310-15(b)(2)(a), 4 "hots" = 0.8 derating factor
40 amperes x 0.8 derating factor = 32 ampacity

Fig. 15-14 Neutral Conductor at Times is
Considered a Current-Carrying Conductor

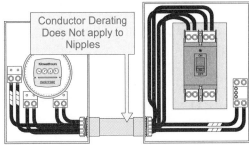

Bundled Conductor Corrections
Table 310-15(b)(2) Exception 3

Conductor Derating
Does Not apply to
Nipples

A nipple is any raceway that is 24 inches or less.

Fig. 15-15 Ampacity Adjustment Factors Do
Not Apply to Bundled Conductors in a Nipple

Question. What is the ampacity of 4 current-
carrying No. 3 THHN conductors installed in
ambient temperature of 114°F?

Answer.   72 ampere
Ampacity at 90°C = 110 amperes
Temperature Multiplier = 0.82
Conductor Bundle Multiplier = 0.80
110 ampere × 0.82 × 0.8 = 72 ampere

(4) Neutral Conductor. All ungrounded conductors (hot wires) are considered current-carrying,
but not all grounded (neutral) conductors are considered current-carrying.

(a) Balanced Circuits. The grounded (neutral) conductor of a *balanced system* is not
considered a current-carrying conductor.

(b) Wye 3-Wire Circuits. The grounded (neutral) conductor of a balanced 3-wire wye system is
considered a current-carrying conductor.

> **Note.** When a 3-wire circuit is supplied
> from a 4-wire, 3-phase, wye-connected sys-
> tem, the grounded (neutral) conductor
> carries approximately the same current as
> the ungrounded conductors.

**(c) Wye 4-Wire Circuits Supplying Nonlin-
ear Loads.** The neutral conductor of a balanced
4-wire wye circuit that is at least 50 percent loaded
with nonlinear loads is considered a current-carrying conductor.

CAUTION: Harmonic current on the neutral conductor from nonlinear loads can be as much as 200 percent of the current on any ungrounded conductor. To measure neutral current for nonlinear loads, you must use a true RMS ammeter, Fig. 15-15. See my comments in Sections 220-22 and 310-13(b).

(5) Grounding Conductors. Grounding and bonding conductors are not considered current-carrying conductors.

(6) Dwelling Unit Feeder/Service Conductors. One-family, two-family, or multifamily dwelling units can size their 3-wire 120/240 volt service-entrance, and feeder conductors that serve as the main power feeder to a dwelling unit using Table 310-15(b)(6).

Table 310-15(b)(6)
120/240-Volt Feeder or Service Conductor

Conductor types
THHW-THW-THWN-THHN-XHHW-USE

Service Rating Amperes	Copper Size Feeder/Service	Aluminum Feeder/ Service
100	4	2
110	3	1
125	2	1/0
150	1	2/0
175	1/0	3/0
200	2/0	4/0
225	3/0	250 kcmil
250	4/0	300 kcmil
300	250 kcmil	350 kcmil
350	350 kcmil	500 kcmil
400	400 kcmil	600 kcmil

WARNING:
Section 310-15(b)(6) does not apply to 208Y/120 volt, 3-wire single-phase systems, because the grounded (neutral) conductor in these systems carry neutral current which generates heat.

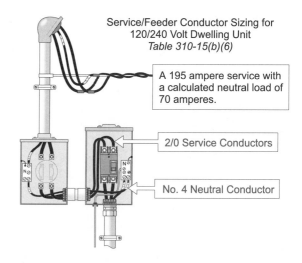

Fig. 15-16   Service/Feeder Conductor Sizing for 120/240 Volt Dwelling Unit

Grounded Conductor Sizing. The **grounded (neutral) conductor** shall not be smaller than required to carry the maximum unbalanced load as determined in Section 220-22. In addition, the grounded (neutral) conductor must be sized to comply with the requirements of Sections 230-42 and 250-24(b).

Question. If the total demand load for a dwelling unit equals 195 ampere and the unbalanced neutral load is 70 ampere, what size feeder conductors would be required, Fig. 15-16?

Answer.   2 - No. 2/0 conductors and 1 - No. 4
The grounded (neutral) conductor only has to be sized to carry the maximum unbalanced load of 70 ampere [220-22].

Note. The grounded (neutral) conductor for the service is used to carry ground-fault current and it cannot be sized smaller than the value listed in Table 250-66 [250-24(b)].

SUMMARY

☐ Article 310 covers the general requirements for single conductors such as the designations, insulation markings, ampacity rating, and uses. ☐ The maximum size solid conductor permitted to be installed in a raceway is No. 10. ☐ Conductors 1/0 and larger can be connected in parallel. ☐ When conductors are run in parallel, each conductor within the parallel set must have the same length, material, circular mil area, insulation, and must terminate in the same manner. ☐ When conductors are paralleled in different raceways, all raceways are required to have the same physical characteristics. ☐ Parallel equipment grounding conductors must be full-size and installed in each raceway. ☐ Parallel conductors are subject to the ampacity derating requirements of Table 310-15(b)(2). ☐ The minimum size copper conductor permitted for general wiring is No. 14. ☐ Conductor insulation must be suitable for any substance that may have a detrimental effect on the insulation. ☐ Conductors shall not be used in any way that the conductor insulation temperature rating can be exceeded. ☐ Conductors shall not be associated in any way that will cause any other conductor's insulation temperature rating to be exceeded. ☐ Grouping of conductors has the dual effect of raising the ambient temperature and preventing the dissipation of heat.

☐ Grounded conductors No. 6 and smaller must be insulated and identified with a white or natural gray insulation along their entire length. ☐ Ungrounded conductors must be insulated and can be identified by any color, stripe, or mark (except white, natural gray, or green). ☐ When two or more ampacities apply to a single conductor length, the lower value must be used. ☐ When more than 3 current-carrying conductors are bundled together, the conductor ampacities must be reduced by the values shown in the Table 310-15(b)(2). ☐ The grounded (neutral) conductor in 240/120 volt, 3-wire; 208Y/120 volt, 4-wire; and 480Y/277 volt, 4-wire circuits is not considered a current-carrying conductor. ☐ The grounded (neutral) conductor of a 3-wire circuit supplied from a 4-wire, 3-phase, wye system is considered a current-carrying conductor and must be counted when applying Table 310-15(b)(2) derating factors. ☐ On a 4-wire, 3-phase, wye (not delta) circuit, where the major portion of the load consists of nonlinear loads, the grounded (neutral) conductor will carry almost 2 times the phase current load and is considered a current-carrying conductor. ☐ A grounding or bonding conductor is not counted when applying Table 310-15(b)(2) derating factors. ☐ Single-phase, 120/240 volt, 3-wire service-entrance and feeder conductors for dwelling units can be sized using Table 310-15(b)(6).

REVIEW QUESTIONS

1. When can uninsulated conductors be used?

2. When is the equipment grounding conductor required to be insulated?

3. When can solid conductors larger than No. 10 be installed in a raceway?

4. What size conductor is required (parallel – 2 raceways) for a 760 ampere load?

5. What is the smallest size conductor for receptacles and lighting for commercial and industrial buildings?

6. What is the color code required for a 480Y/277 volt system?

7. What is the maximum operation temperature of THW, when installed within 3 inches of a ballast in fluorescent fixtures?

8.   What is the ampacity of three No. 14 THHN in an ambient temperature of 79°C?

9.   A raceway contains 2 No. 12 THHN and 3 No. 14 THHN. What is the conductor ampacity?

10. A raceway contains 2 No. 12 THHN and 3 No. 14 TW, and the ambient temperature is 140°F. What is the ampacity of the No. 12 conductors?

11. In your own words explain the meaning of the following Article 100 definitions:

Ampacity
Equipment grounding conductor
Grounded (neutral) conductor
Nonlinear load

12. In your own words explain the meaning of the following Glossary terms:

Balanced system
Circular mils
Harmonic current
Nipples
Wye-connected

Unit 16

Article 328
Flat Conductor Cable (Type FCC)

OBJECTIVES

After studying this unit, the student should be able to understand:
- the primary uses of flat conductor cable.
- where flat conductor cable can be installed.
- what the installation requirements are for the carpet squares.

Definitions and Glossary Terms

To better understand the *NEC* rules contained in this unit, review the following:

Definitions Article 100 - Unit 2
Branch circuit
Equipment grounding conductor
Identified

PART A. GENERAL

328-1 Scope

Flat conductor cable (FCC) is a **branch circuit** wiring system designed for installation under adhesive carpet squares. FCC is constructed of 3 or more flat copper conductors laid edge-to-edge in an insulating material. This flexible ribbon of flat conductors is installed between a bottom and top protective shield. When the installation is complete, the cable is less than 1/8 inch thick; then it is cov-

ered with carpet squares. This branch circuit wiring method is used primarily in banks and office buildings for wiring receptacles in open floor areas.

328-2 Definitions

FCC Cable. Cable consisting of three or more flat copper conductors placed edge-to-edge, enclosed, and separated within an insulation assembly.

Top Shield. A grounded metal shield covering FCC cable for the purpose of protection against physical damage.

Bottom Shield. A shield installed between the floor and FCC cable to protect the cable from physical damage.

Transition Assembly. An assembly to connect the FCC cable to another wiring method, providing for electrical connections and a box or covering to protect against physical damage.

328-4   Uses Permitted

FCC can be used for general purpose, appliance, and individual branch circuits. The cable can only be installed on dry or damp floor surfaces that are sound, smooth, and continuous, such as concrete and wood. When installed on wall surfaces, the cable must be installed in surface metal raceways [Article 352, Part A].

328-5   Uses Not Permitted

FCC cannot be installed in residential locations, schools, hospitals, wet locations, outdoors, or where corrosive vapors are present, or in any hazardous (classified) location.

> **Note.** Although schools and hospitals have office areas, FCC cannot be installed in any part of these buildings.

328-6   Branch Circuit Ratings

FCC can be used only for 3-wire, 120/240 volt, single-phase; 3-wire, 208Y/120 volt, single-phase; or 4-wire, 208Y/120 volt, 3-phase systems. The branch circuit ampere rating shall not exceed 30 ampere for individual branch circuits and 20 ampere for general purpose and appliance branch circuits.

PART B.   INSTALLATION

328-10   Coverings

Flat conductor cable, connectors, and insulating ends must be covered with carpet squares. The carpet squares are required to have a release type adhesive and must not be larger than 36 inches in any dimension. The purpose of limiting the carpet squares to 36 inches, with release type adhesive, is to permit access to the wiring method to minimize damage to the cable by cutting through the carpet.

328-12   Shields

A protective top (metal) and bottom shield shall completely cover all cable runs, corners, connectors, and ends. Metal shields must be connected to the **equipment grounding conductor** at each receptacle [328-14].

328-13   Enclosures and Shield Connections

All metal components (shields, boxes, receptacles, housings, and self-contained devices) shall be electrically continuous to the branch circuit equipment grounding conductor. All grounding connections shall be made with fittings **identified** for the use.

328-14   Receptacles

Receptacles, receptacle housings, and self-contained devices used with FCC systems shall be identified for the use. These items must be electrically connected to the FCC and to the metal shield. The grounding conductor of the FCC shall be connected to the protective metal shield at each receptacle.

328-16   Anchoring

FCC systems shall be firmly anchored with adhesive or mechanical anchoring systems identified for the use.

328-17   Crossings

No more than two layers of FCC can cross at any one point, and a grounded metal shielding must separate the cables. FCC can cross over or under flat communications or signal cable if a grounded metal shield separates the cables.

SUMMARY

Flat conductor cable (FCC) is a branch circuit wiring system designed for installation under carpet squares. ☐ This wiring method can be used only on dry or damp floor surfaces that are smooth and hard. ☐ FCC is permitted to supply 125 volt, 15 and 20 ampere general purpose and appliance branch circuits and 30 ampere individual branch circuits. ☐ A grounded protective metal shield shall completely cover all cable runs, corners, connectors, and ends. ☐ FCC systems shall be firmly anchored with adhesive or mechanical anchoring systems identified for the use. ☐ No more than two layers of FCC can cross at any one point, separated by a grounded metal shield.

REVIEW QUESTIONS

1. What are the primary uses of flat conductor cable?

2. Where is FCC permitted to be installed?

3. After the cable has been installed, what are the installation requirements for the carpet squares?

4. In your own words explain the meaning of the following Article 100 definitions:

    Branch circuit
    Equipment grounding conductor
    Identified

Unit 17

Article 331
Electrical Nonmetallic Tubing (ENT)

OBJECTIVES

After studying this unit, the student should be able to understand:
- when electrical nonmetallic tubing must be installed concealed.
- when ENT can be installed exposed.
- where electrical nonmetallic tubing is not permitted.
- the mechanical installation requirements of ENT, such as trimming, termination, supports, bushings, and through framing members.

Definitions and Glossary Terms

To better understand the *NEC* rules contained in this unit, review the following:

Definitions Article 100 - Unit 2
Approved
Concealed
Exposed

PART A. GENERAL

331-1 Definition

Electrical nonmetallic tubing is a pliable, corrugated, circular raceway made of polyvinyl chloride (PVC). In some parts of the country, the trade name for electrical nonmetallic tubing is "smurf pipe," because originally it was only available in blue when it came out at the height of popularity of the children's character, the Smurfs.

The conduit now is available in a rainbow of colors such as white, yellow, red, green, and orange and is sold in both fixed lengths and on reels.

331-3 Uses Permitted

Electrical nonmetallic tubing and fittings shall be permitted:

(1) Concealed or exposed, in dry or damp locations, in buildings not exceeding 3 floors when not subject to physical damage.

(2) In buildings over 3 floors, ENT can be installed only **concealed** in floors, walls, and ceilings, and must have a thermal barrier identified as

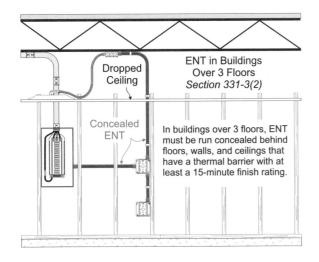

Fig. 17-1 ENT in Buildings over 3 Floors Must Be Concealed Behind a 15-Minute Finish Barrier

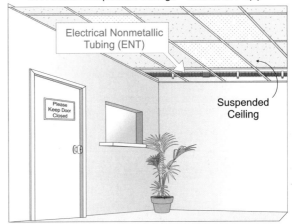

ENT is permitted above a suspended ceiling not used for environmental air. The suspended ceiling must provide a thermal barrier having at least a 15-minute finish rating.

Fig. 17-2 ENT above Suspended Ceilings Must Be Behind a 15-Minute Finish Rated Ceiling

having a minimum 15-minute finish rating, as listed for fire rated assemblies, Fig. 17-1.

(3) Electrical nonmetallic tubing can be installed in severe corrosive and chemicals locations when identified for this use.

(4) ENT can be installed in dry and damp concealed locations where not prohibited by Section 331-4.

(5) Electrical nonmetallic tubing can be installed above suspended ceilings not used for environmental air handling purposes [300-22(c)] in buildings of any height, if the ceiling provides a thermal barrier having a 15-minute finish rating, Fig. 17-2.

(6) Electrical nonmetallic tubing can be encased or embedded in a concrete slab on grade where the tubing is placed upon sand or **approved** screenings, provided fittings identified for the purpose are used.

Note. Electrical nonmetallic tubing cannot be buried in the earth, see 331-4(5).

(7) ENT can be installed in wet locations indoors as permitted in this Section or in a concrete slab on or below grade, with fittings listed for the purpose.

(8) Listed prewired ENT in sizes from ½ inch to 1 inch shall be permitted.

331-4 Uses Not Permitted

Electrical nonmetallic tubing is not permitted:

(1) In hazardous (classified) locations, except in *intrinsically safe systems* [Article 504].

(2) For the support of lighting fixtures or equipment [370-23].

(3) Where the ambient temperature is in excess of 50°C.

(4) To contain conductors that operate at a temperature higher than the raceway temperature rating.

(5) For direct earth burial, but it can be encased in concrete [331-3(6)].

(6) As a wiring method for systems over 600 volts.

(7) To be installed **exposed** in buildings over 3 floors [331-3(1)], but can be installed concealed in buildings over 3 floors [331-3(2)].

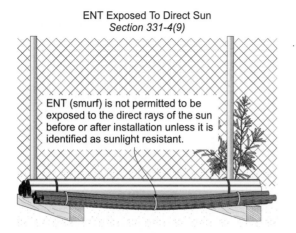

ENT Exposed To Direct Sun
Section 331-4(9)

ENT (smurf) is not permitted to be exposed to the direct rays of the sun before or after installation unless it is identified as sunlight resistant.

Fig. 17-3 ENT Not Permitted to Be Exposed to Direct Sunlight for Extended Period of Time

(8) In places of assembly or theaters, except as permitted in Sections 518-4 and 520-5.

(9) To be exposed to the direct rays of the sun for an extended period of time, before or after installation, unless listed as sunlight resistant, Fig. 17-3.

Note. Exposing ENT to direct rays of the sun for an extended period of time may result in the product becoming brittle, unless it has specific compounds to resist the effects of ultraviolet (UV) radiation.

> **WARNING:**
> Electrical nonmetallic tubing is also prohibited for ducts, plenums, and other spaces used for environmental air [300-22], and patient care area receptacles in health care facilities [517-13(b)].

PART B.   INSTALLATION

331-5   Sizes

Electrical nonmetallic tubing is manufactured in trade sizes from ½ inch through 2 inches and comes in various colors.

331-6   Number of Conductors

When determining the number of conductors permitted in ENT, the cross-sectional area of the raceway and the conductors according to Chapter 9, Table 1, see Section 300-17.

331-7   Trimming

When ENT is cut, the tubing ends must be trimmed both inside and outside to remove the rough edges. Trimming is very easy; most of the burrs will rub off with fingers, and a knife can be used to smooth the rough edges.

331-8   Joints

Joints (couplings and connectors) must be made in an approved manner. This means the tubing must be cut straight, and if the raceway is to be glued, the installation must comply with the instruction on the glue. Rigid nonmetallic conduit fittings (PVC) can be used with ENT according to testing laboratory publications.

> CAUTION: When using glue for ENT, the glue must be approved for the tubing and the fittings. PVC glue generally cannot be used with ENT, because the solvent will damage the raceway tubing; check the glue instructions.

331-9   Bends

When ENT is bent, which can be done easily by hand, the raceway must not be damaged and the internal diameter of the raceway must not be significantly changed (no kinks). The radius of the curve of the inner edge shall not be less than the dimensions listed in Table 346-10.

331-10   Number of Bends (360°)

The maximum number of bends between pull points (boxes and conduit bodies) cannot exceed

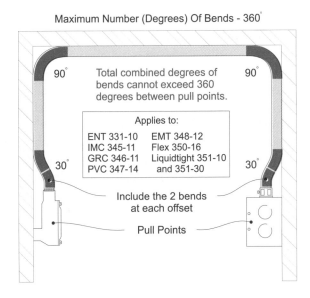

Maximum Number (Degrees) Of Bends - 360°

Total combined degrees of bends cannot exceed 360 degrees between pull points.

Applies to:

ENT 331-10   EMT 348-12
IMC 345-11   Flex 350-16
GRC 346-11   Liquidtight 351-10
PVC 347-14   and 351-30

Include the 2 bends at each offset

Pull Points

Fig. 17-4 No More Than 360° of Bends between Pull Points, Including Offsets

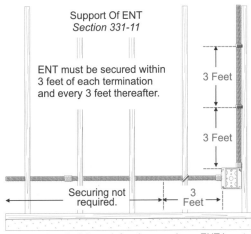

Support Of ENT
Section 331-11

ENT must be secured within 3 feet of each termination and every 3 feet thereafter.

3 Feet

3 Feet

Securing not required.   3 Feet

Where installed through framing members, ENT is not required to be secured except within 3 feet of termination.

Fig. 17-5 ENT Must Be Supported Every 3 Feet Except When Run Horizontally

360°, including any offsets. The 360° limitation is to reduce stress and friction on conductor insulation during installation, Fig. 17-4.

331-11 Secured and Supported

Electrical nonmetallic tubing must be securely fastened in place within 3 feet of each termination. In addition the raceway must be supported every 3 feet thereafter, Fig. 17-5.

Horizontal Runs. When installed horizontally through framing members, the raceway is not required to be supported, but it must be secured within 3 feet of the termination, Fig. 17-5.

Exception: Fixture Whips. When ENT is installed according to Section 410-67(c) as a fixture whip between an outlet box and the fixture, the tubing is not required to be secured or supported.

331-14 Bushings

Bushings or adapters are required at the raceway terminations to protect the conductors from abrasion. ENT and rigid nonmetallic conduit fittings (adapters) provide the protection required in this Section [300-4(f)].

331-15 Construction Specifications

The type, size, and quantity of conductors used in prewired manufactured assemblies shall be identified by means of a printed tag or label attached to each end of the manufactured assembly. ENT, as a prewired manufactured assembly, shall be provided in continuous lengths capable of being shipped in a coil, reel, or carton without damage.

SUMMARY

☐ Electrical nonmetallic tubing is a pliable (can be bent by hand), corrugated, circular raceway made of polyvinyl chloride (PVC). ☐ ENT can be installed exposed in buildings not over three floors, if the raceway is not subject to physical damage. ☐ Electrical nonmetallic tubing can be concealed in walls, floors, and ceilings in buildings of any height, if the finish of these surfaces has a thermal barrier of at least 15 minutes. ☐ ENT can be installed embedded in concrete and installed in wet locations, but not directly buried in the earth. ☐ Electrical nonmetallic tubing is not permitted for hazardous (classified) locations, support of lighting fixtures or boxes, exposed in buildings over 3 floors, services, and in ducts, plenums, or other spaces used for environmental air. ☐ ENT must have no more than 360° of bends between pull points and must be securely fastened in place every 3 feet.

REVIEW QUESTIONS

1. Can electrical nonmetallic tubing be installed concealed behind wood paneling in a 3 story building?

2. Can ENT be installed exposed for the first 3 floors of a 5 story building?

3. Can electrical nonmetallic tubing be installed exposed behind dropped ceiling panels that give access to the raceway in buildings with over 3 floors above ground?

4. Where is ENT not permitted?

5. What are the mechanical installation requirements of ENT, such as trimming, termination, supports, bushings, and through framing members?

6. In your own words explain the meaning of the following Article 100 definitions:

    Approved
    Concealed
    Exposed

Unit 18

Article 333
Armored Cable (Type AC)

OBJECTIVES

After studying this unit, the student should be able to understand:
- the installation requirements for armored cables with protective shields.
- where armored cable can be installed.
- where armored cable is prohibited.
- the requirements for securing armored cable.
- the termination requirements of armored cable.
- the equipment grounding conductor requirements for armored cable.

Definitions and Glossary Terms

To better understand the *NEC* rules contained in this unit, review the following:

Definitions Article 100 - Unit 2
Approved
Concealed
Exposed

Glossary Terms
Studs
Reactance

333-1 Definition

Armored cable (*BX®*) is an assembly of insulated conductors, No. 14 through No. 1, that are individually wrapped within waxed paper. The conductors are contained within a flexible metal (steel or aluminum) sheath that interlocks at the edges. Armored cable has an outside appearance like flexible metal conduit, Fig. 18-1.

The advantage of any cable system is that there is no limit to the number of bends between boxes; this permits the cable to be routed to its needed location quickly and efficiently.

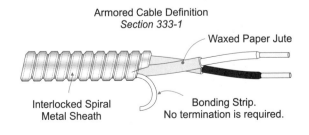

Armored Cable Definition
Section 333-1

Waxed Paper Jute

Interlocked Spiral
Metal Sheath

Bonding Strip.
No termination is required.

Fig. 18-1 Armored Cable Definition

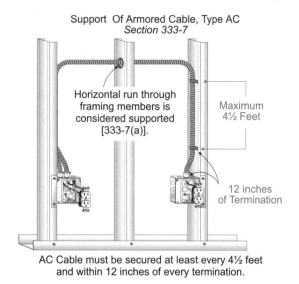

AC Cable must be secured at least every 4½ feet
and within 12 inches of every termination.

Fig. 18-2 AC Cable Must Be Secured Within 12 of
Termination, and Every 4 1/2 Feet Thereafter

333-3 Uses Permitted

Armored cable can be used in **dry locations** where not subject to physical damage for branch circuits, feeders, and services. The cable can be run exposed or concealed [333-11], *fished* in air voids of masonry walls, embedded in plaster, brick, or other masonry locations not exposed to excessive moisture or dampness.

> CAUTION: The NEC does not contain any requirement on cutting of armored or metal clad cables. When cutting cables with a hacksaw, be sure to cut only one spiral of the cable and be careful not to nick the conductors; this is done by cutting the cable at an angle. Some electricians break the cable spiral (bending the cable very sharply), then cut the cable with a pair of dikes. Some local Codes prohibit the use of hacksaws or dikes, and require the use of a tool specially designed and available to cut armored as well as metal clad cable.

333-4 Uses Not Permitted

Armored cable is limited or not permitted in commercial garages [511], hazardous (classified) locations [501-4(b)], motion picture studios [530-11], places of assembly [518-4], pool wiring [680-25], and theaters [520-5].

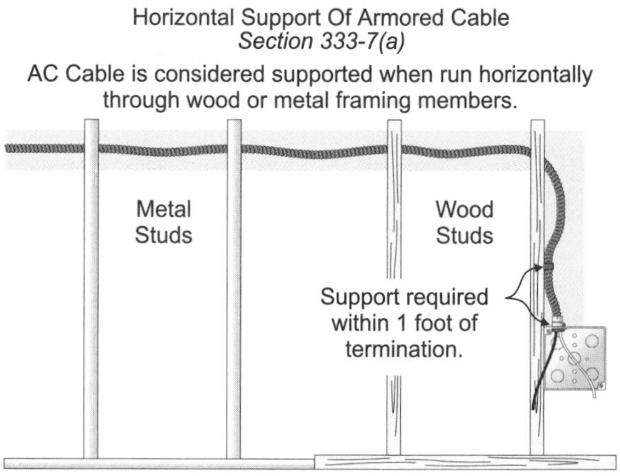

Horizontal Support Of Armored Cable
Section 333-7(a)

AC Cable is considered supported when run horizontally
through wood or metal framing members.

Fig. 18-3 Horizontal Runs Must Be Secured Within
12 Inches of Termination

333-7 Secured and Supported

Type AC cable shall be secured by staples, cable ties, straps, hangers, or fittings designed and installed so the cable is not damaged. In addition, AC cables must be supported at intervals not exceeding 4½ feet and the cable shall be secured within 12 inches of the cable termination, Fig. 18-2.

(a) Horizontal Runs. AC cable run horizontally through framing members, or notches in wooden members is considered supported. However the cable must be securely fastened within 12 inches of termination, Fig. 18-3.

> **Note.** Armored cable through framing members must be protected from nails and screws in accordance with Section 300-4.

(b) Unsupported. AC cable can be unsupported where the cable is:

(1) Fished. When the cable is installed by fishing, that portion of the cable that is concealed is not required to be secured.

(2) Where Flexibility Is Required. Where flexibility is required, for example connections to moving or vibrating equipment, securing of the cable is not required for lengths up to 2 feet at the equipment termination.

(3) Lengths Not More than 6 Feet. Armored cable in an **accessible** ceiling used for the connection of lighting fixtures or equipment is not

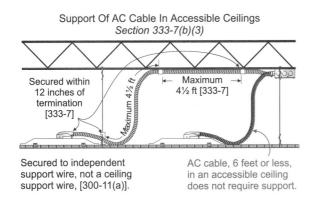

Support Of AC Cable In Accessible Ceilings
Section 333-7(b)(3)

Secured within 12 inches of termination [333-7]

Maximum 4½ ft

Maximum 4½ ft [333-7]

Secured to independent support wire, not a ceiling support wire, [300-11(a)].

AC cable, 6 feet or less, in an accessible ceiling does not require support.

Fig. 18-4 AC Cable Used for Fixture Whips Are Not Required to Be Secured or Supported

required to be secured or supported for lengths up to 6 feet, Fig. 18-4.

Note. Armored cable can be secured to independent support wires within suspended ceiling [300-11(a)].

333-8 Bends

Armored cable shall not be bent in a manner that will damage the cable. This is accomplished by limiting the bending radius of the inner edge of the cable to not less than 5 times the internal diameter of the cable.

333-9 Boxes and Fittings

Armored cable must terminate in boxes or fittings listed specifically for armored cable. The fittings and boxes must protect the conductors from abrasion. An approved insulating anti-short bushing, also referred to in the trade as a "redheads" or "red devils," must be installed at the cable terminations between the metal armor and the conductors, Fig. 18-5.
The termination fitting must permit the visual inspection of the anti-short bushing once the cable has been installed.

Note. Many electricians use the aluminum bonding strip contained within the cable to secure the anti-short bushing. This is not a *Code* requirement.

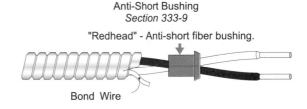

Anti-Short Bushing
Section 333-9

"Redhead" - Anti-short fiber bushing.

Bond Wire

Fig. 18-5 Anti-Short Bushing Required for AC Cable, but Not Required for MC Cable

333-10 Through or Parallel to Framing Members

When running armored cable through or parallel to framing members, the cable must comply with the physical protection requirements contained in Section 300-4.

333-11 Exposed Work

Armored cable must closely follow the surface of the building finish or running boards when exposed. Armored cable can be run on the bottom of floor or ceiling joists if secured strapped at every joist and located so as not to be subject to physical damage.

333-12 In Accessible Attics or Roof Spaces

(a) On the Surface of Floor Joist, Rafters, or Studs.

Attic or Roof Space Accessible by Permanent Ladder or Stairs. Armored cable run on the surface of attic or roof space floor joists, rafters, or *studs* within 7 feet of the attic floor must be protected by guard strips.

Attic or Roof Space with No Permanent Ladder or Stairs. Armored cable run on the surface of attic or roof space floor joists must be protected by guard strips when located within 6 feet of the nearest edge of the attic or roof space entrance. Protective wood strips are generally required only on one side of the cable.

(b) Along the Side of Floor Joists. When armored cable is run on the side of rafters, studs, or

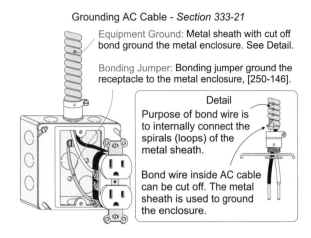

Grounding AC Cable - *Section 333-21*

Equipment Ground: Metal sheath with cut off bond ground the metal enclosure. See Detail.

Bonding Jumper: Bonding jumper ground the receptacle to the metal enclosure, [250-146].

Detail
Purpose of bond wire is to internally connect the spirals (loops) of the metal sheath.

Bond wire inside AC cable can be cut off. The metal sheath is used to ground the enclosure.

Fig. 18-6 AC Cable Bonding Strip Can Be Cut Off at Termination

floor joists, no protection is required for the cable, however the cable must be installed in accordance with Section 300-4(d).

333-20 Conductor Ampacities

The ampacities of armored cable shall be as listed in Section 310-15, which basically means Table 310-16. When armored cable is installed in thermal insulation, the conductor insulation must be rated 90°C, but the ampacity of the conductors must be based on 60°C temperature ratings as listed in Table 310-16.

333-21 Grounding

The combination of the armor and No. 18 aluminum bonding strip provides an adequate path for equipment grounding as required by Section 250-2(d). The equipment grounding path must be maintained by the use of fittings that are specifically listed for armored cable [300-15(a), 333-9], Fig. 18-6.

Note. Armored cable is manufactured with an internal bonding strip of copper or aluminum in direct contact with the metal armored. The internal bonding strip is not an equipment grounding conductor, but it is used to reduce the inductive *reactance* of the armored spirals (decreases the resistance of the armored cable). Once the bonding strip exits the cable, it can be cut off because it no longer serves any purpose.

SUMMARY

☐ Armored cable is an assembly of insulated conductors, individually wrapped within a waxed paper jute, within a flexible metal (steel or aluminum) sheath that interlocks at the edges. ☐ Armored cable can be used exposed or concealed in dry locations where not subject to physical damage.   ☐ Armored cable shall be secured by approved staples, straps, hangers, or similar fittings every 4½ feet, and within 12 inches of every box, cabinet, enclosure, or termination fitting. ☐ Boxes and fittings shall be specifically listed for use with armored cable. ☐ Exposed armored cable must be protected from physical damage. ☐ The grounding continuity of the armored cable is maintained through the use of fittings and boxes that are approved for armored cable. ☐ The bonding strip is not an equipment grounding conductor.

REVIEW QUESTIONS

1. Give a brief description of the construction requirements for armored cable.

2. Where can armored cable be installed?

3. Where is armored cable prohibited?

4. What is the ampacity of No. 8 THHN armored cable?

5. What are the requirements for securing armored cable?

6. Briefly describe the termination requirements of armored cable.

7. Is armored cable required to have an equipment grounding conductor?

8. In your own words explain the meaning of the following Article 100 definitions:

   Accessible
   Location, dry

9. In your own words explain the meaning of the following Glossary terms:

   Reactance
   Studs

Unit 19

Article 334

Metal-Clad Cable(Type MC)

OBJECTIVES

After studying this unit, the student should be able to understand:
- that an equipment grounding conductor must be installed in metal-clad cable.
- the advantages of MC cable as compared to AC cable.

PART A. **GENERAL**

334-1 Definition

Metal-clad cable (Type MC) encloses one or more insulated conductors in a metal sheath of either corrugated or smooth copper or aluminum tubing. The physical characteristics of MC cable make it a versatile wiring method. It can be used in almost any location and for almost any application.

> **Note.** The most common type of MC cable is interlocking type, which has an appearance similar to armored cable or flexible metal conduit.

334-3 Uses Permitted

Metal-clad cable is permitted for:

(1) Branch circuits, feeders, and services [230-43,334-10(e)]
(2) Power lighting, control, and signal circuits

(3) Indoors or outdoors [225-10]
(4) Exposed or concealed
(5) Directly buried (if identified) [334-10(d)]
(6) Cable Tray
(7) In a raceway
(8) Open runs
(9) Aerial Cable on a messenger [334-10(f)]
(10) Hazardous locations
(11) Embedded in dry plaster
(12) Wet locations, where the cable is impervious to moisture and the conductors are listed for use in a wet location.

> **Note.** MC cable can be installed in plenums [300-22(b)], places of assembly [514-4], and theaters [520-5].

334-4 Uses Not Permitted

Metal-clad cable is limited or prohibited in commercial garages [511], exposed to corrosive fumes or vapors, unless the metallic sheath is suitable for the condition or the sheath is protected for the condition. In addition MC cable must not be

228

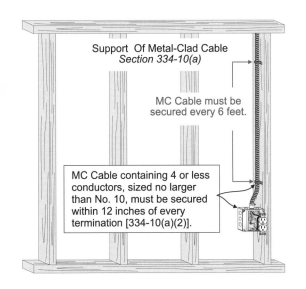

Fig. 19-1 MC Cable Must Be Secured Every 6 Feet, and Within 12 Inches of Termination

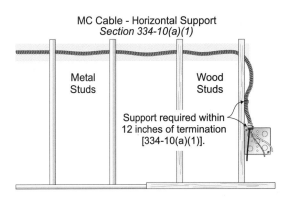

Fig. 19-2 Horizontal Runs Must Be Secured Within 12 Inches of Termination

used where restricted in hazardous (classified) locations [501-4(b), 502-4(b), and 503-3], patient care area receptacles [517-13(a)], or for swimming pool, fountain, spa or hot tub wiring [680-25].

PART B. INSTALLATION

334-10 Installation Requirements

MC cable must be installed in accordance with the requirements of Articles 300, 490, 725 and Section 770-52 and in accordance with the following:

(a) Support. MC cable must be supported and secured at intervals not exceeding 6 feet, Fig. 19-1.

Note. MC cable installed within suspended ceiling must be supported in accordance with Section 300-11(a).

(1) Horizontal Runs. MC cable installed horizontally in bored or punched holes in wood or metal framing members, or notches in wooden members is considered to be adequate, Fig. 19-2.

(2) Termination. MC cable with conductors no larger than No. 10 must be securely fastened within 12 inches of of termination to prevent stress onto terminal fittings, Fig. 19-1 and 19-2.

(b) Unsupported Cable. Type MC cable can be unsupported where the cable meets the following:

Lengths Not More than 6 Feet. MC cable in lengths of not more than 6 feet used as a fixture whip to supply recessed lighting fixtures [410-67(c)] is not required to be secured or supported when, Fig. 19-3.

Fished. When MC cable is installed by fishing, that portion of the cable that is fished is not required to be secured or supported.

(d) Direct Burial. Only MC cable marked as suitable for direct burial can be buried directly in the earth. When MC cable is directly buried, it must comply with the burial requirements contained in Section 300-5.

(e) Installed for Services. MC cable used for service-entrance conductors must comply with the installation requirements of Article 230, particularly Sections 230-50 and 51.

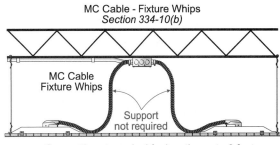

Fig. 19-3 MC Cable Used for Fixture Whips Are Not Required to Be Secured or Supported

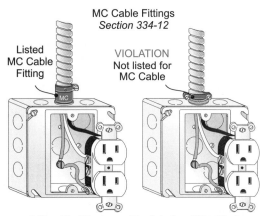

MC cable fittings must be listed and identified
for the connection of Type MC Cable.

Fig. 19-4 MC Cable Fittings and Boxes Must
Be Listed for Use With MC Cable

(f) Outside. Metal-clad cable installed out-doors must be installed in accordance with Article 225 and must of the type that is impervious to moisture.

(g) Framing Members. When running metal-clad cable through framing members, the cable must be installed in accordance with Section 300-4.

(h) In Accessible Attics or Roof Spaces. MC cable installed in accessible attics or roof spaces must comply with the protection requirement of Section 333-12.

334-11 Bends

(a) Smooth Sheath Cables. Smooth sheath cables cannot be bent in a manner that will damage the cable. This is accomplished by limiting the bending radius of the inner edge of the cable to:

(1) Ten times the external diameter of the metallic sheath for cable not more than ¾ inch in external diameter;

(2) Twelve times the external diameter of the metallic sheath for cable more than ¾ inch but not more than 1½ inch in external diameter; and

(3) Fifteen times the external diameter of the metallic sheath for cable more than 1½ inch in external diameter.

(b) Interlocked or Corrugated Sheath. Inter-locked or corrugated sheath cables shall not be bent in a manner that will damage the cable. This is accomplished by limiting the bending radius of the inner edge of the cable to not less than 7 times the diameter of the cable.

334-12 Fittings

Fittings used with metal-clad cable must be listed and identified for use with MC cable. This means that fittings and boxes intended for armored cable, nonmetallic-sheathed cable, and flexible metal conduit cannot be used with MC cable unless identified specially for MC cable [300-15(a)], Fig. 19-4.

It's been reported that metal-clad cable installations have experienced pull-out problems related to the use of improper fittings.

> **Note.** Anti-short bushings are not required for MC cable; but they are supplied with the cable to prevent aggravation in case some inspectors insist they be used.

334-13 Conductor Ampacity

The ampacity of MC cable conductors shall be based on the insulation rating of the conductors, which is generally THHN. However, conductors must be size to terminal temperature rating in accordance with Section 110-14(c).

PART C. CONSTRUCTION

334-23 Grounding

Interlocking MC cable is not listed as an equipment grounding conductor, must have an equipment grounding conductor and it requires a separate equipment grounding conductor to be within the metal sheath [250-118(11)].

> **Note.** Because MC cable is not listed as an equipment grounding conductor, it cannot be used to supply receptacles and switches in patient care areas [517-13(a)].

SUMMARY

☐ MC cable comes in corrugated or smooth tubing. ☐ Where not subjected to physical damage, MC cable can be installed concealed or exposed, in wet or dry locations, and can be directly buried (if identified). ☐ MC cable shall be secured and supported at intervals not to exceed 6 feet and within 12 inches of termination. ☐ Fittings for metal-clad must be identified for use with MC cable. ☐ The interlocking metal of MC cable is not approved as an equipment grounding conductor, and a separate equipment grounding conductor must supplement the metal armored sheath.

REVIEW QUESTIONS

1. What are the installation requirements for metal-clad cables?

2. Where can MC cable be installed?

3. Where is metal-clad cable prohibited?

4. What are the requirements for securing metal-clad cable?

5. What are the termination requirements of metal-clad cable?

6. What are the advantages of MC cable versus AC cable?

7. Is an equipment grounding conductor required to be installed in metal-clad cable?

Unit 20

Article 336

Nonmetallic-Sheath Cable (Types NM and NMC)

OBJECTIVES

After studying this unit, the student should be able to understand:
- where nonmetallic-sheathed cable can be installed.
- the disadvantages of nonmetallic-sheathed cable.
- when NM cable must be installed in a raceway.
- the installation requirements of nonmetallic-sheathed cable when installed through or parallel to framing members.
- the installation requirements for installing nonmetallic-sheathed cable in basements.
- the installation requirements for installing NM cable in accessible attics or roof spaces.
- the requirements for securing nonmetallic-sheathed cable.

PART A. GENERAL

336-2 Definition

Nonmetallic-sheathed cable is a wiring method enclosing 2 or 3 insulated conductors, No. 14 through No. 2, within a nonmetallic outer cover. Because this cable is nonmetallic, it contains a separate equipment grounding conductor. Non-metallic-sheathed cable is a common wiring method used for residential and commercial branch circuits. It is called *Romex*® by most electricians.

336-4 Uses Permitted

Nonmetallic-sheathed cable shall be permitted in:

(1) One- and two-family dwellings of any height, Fig. 20-1.

NM Cable - Over 3 Floors For Dwelling Units
Section 336-4(1)

1 or 2 Family Dwelling

1 or 2 Family Dwelling

VIOLATION
Multifamily
Dwelling

NM cable is permitted for one- and two-family dwelling units of any height.

Fig. 20-1 NM Cable over 3 Floors for Dwelling Units

232

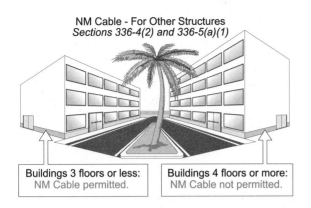

NM Cable - For Other Structures
Sections 336-4(2) and 336-5(a)(1)

Buildings 3 floors or less:
NM Cable permitted.

Buildings 4 floors or more:
NM Cable not permitted.

Fig. 20-2 NM Cable Not Permitted in Buildings over 3 Floors

(2) Other buildings not over 3 floors, unless prohibited in Section 336-5, Fig. 20-2.

(a) NM Cable. The outer covering of NM cable is flame retardant and moisture resistant. NM cable is intended for use in dry locations only, but can be fished into masonry air voids when the wall is not exposed to excessive moisture or dampness.

(b) NMC Cable. The outer cover of NMC cable is similar to NM cable, except that it is corrosion resistant, as shown by the letter C. This cable can be installed in:

(1) Exposed or concealed in dry, moist, damp, and corrosive locations [300-6].

(2) Inside or outside masonry or tile walls.

(3) In shallow chases of concrete, masonry, or adobe when protected from nails and screws with a metal plate.

336-5 Uses Not Permitted

(a) Type NM and NMC.

(1) Nonmetallic-sheathed cable cannot be installed in multifamily buildings and other structures of more than 3 floors, Fig. 20-2.

(2) Services [230-43].

(3) Commercial garages [511-3] and hazardous (classified) locations areas [501, 502, 503].

(4) Places of assembly [518-4] and theaters [520-5].

(5) Motion picture studios [Article 530].

(6) Storage battery rooms [Article 480].

(7) Hoistways [Article 620].

(8) Embedded in concrete.

(9) Hazardous (classified) locations.

Note. In addition, NM cable is not permitted in ducts, plenums, or other spaces used for environmental air [300-22], some swimming pool wiring [680-25], or for receptacles and switches in patient care areas [517-13(a)].

PART B. INSTALLATION

336-6 Exposed

(a) Surface of the Building. Nonmetallic-sheathed cable when run exposed, must closely follow the surface of the building finish or running boards when run exposed. This means that loose cable is not allowed. Some inspectors require the cable to be secured more often than 4½ feet.

(b) Protected from Physical Damage. Non-metallic sheath cable must be protected from physical damage by rigid metal conduit, intermediate metal conduit, Schedule 80 rigid nonmetallic conduit (PVC) [347-3(c)], electrical metallic tubing (EMT), guard strips, or other means. Where nonmetallic-sheathed cable is installed in metal raceway for protection, the raceway is not required to be grounded [250-86 Ex 2, 300-12 Ex.]

CAUTION: Some local codes require NM cable around windows and doors to be installed in electrical metallic tubing.

Note. When an NM cable is installed in a raceway, the cable must be protected from abrasion by a fitting installed on the end of

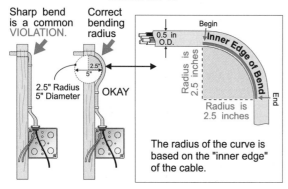

Fig. 20-3 Nonmetallic Sheathed Cable Must Be Carefully Bent

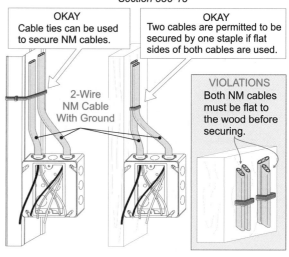

Fig. 20-4 NM Cable Can be Secured With Cable Ties, and Flat Cable Cannot Be Stapled on Edge

the raceway [300-15(c)]. Most inspectors permit raceway connectors and couplings to meet this requirement.

(c) Unfinished Basements. 6/2, 8/3 and larger NM cables can be mounted directly to the bottom of the joist without protection of running boards. Smaller cables must be run through bored holes in the joist or mounted on running boards. Cables run parallel with the joist must be secured to the side of the joist.

(d) Attics and Roof Spaces. Nonmetallic-sheathed cable installed in accessible attics or roof spaces must comply with the physical protection requirements contained in Section 333-12.

336-9 Through or Parallel to Framing Members

Nonmetallic-sheathed cables run through or parallel to framing members must comply with the physical protection requirements contained in Section 300-4.

336-16 Bends

Nonmetallic-sheathed cable must not be bent in any manner that will result in damage to the cable or conductors. To accomplish this, the bending radius of the inner edge of the cable must not be less than 5 times the diameter of the cable, Fig. 20-3.

Note. NM cable is often installed with a sharp short radius bend in violation of this requirement.

336-18 Secured or Supported

Nonmetallic-sheathed cable must be secured by **approved** staples, straps, or cable ties in such a manner that the cable will not be damaged. Flat nonmetallic-sheathed cable cannot be stapled on edge, Fig. 20-4.

Nonmetallic-sheathed cable must be secured within 12 inches of every box, cabinet, enclosure, or termination fitting, except as required in Section 370-17(c) Exception, or as permitted in Section 373-5(c) Exception. In addition the cable must be secured in intervals not to exceed 4½ feet.

Cables installed horizontally through framing members are considered to be adequately supported, Fig. 20-5.

Exception No. 1: Fished. That portion of non-metallic-sheathed cable that is fished is not required to be secured.

Exception No. 3: Fixture Whips. Nonmetallic-sheathed cable installed within an accessible ceiling for the connections to lighting fixtures and equipment is not required to be secured within 12 inches of termination where the free length does not exceed 4½ feet.

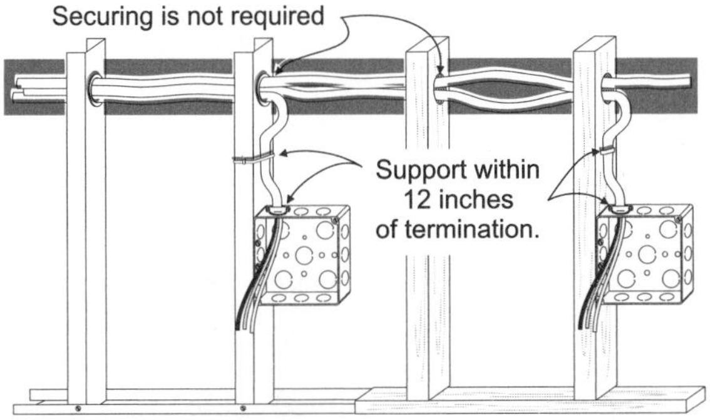

Framing Members Securing NM Cable
Section 336-18

Securing is not required

Support within
12 inches
of termination.

NM cable must be secured at intervals not more than
4½ feet but are considered supported where run through
holes in wood or metal framing members.

Fig. 20-5 Horizontal Runs Must Be Secured
Withing 12 Inches of Termination

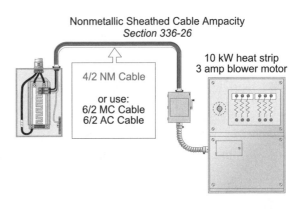

Nonmetallic Sheathed Cable Ampacity
Section 336-26

10 kW heat strip
3 amp blower motor

4/2 NM Cable

or use:
6/2 MC Cable
6/2 AC Cable

Fig. 20-6 Nonmetallic Sheathed Cable Must Be
Sized Based on 60°C Ampacity of Table 310-16

336-26 **Ampacity**

(b) Ampacity. Nonmetallic-sheathed cable conductors must be rated 90°C, but the ampacity of the conductors must be based on the 60°C temperature ratings as listed in Table 310-16.

However, the 90°C ampacity rating can be used for ampacity derating purposes, provided the final derated ampacity does not exceed the value of a 60°C rated conductor. See Sections 110-14(c), 310-10 and 310-15 for conductor sizing and ampacity adjustment.

Question. What size THHN conductor is required to supply a 10 kW, 230 volt, single-phase fixed space heater that has a 3 ampere blower motor [424-3(b)]? Terminal rating of 75°C, Fig. 20-6.

Answer. No. 4

According to Section 424-3(b) the conductors and overcurrent protection device to electric space heating equipment shall be sized no less than 125 percent of the total load (heat plus motors).

10,000 watts/230 volt + 3 ampere, = 46 ampere
46 ampere × 1.25 = 58 ampere
No. 4 THHN rated 70 ampere at 60°C

Note. Article 333 – Type AC cable and Article 334 – Type MC cable do not have a 60°C limitation for conductor sizing; therefore, 6/2 Type AC or MC cable (rated 65 amperes) could be used. I know this doesn't make sense, but it's the *Code.*

SUMMARY

☐ The outer cover of NM cable is flame retardant and moisture resistant intended for use in dry locations only. ☐ NMC cable is similar to NM cable, except its outer cover is corrosion resistant as indicated by the letter C, and is permitted in dry, moist, damp, or corrosive locations. ☐ Nonmetallic-sheathed cables are permitted in residential, commercial, and industrial locations, except where prohibited by Section 336-5. ☐ Nonmetallic-sheathed cables must be protected from physical damage. ☐ The cable shall be secured in place within 12 inches of each box, cabinet, or fitting, and at intervals not exceeding 4 feet. ☐ When nonmetallic-sheathed cable is installed in a 2¼ × 4 inch nonmetallic box without cable clamps, the cable must be secured within 8 inches of the box. ☐ Nonmetallic-sheathed cable must have 90°C insulated conductors, but the ampacity shall be based on 60°C insulation.

REVIEW QUESTIONS

1. Where can nonmetallic-sheathed cable be installed?

2. What is the greatest disadvantage of nonmetallic-sheathed cable?

3. When is NM cable required to be installed in a raceway?

4. When nonmetallic-sheathed cable is installed through framing members, what are the installation requirements?

5. What are the installation requirements for installing nonmetallic-sheathed cable in basements?

6. What are the installation requirements for installing NM cable in accessible attics or roof spaces?

7. What are the requirements for securing nonmetallic-sheathed cable?

Unit 21

Article 338

Service-Entrance Cable (Types SE and USE)

OBJECTIVES

After studying this unit, the student should be able to understand:
- when SE cable can be used for interior wiring.
- which rules in Articles 230, 250, 300, and 336 particularly apply to service cables.

338-1 Definition

Service-entrance cable can be a single conductor or multiconductor assembly with an overall covering. This cable is used primarily for services not over 600 volt, but can also be used for feeders and branch circuits, Fig. 21-1.

(a) Service-Entrance Cable (Type SE). Type SE cable has a flame-retardant, moisture-resistant covering and is only permitted to be used in aboveground installations. This cable can be used for branch circuits or feeders when installed according to Section 338-3 and 338-4.

(b) Underground Service-Entrance Cable (Type USE). Underground service-entrance cable Type USE can be used for below and above ground installations. When the cable is directly buried, the cable must comply with the burial requirements of Section 300-5.

338-2 Used for Services

Type SE and USE cable used for service-entrance conductors must comply with the applicable requirement of Article 230, particularly Sections 230-50 and 51.

Type USE cable used for service-laterals are permitted to emerge above ground outside at terminations in meter bases or other enclosures where protected in accordance with Section 300-5(d).

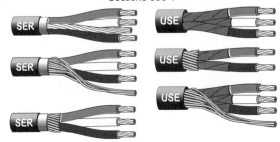

Service-Entrance Cable - Definition
Sections 338-1

Service-entrance cable is a single conductor or multiconductor assembly with or without an overall outer cover. It is mostly used for services but can be used for feeders and branch circuits [338-4].

Fig. 21-1 Service-Entrance Cable Types

237

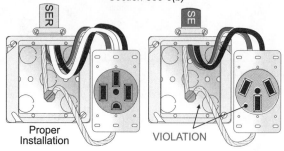

Neutral Not Permitted As Grounding Conductor
Section 338-3(b)

Proper Installation

VIOLATION

Section 250-60 does not permit the grounded conductor to ground the frames of ranges, ovens, and dryers. These circuits require a separate equipment ground.

Fig. 21-2 Uninsulated Conductor Not Permitted to be Used for the Neutral Conductor

338-3   Type SE Cable Used for Branch Circuits or Feeders

(a) **Insulated Grounded Conductor.** Service entrance cable Type SER (round) can be used for branch circuits and feeders [338-3(b)], but the uninsulated conductor within the cable can only be used for equipment grounding purposes [250-118].

(b) **Uninsulated Grounded Conductor.** Service-entrance cable (Type SE or USE) with an uninsulated grounded (neutral) conductor cannot be used for branch circuits or feeders, except where it is used as a feeder to a separate building or structure in accordance with Section 250-32(b)(2), Fig. 21-2.

> **Note.** The 1996 *Code* revised Section 250-140 to prohibit the grounding of ranges, ovens, and dryers to the grounded (neutral) conductor. Therefore Type SE or USE cable cannot be used to supply power to these appliances, Fig. 21-2.

338-4   SE Used for Interior Wiring

(a) **Interior Installations.** When Type SE cable is used for interior wiring, it must comply with the installation requirements for Parts A and B of Article 336. This means that it must be installed according to the same requirements as nonmetallic-sheathed cable.

(b) **Exterior Installations.** SE cable installed outdoors for branch circuits and feeders shall be installed in accordance with the requirements of Article 225 – Outside Wiring. When installed underground the cable shall comply with the underground requirements in Article 339 – Underground Feeder Cable (Type UF).

SUMMARY

☐ Service-entrance cable can be a single conductor or multiconductor assembly with an overall covering. ☐ SE cable has a flame-retardant, moisture-resistant covering and is listed for above-ground installations only. ☐ Underground service-entrance cable (Type USE) can be used below and above ground. ☐ SE cable with an insulated grounded (neutral) conductor can be used for interior or exterior wiring for feeders and branch circuits. ☐ SE cable with an uninsulated grounded (neutral) conductor cannot be used for branch circuits. ☐ SE cable with an uninsulated grounded (neutral) conductor cannot be used for feeders, except when used to supply power to another building on the same premises [250-32]. ☐ When SE cable is used for interior, the cable must comply with the requirements of nonmetallic-sheathed cable, Article 336 Parts A and B.

REVIEW QUESTIONS

1. SE cable is used primarily for services. When can SE cable be used for interior wiring?

2. Which rules in Articles 230, 250, 300, and 336 apply to service-entrance cables?

Unit 22

Article 339

Underground Feeder and Branch Circuit Cable (Type UF)

OBJECTIVES

After studying this unit, the student should be able to understand:
- underground installation requirements for UF cable.
- where UF cable cannot be used.
- the requirement for UF cable installed indoors

339-1 Description

Underground feeder and branch circuit cable is a moisture, fungus, and corrosion-resistant cable system suitable for direct burial in the earth. It comes in sizes No. 12 through No. 4/0. The covering of multiconductor UF cable is molded plastic that encapsulates the insulated conductors.

This makes it more difficult to strip off the outer jacket, but it provides excellent corrosion protection.

339-3 Uses

(a) Uses Permitted.

Underground. When run underground, Type UF cable must comply with the underground cable requirements of Section 300-5.

Interior Wiring. When UF cable is to be used for interior wiring, it must be installed to the same requirements as nonmetallic-sheathed cable, Article 336 Parts A and B.

(b) Uses Not Permitted. UF cable cannot be installed in commercial buildings of more than 3 floors [336-5(a)(1)], in ducts, plenums or other spaces used for environmental air [300-22]. In addition, Type UF cable is not permitted in:

(1) Services [230-43]

(2) Commercial garages [511-3]

(3) Theaters [520-5]

(4) Motion picture studios [530-11]

(5) Storage battery rooms [Article 480]

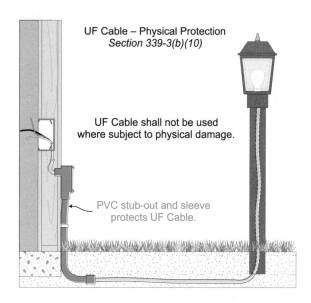

UF Cable – Physical Protection
Section 339-3(b)(10)

UF Cable shall not be used
where subject to physical damage.

PVC stub-out and sleeve
protects UF Cable.

Fig. 22-1 UF Cable Must Be Protected from
Physical Damage

(6) Hoistways [Article 620]

(7) Hazardous (classified) locations areas

(8) Embedded in concrete

(9) Exposed to direct sunlight unless identified

(10) Type UF cable shall not be used where subject to physical damage, Fig. 22-1.

When UF cable is subject to physical damage, it must be protected by a suitable method such as a raceway [250-33 Ex., 300-12 Ex.], Fig. 22-1.

Other Restrictions Include:

Location	Section
Places of assembly	518-4
Pool wiring	680-25
Patient care receptacles	517-13(a)

339-5 **Ampacity**

The ampacity of conductors in UF cable must be based on 60°C ratings as listed in Table 310-16.

SUMMARY

☐ Underground feeder and branch circuit cable (UF) is moisture, fungus, and corrosion-resistant cable suitable for direct burial in the earth. ☐ When UF cables are run underground, they must comply with the underground requirements of Section 300-5. ☐ UF cable used for interior wiring must comply with the requirements of nonmetallic-sheathed cable, Article 336 Parts A and B.

REVIEW QUESTIONS

1. Describe the underground installation requirements for UF cable.

2. When UF cable is installed indoors, it must comply with the installation requirements for AC cable in Article 333. True or False?

Unit 23

Articles 345 and 346
Intermediate Metal Conduit and Rigid Metal Conduit

OBJECTIVES

After studying this unit, the student should be able to understand:
- the advantages of intermediate metal conduit over rigid metal conduit.
- the advantages and disadvantages of aluminum rigid metal conduit as compared to IMC or rigid.
- why a running thread cannot be installed at couplings.
- the requirements of bushings for IMC and rigid metal conduit containing conductors No. 6 and smaller.
- the requirements of bushings for IMC and rigid metal conduit containing conductors No. 4 and larger.

Since the requirements for intermediate metal conduit and rigid metal conduit are identical, both raceway systems have been combined into one unit in this book.

PART A. GENERAL/INSTALLATION

Intermediate metal conduit (IMC) [Article 345] is a circular metal raceway with an outside diameter the same as rigid metal conduit (rigid) [Article 346]. The wall thickness of IMC is thinner than rigid conduit, therefore, it has a greater interior cross-sectional area.

Intermediate metal conduit is lighter and less expensive than rigid metal conduit and it can be used in all of the same locations as rigid conduit.

The type of steel from which IMC is manufactured, the process by which it is made, and the corrosion protection applied is equal to or superior to that of rigid conduit. Impact, crush, and hydrostatic tests on both intermediate metal conduit and rigid conduit show that IMC to be equal to or better than rigid conduit.

345 / 346-3 Uses Permitted

(a) Listed IMC or Rigid. Listed IMC or rigid conduit can be installed in all atmospheric conditions and occupancies.

Severe Corrosive Areas. In severe corrosive areas where the mechanical and physical properties of metal conduit are required, plastic-coated rigid metal conduit must be used. This type of raceway is used in the petrochemical industry and the common trade name for this material is "Plasti-bond."

241

Dissimilar Metals. Where practical, contact with dissimilar metals should be avoided to prevent the deterioration of the metal as a result of galvanic action (corrosion).

Aluminum Metal Conduit. Aluminum rigid metal conduit fittings and enclosures can be used with IMC or rigid metal conduit. Rigid steel fittings can be used with aluminum rigid metal conduit.

Aluminum rigid metal conduit is much lighter, has better grounding capability, and is easier to thread and install. In addition, aluminum raceways have lower *induction* losses as compared to rigid metal steel, and it's excellent to be used for high-frequency circuits (415 Hz computer power).

However, aluminum conduit must not be installed underground because it corrodes at an accelerated rate.

(b) Corrosion Protection: Concrete or Earth Burial. Supplementary corrosion protection is not required for IMC or rigid conduit when installed in concrete or buried in the earth, unless a severe corrosion condition exists.

Aluminum Metal Conduit. However, testing laboratory publications indicate aluminum rigid metal conduit requires additional corrosion protection when encased in concrete or directly buried [300-6]. According to UL, supplemental corrosion protection is necessary for metal raceway at the location where the raceway leaves the concrete or soil.

Note. Supplementary nonmetallic coatings have not been investigated for resistance to corrosion and these coatings are also known to cause cancer in laboratory animals. I know of a case where an electrician was taken to the hospital for lead poisoning, after using this product in a poorly ventilated area.

PART B.   INSTALLATION

345 / 346-5   Wet Locations

All materials such as screws, straps, supports, etc. installed in a wet location must be made of cor-

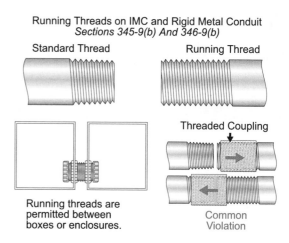

Fig. 23-1 Running Threads Permitted at Enclosures, but Not at Couplings

rosion-resistant material or it must be protected by corrosion-resistant coatings [300-6].

345 / 346-6   Size

Intermediate metal conduit is manufactured in sizes from ½ inch to 4 inches, and rigid metal conduit is made in sizes from ½ inch to 6 inches.

345 / 346-7   Conductor Fill

When determining the number of conductors in a raceway, the area of the raceway must be used, according to Chapter 9, Table 1, see Section 300-17.

345 / 346-8   Reaming

Reaming to remove the burrs and rough edges is required when the raceway is cut. Small raceways cut with a hacksaw or band saw are often reamed by the use of a screw driver or pliers. When the raceway is cut with a three-wheel pipe cutter, a reaming tool (reamer) must be used to remove the indented burrs.

345 / 346-9   Couplings and Connectors

(a) Installation. Threadless couplings and connectors must be made up wrenchtight to main-

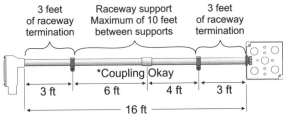

IMC and Rigid Metal Conduit Support
Sections 345-12(a) and 346-12(a)

Note: Support is not required within 3 feet of a coupling.

Fig. 23-2 EMT Must Be Secured Within 3 Feet of Termination, and Every 10 Feet Thereafter

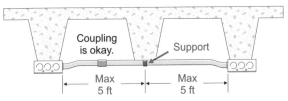

Support of Rigid Metal and IMC on Structural Members
Sections 345-12(a) and 346-12(a)

Where the structural members do not permit fastening within 3 feet of the conduit terminations, the conduit is permitted to be supported within 5 feet.

Fig. 23-3 EMT Support Permitted to Be up to 5 Feet from Termination, If Necessary

tain an effective low impedance path to conduct safely fault-current [250-2(d), 250-96(a), and 300-10]. When IMC or rigid metal conduit is installed in concrete, the fittings must be concretetight, and when installed in a wet location, the raceway must be **raintight** and arranged to drain [225-22, 230-53].

(b) Running Threads. Running threads cannot be used on conduits for the connection of couplings, but they can be used between enclosures, Fig. 23-1.

345 / 346-10 Bends

Raceway bends shall not be made in any manner that would damage the raceway or significantly change the internal diameter of the raceway (no kinks). This is accomplished by complying with the bending radius requirements contained in Table 346-10. This is usually not a problem because most benders are made to comply with this Table; however, when using a *hickey* bender, be careful not to overbend the raceway.

345 / 346-11 Number of Bends (360°)

To reduce the stress and friction on the conductor insulation during installation bends (including offsets) between pull points such as boxes and conduit bodies cannot exceed 360°.

345/346-12 Secured and Supported

(a) Securely Fastened. IMC or rigid conduit must be securely fastened within 3 feet of every

box, cabinet, or termination fitting, Fig. 23-2.

When structural members do not permit the raceway to be secured within 3 feet of a box or termination fitting, the raceway is permitted to be secured within 5 feet of a box or termination fitting, Fig. 23-3.

Note. Support is not required within 3 feet of each coupling.

(1) Support. IMC and rigid must be supported at intervals not exceeding 10 feet, Fig. 23-2.

(2) Support. Straight horizontal runs made with threaded couplings can be supported according to dimensions listed in Table 346-12.

Table 346-12 Supports for Rigid Metal Conduit

Conduit Size	Maximum Distance between Supports
½-¾	10
1	12
1¼-1½	14
2-2½	16
3 & larger	20

(3) Industrial Machinery. For industrial machinery, straight exposed vertical risers with threaded couplings are permitted if supported at the top and bottom no more than 20 feet apart.

(4) Horizontal Runs. Conduits installed horizontally in bored or punched holes in wood or metal framing members, or notches in wooden members is considered to be adequately supported, but the raceway must be secured within 3 feet of termination.

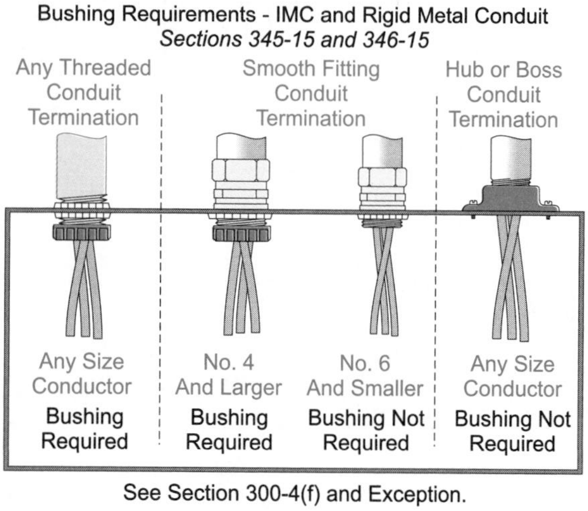

Bushing Requirements - IMC and Rigid Metal Conduit
Sections 345-15 and 346-15

Any Threaded Conduit Termination	Smooth Fitting Conduit Termination	Hub or Boss Conduit Termination

Any Size Conductor	No. 4 And Larger	No. 6 And Smaller	Any Size Conductor
Bushing Required	Bushing Required	Bushing Not Required	Bushing Not Required

See Section 300-4(f) and Exception.

Fig. 23-4 Insulated Bushing Requirements for Rigid Conduit and IMC

345 / 346-15   Bushings

A metal or plastic bushing must be on the conduit termination threads to protect conductors from abrasion. A bushing is not required if the design of the box or fitting (connectors) will provide the equivalent protection, Fig. 23-4.

Note. Section 300-4(f) requires No. 4 and larger to be protected with a fitting having a smooth, rounded insulating surface.

SUMMARY

□ Rigid metal conduit is called "rigid" or "heavy wall." □ Intermediate metal conduit is a circular metal raceway with a wall thickness that is thinner than rigid metal conduit. □ Where practical, contact with dissimilar metals should be avoided to prevent the deterioration of the metal, but aluminum and steel fittings can be intermixed. □ IMC or rigid metal conduit can serve as the circuit equipment grounding conductor [250-118]. □ The maximum number of bends between two pull points, including offsets, must not exceed 360°; and the raceway must be secured within 3 feet of each box or termination fitting. □ A metal or plastic bushing must be on the conduit termination threads to protect the conductors from abrasion for conductors No. 6 and smaller. □ Raceways containing conductors No. 4 and larger require a fitting having a smooth, rounded insulating surface to protect the conductors from abrasion.

REVIEW QUESTIONS

1. What are the advantages of intermediate metal conduit over rigid metal conduit?

2. What are the advantages and disadvantages of aluminum rigid metal conduit compared to IMC or rigid metal conduit?

3. Explain the requirements of bushings for IMC and rigid metal conduit containing conductors No. 6 and smaller.

4. Explain the requirements of bushings for IMC and rigid metal conduit containing conductors No. 4 and larger.

5. How long can a raceway be before a junction/pull box is required?

6. Define the glossary term hickey.

Unit 24

Article 347
Rigid Nonmetallic Conduit

OBJECTIVES

After studying this unit, the student should be able to understand:
- the different types of rigid nonmetallic conduits to which Article 347 applies.
- where rigid nonmetallic conduit can be used.
- where rigid nonmetallic conduit is prohibited.
- the mechanical installation requirements of rigid nonmetallic conduit, such as trimming, termination, supports, and expansion fitting.
- the major differences between Schedule 40 and Schedule 80 rigid nonmetallic conduits.

347-1 Description

Rigid nonmetallic conduit and fittings are moisture and chemical resistant. When installed above-ground, the raceways and fittings must be resistant to impact, crushing, distortion from heat that occurs in normal use, and must be resistant to low temperatures and sunlight effects.

Types of Rigid Nonmetallic Conduits

Most of us in the electrical trade think of rigid nonmetallic conduit as PVC (polyvinyl chloride). Though this is the most common type installed, there are other kinds such as Type A or Type EB, HDPE (high density polyethylene) Schedule 40, asbestos, cement, soapstone, and nonmetallic fiber.

347-2 Uses Permitted

Only listed rigid nonmetallic conduit can be installed as contained in (a) through (h).

FPN: Rigid nonmetallic conduit in extreme cold can become brittle and is more susceptible to physical damage.

(a) Concealed. Rigid nonmetallic conduit can be concealed within walls, floors and ceilings, directly buried, or embedded in concrete in buildings of any height.

(b) Corrosive Influences. Rigid nonmetallic conduit can be installed in locations subject to severe corrosion or chemicals for which the material is specifically approved [300-6].

(d) Wet Locations. Rigid nonmetallic conduit can be installed in wet locations such as dairies, laundries, canneries, car washes, and other areas that are frequently washed or outdoors.

Supporting equipment such as straps, screws, and bolts must be made of corrosion-resistant materials, or it must be protected with a corrosion-resistant coating [300-6].

(e) Dry and Damp Locations. Rigid non-metallic conduit can be installed in dry and damp locations, except where limited in Section 347-3.

(f) Exposed. Schedule 40 rigid nonmetallic conduit is approved for exposed locations where not subject to physical damage [347-3(c)]. In areas exposed to physical damage, Schedule 80 rigid nonmetallic conduit or rigid metal conduit must be used.

Note. Rigid nonmetallic conduit can be installed exposed throughout a building of any height, but ENT (made of PVC, but thinner) cannot be installed exposed in buildings over 3 floors [331-3(1)]. Crazy, but that's the *Code*.

(g) Underground. Rigid nonmetallic conduit installed underground must comply with the burial requirements contained in Section 300-5.

(h) Conduit Bodies. Rigid nonmetallic conduit can be used to support nonmetallic conduit bodies, but the conduit bodies must not contain any devices, lighting fixtures, or other equipment such as a photo-cell.

347-3 Uses Not Permitted

(a) Hazardous Locations. Rigid nonmetallic conduit cannot be installed in hazardous (classified) locations, except as permitted for service stations [514-8 Ex. 2], bulk storage plants [515-5], Class I locations [501-4(a)(1) Ex. 1], or Class III locations [503-3(a)].

(b) Support of Equipment. Rigid nonmetallic conduit cannot be used to support lighting fixtures or equipment such as boxes [347-2(h), 370-23].

(c) Physical Damage. Schedule 40 rigid nonmetallic conduit cannot be installed where subject to physical damage, but Schedule 80 rigid nonmetallic conduit is listed for this purpose.

(d) Ambient Temperature. Rigid nonmetallic conduit cannot be used where the ambient temperature exceeds 50°C.

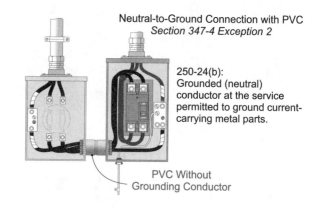

Neutral-to-Ground Connection with PVC
Section 347-4 Exception 2

250-24(b): Grounded (neutral) conductor at the service permitted to ground current-carrying metal parts.

PVC Without Grounding Conductor

Fig. 24-1   Rigid Nonmetallic Conduit Does Not Always Need an Equipment Grounding Conductor

(e) Conductor Insulation. Rigid nonmetallic conduit cannot contain conductors whose operating temperature would exceed the conduit temperature rating.

(f) Places of Assembly. Because of the toxicity of rigid nonmetallic conduit when heated or burned, it cannot be installed in places of assembly and theaters, except as permitted in Sections 518-4(b) and 520-5(c).

Note. Rigid nonmetallic conduit is also prohibited for patient care area receptacles in health care facilities [517-13(b)], ducts, plenums, and other spaces used for environmental air handling [300-22].

347-4 Grounding

An equipment grounding conductor shall be installed in rigid nonmetallic conduit.

Exception No. 2: Neutral-to-Ground Connections. An equipment grounding conductor is not required in rigid nonmetallic conduit if the grounded (neutral) conductor is used to ground equipment as permitted in Section 250-142(a) for, Fig. 24-1:

- Service equipment   250-24(b)
- Separately derived system   250-30(a)
- Separate buildings/structures 250-32(a)

347-5 Trimming

The cut ends of rigid nonmetallic conduit must be trimmed (inside and out) to remove the burrs and rough edges. Trimming nonmetallic raceway is very easy; most of the burrs will rub off with fingers, and a knife can be used to smooth the rough edges.

347-6 Joints

Joints (couplings and connectors) must be made in an approved manner. This means following the instructions of the raceway, fitting, and glue.

> **Note.** Rigid nonmetallic conduit glue is a solvent cement approved for use on rigid nonmetallic conduit only. Some PVC solvents require the surface to be cleaned with a cleaning solvent before the application of the glue. Other solvents are a one-process and no cleaning is required. After the application of the cement to both surfaces, a quarter turn of the fitting is required.

347-8 Secured and Supports

(a) **Secured.** Rigid nonmetallic conduit must be installed as a complete system before conductors can be installed [300-18(a)]. Rigid nonmetallic conduit must be secured within 3 feet of every box, cabinet, or termination fitting, such as a conduit body.

(b) **Supports.** Rigid nonmetallic conduit must also be supported at intervals not exceeding the values listed in Table 347-8 and the raceway must be fastened in a manner that permits movement from thermal expansion or contraction.

Table 347-8 Supports for Rigid Metal Conduit

Conduit Size	Maximum Spacing
½-1	3 feet
1¼-2	5 feet
2½-3	6 feet
3½-5	7 feet
6	8 feet

Some types of rigid nonmetallic conduits require less support than specified in Table 347-8,

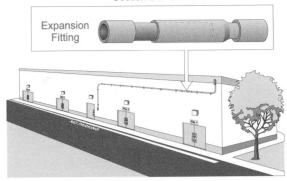

PVC Expansion-Contraction
Section 347-9

Expansion Fitting

Table 347-9 provides thermal expansion characteristics of rigid nonmetallic conduit for using expansion fittings.

Fig. 24-2 Expansion Fittings Generally Required for Runs More than 50 Feet in Length Outdoors

unless the raceways instructions indicated other support requirements.

Horizontal runs of rigid nonmetallic conduit through framing members and securely fastened within 3 feet of terminations are considered adequately supported.

347-9 Expansion Fittings

Expansion fittings must be provided to compensate for thermal expansion and contraction of the raceway or building structure where the length change is expected to be ¼ inch or greater in a straight run between securely mounted items such as boxes, cabinets, elbows, or other conduit terminations, Fig. 24-2.

Temperature Change	Length Change
25 degrees	1 inch
50 degrees	2 inch
75 degrees	3 inch
100 degrees	4 inch

> **Note.** Listing instructions state that you should add 30°F to the ambient temperature when the raceway is in direct sunlight.

CAUTION: If the ambient temperature during installation is high, you must realize that the conduit is at its expanded range and it will contract when the temperature drops. Of course the opposite applies if the ambient temperature is low.

347-10   Size

Rigid nonmetallic conduit is manufactured in sizes from ½ inch to 6 inches.

347-11   Number of Conductors

When determining the number of conductors in rigid nonmetallic conduit, the area of the raceway must be used according to the percentages as listed in Chapter 9 Table 1, see Section 300-17.

> CAUTION: Schedule 80 rigid nonmetallic conduit has the same outside diameter as Schedule 40 rigid nonmetallic conduit, but the wall thickness of Schedule 80 is greater. This results in a reduced interior area for conductor fill. The area of Schedule 80 PVC is marked on the surface of the raceway.

347-12   Bushings

Where a No. 4 or larger conductor conduit enters a box, fitting, or other enclosure, a fitting that provides a smooth rounded insulating surface a bushing or adapter shall be provided to protect the wire from abrasion during and after installation [300-4(f)]. Some rigid nonmetallic conduit adapters (connectors) and all bell-ends provide the conductor protection required in this Section, Fig. 24-3.

Note. When rigid nonmetallic conduit is stubbed into an open bottom switchboard or other apparatus, the raceway, including the end fitting (bell-ends), must not rise more

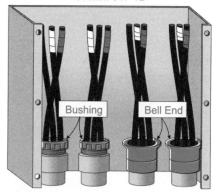

Bushings On Nonmetallic Conduit
Section 347-12

Bushing   Bell End

A bushing or similar fitting is required where nonmetallic conduit enters any enclosure for the protection of conductors. See Sections 300-4(f), 300-16(b), and 384-10.

Fig. 24-3 Rigid Nonmetallic Conduit Can Terminate to a Bushing

than 3 inches above the bottom of the switchboard enclosure [300-16(b), 384-10].

347-13   Bends, How Made

The bending radius requirements for rigid nonmetallic conduit shall be according to Section 346-10 for rigid metal conduit. Field bends must be made with heaters approved for the purpose. Blow torches and open flames should not be used since they may damage the nonmetallic raceway.

347-14   Number of Bends (360°)

To reduce the stress and friction on the conductor insulation during installation bends (including offsets) between pull points such as boxes and conduit bodies cannot exceed 360°.

SUMMARY

☐ Rigid nonmetallic conduit and fittings are moisture- and chemical-resistant. ☐ Rigid nonmetallic conduit can be concealed within walls, floors, and ceilings, embedded in concrete, or directly buried in the earth or run exposed. ☐ Rigid nonmetallic conduit cannot be installed in hazardous (classified) locations, except as permitted for service stations [514-8 Ex. 2], bulk storage plants [515-5], and Class III locations [503-3(a)]. ☐ Rigid nonmetallic conduit cannot be installed in ducts, plenums, and other spaces used for environmental air. ☐ Rigid nonmetallic conduit cannot be used for the wiring of patient care area receptacles in health care facilities [517-13(b)]. ☐ Grounding: When an equipment grounding conductor is required, a separate equipment grounding conductor must be installed in the raceway [300-3(b)].

REVIEW QUESTIONS

1. List the different types of rigid nonmetallic conduits.

2. Where can rigid nonmetallic conduit be used?

3. Where is rigid nonmetallic conduit prohibited?

4. What are the mechanical installation requirements of rigid nonmetallic conduit such as trimming, termination, supports, and expansion fittings?

5. What are the major differences between Schedule 40 and Schedule 80 rigid nonmetallic conduits?

Unit 25

Article 348
Electrical Metallic Tubing

OBJECTIVES

After studying this unit, the student should be able to understand:
- the limitations of electrical metallic tubing as compared to IMC or rigid metal conduit.
- support requirements for electrical metallic tubing.
- why bushings are not required for conductors No. 6 and smaller installed in electrical metallic tubing.

Electrical metallic tubing is commonly called EMT or thin wall. It is available as either galvanized steel or aluminum. Electrical metallic tubing is to bend, cut, and ream and because it is threaded, all connectors and couplings are of the threadless type.

348-4 Use

(a) Exposed and Concealed. EMT can be installed exposed, concealed where not subject to severe physical damage, or buried in concrete without supplemental protection.

(b) Corrosion Protection. When installed in areas subject to severe corrosive influences, supplementary corrosion protection is required. However, because EMT is thinner than IMC or rigid metal conduit, corrosion can eat through the thinwall raceway much quicker.

Supplemental Corrosion Protection. At the point where EMT leaves the concrete or the soil, supplemental corrosion protection is necessary.

348-5 Uses *Not Permitted*

Electrical metallic tubing cannot be installed:

(1) Where, during installation or afterward, it will be subject to severe physical damage;

(2) Where protected from corrosion solely by enamel;

(3) In cinder concrete or cinder fill where subject to permanent moisture unless protected on all sides by a layer of noncinder concrete at least 2 in. thick or unless the tubing is at least 18 in. under the fill;

(4) In any hazardous (classified) location except as permitted by Sections 502-4, 503-3, and 504-20; or

(5) For the support of fixtures or other equipment except conduit bodies no larger than the largest trade size of the tubing, Fig. 25-1.

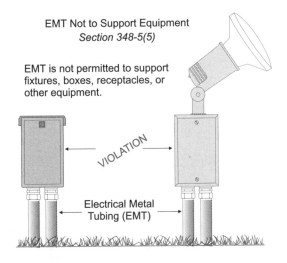

Fig. 25-1 EMT Can Not Be Used to Support Equipment

Where practicable, dissimilar metals in contact anywhere in the system shall be avoided to eliminate the possibility of galvanic action.

Exception. Aluminum fittings on steel electrical metallic tubing and steel fittings on aluminum electrical metallic tubing are permitted.

348-7 Size

Electrical metallic tubing is manufactured in trade sizes from ½ inch to 4 inches.

348-8 Number of Conductors

When determining the number of conductors in electrical metallic tubing, the area of the raceway for raceway fill must be limited to the percentages listed in Chapter 9, Table 1, see 300-17.

348-9 Reaming

Reaming is required to remove burrs and rough edges after the raceway has been cut. The raceway is often reamed with the back of pliers or with a screwdriver.

348-10 Coupling and Connectors

Threadless couplings and connectors must be made up tight to safely maintain an effective low

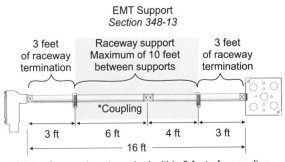

Fig. 25-2 EMT Must Be Secured within 3 Feet of Termination, and Every 10 Feet Thereafter

impedance ground-fault path [250-2(d), 250-96(a), and 300-10]. When EMT is installed in masonry or concrete, the installation must be concrete-tight. When used in wet locations or outdoors, the raceway must be raintight and arranged to drain [225-22, 230-53].

348-11 Bends

When bending EMT, it must not be damaged and there should not be a significant change to the internal diameter of the raceway (no kinks). The bending radius must comply with the requirement of Section 346-10, and a short radius bender is not permitted.

348-12 Number of Bends (360°)

To reduce the stress and friction on the conductor insulation during installation bends (including offsets) between pull points such as boxes and conduit bodies cannot exceed 360°.

348-13 Secured and Supports

Electrical metallic tubing must be securely fastened in place within 3 feet of every box, cabinet, or termination fitting. In addition the raceway must be supported at intervals not exceeding 10 feet, Fig. 25-2.

Electrical metallic tubing run horizontally through holes in wood or metal joists, rafters, or studs is considered supported when securely fastened within 3 feet of termination.

Note. Electrical metallic tubing is not required to be secured within 3 feet of each coupling.

Exception No. 1: No Structural Members. When structural members do not permit the raceway to be secured within 3 feet of a box or termination fitting, the raceway must be secured within 5 feet of the box or termination fitting. This is permitted only where the raceway is continuous from the box or termination fitting to the securing fitting, Fig. 25-3.

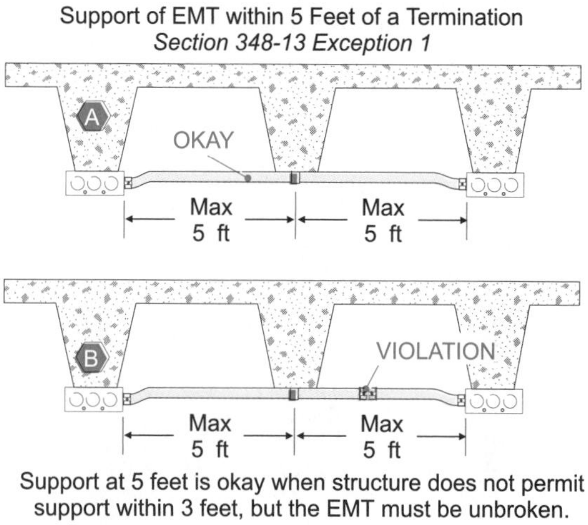

Support of EMT within 5 Feet of a Termination
Section 348-13 Exception 1

Support at 5 feet is okay when structure does not permit support within 3 feet, but the EMT must be unbroken.

Fig. 25-3   EMT Support Permitted to Be up to 5 Feet from Termination, If Necessary

SUMMARY

☐ Electrical metallic tubing is commonly called EMT or thin wall and is available as galvanized steel or aluminum. ☐ Electrical metallic tubing can be installed exposed, concealed, and in concrete without supplemental protection but cannot be installed in the earth. ☐ When installed in areas subject to severe corrosive influences, supplementary corrosion protection is required. ☐ Where practical, contact with dissimilar metals should be avoided to prevent galvanic action. ☐ Electrical metallic tubing cannot be used to support lighting fixtures or boxes. ☐ Electrical metallic tubing shall have a maximum of 360° between pull points, including all offsets. ☐ Electrical metallic tubing must be secured within 3 feet of every termination point and at intervals not exceeding 10 feet.

REVIEW QUESTIONS

1. What is galvanic action?

2. What are the limitations of electrical metallic tubing as compared to IMC or rigid?

3. What are the support requirements for electrical metallic tubing?

Unit 26

Article 350
Flexible Metal Conduit

OBJECTIVES

After studying this unit, the student should be able to understand:
- permitted uses for flexible metal conduit.
- support requirements for flexible metal conduit.
- when a separate equipment grounding conductor is required in flexible metal conduit.

350-2 Definition

Flexible metal conduit is a single spiral of interlocking metal, either steel or aluminum and it is commonly called "Greenfield" (the inventor), or simply "Flex."

350-4 Uses

Listed flexible metal conduit can be installed exposed or concealed where not subject to physical damage [350-5(7)].

Note. Flexible metal conduit is permitted in any length except for service raceways [230-43], fixture whips [410-67(c)], when used as an equipment grounding conductor [250-119(6) and 350-14], or where installed in ducts and plenums [300-22(b)].

350-5 Uses Not Permitted

Flexible metal conduit cannot be installed:

(1) In wet locations, unless the conductors are THWN, XHHW, THW, or TW and the raceway is installed in a way that will prevent water from entering the enclosure [225-10 and 225-22 Ex.].

(2) In hoistways, other than as permitted in Section 620-21(a)(1).

(3) In storage-battery rooms.

(4) In hazardous locations, except as permitted in Section 501-4(b).

(5) Where exposed to material having a deteriorating effect, such as oil or gasoline, unless the conductors are approved for the condition.

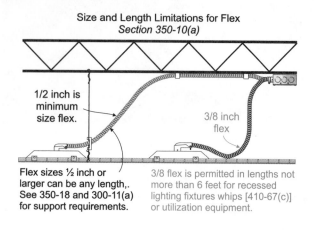

Fig. 26-1 Flex Limited to 6 Feet for the 3/8 Inch Size, No Length Limit for 1/2 Inch and Larger Flex

(6) Installations underground or embedded in poured concrete.

(7) Where subject to physical damage.

350-10 Size

(a) Minimum. Flexible metal conduit is manufactured in sizes from ½ inch to 4 inches.

3/8 Inch. 3/8 flexible metal conduit is permitted in lengths not to exceed 6 feet for recessed lighting fixture whips when installed according to Section 410-67(c), Fig. 26-1.

Note. There is no maximum length for 1/2 flex metal conduit, Fig. 26-1.

350-12 Number of Conductors

When determining the number of conductors in ½ inch through 4 inch flexible metal conduit, the area of the raceway must be used according to Chapter 9, Table 1, see Section 300-17.

Table 350-12 Fittings

Size AWG	Inside	Outside
No. 18	5	8
No. 16	4	6
No. 14	3	4
No. 12	2	3

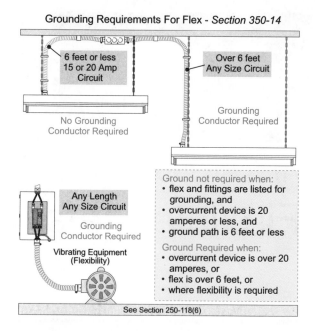

Fig. 26-2 Flexible Metal Conduit Can Be Used without an Equipment Grounding Conductor

3/8 Inch Flexible Metal Conduit. The number of conductors in 3/8 inch flexible metal conduit shall be according to Table 350-12.

350-14 Grounding

According to Section 250-118, flexible metal conduit is not listed as a safe grounding path for ground-fault current, except if, Fig. 26-2:

(1) The ground return path of the flexible metal conduit does not exceed 6 feet,

(2) The circuit conductors in the flexible metal conduit are protected by overcurrent protection devices rated 20 ampere or less, and

(3) The flexible metal conduit termination fittings are listed for grounding.

Note. All listed flexible metal conduit fittings sized from 3/8 inch through ¾ inch are listed for grounding.

Moving or vibrating equipment should have a separate equipment grounding conductor installed because vibrating and moving equipment tends to loosen fittings and locknuts which impairs the equipment grounding path, Fig. 26-2.

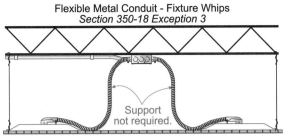

Flexible Metal Conduit - Fixture Whips
Section 350-18 Exception 3

Support
not required.

Support is not required for flex in lengths not exceeding
6 feet from a fixture terminal connection for tap connections
to light fixtures, [410-67(c)].

Fig. 26-3 Flexible Metal Conduit Used for Fixture
Whips Not Required to Be Secured or Supported

350-16 Bends

Bends between pull points such as boxes and conduit bodies cannot exceed 360° and the raceway must be installed so the conductors can be pulled or removed without damaging the conductor insulation.

Note. Article 350 contains no requirements on minimum bending radius as contained in the other raceway Articles.

350-18 Secured and Supports

Flexible metal conduit must be securely fastened within 12 inches of every box, cabinet, or termination fitting. In addition flex must be secured and supported at intervals not exceeding 4½ feet.

Exception No. 1: Fished. Where flex is fished, that portion of the raceway that cannot be reached is not required to be secured.

Exception No. 2: Where Flexibility Required. Securing the raceway is not required for lengths up to 3 feet from the last support to the termination where flexibility is required, such as for the connection to moving or vibrating equipment.

Note. An equipment grounding conductor should be installed in this application, see my note in Section 350-14.

Exception No. 3: Fixture Whips. Flexible metal conduit used as a fixture whip as permitted in Section 410-67(c) is not required to be secured or

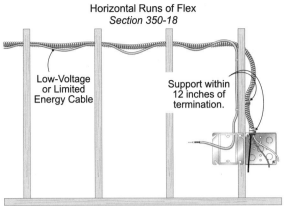

Horizontal Runs of Flex
Section 350-18

Low-Voltage
or Limited
Energy Cable

Support within
12 inches of
termination.

Flexible metal conduit run horizontally through framing members
is considered supported if the raceway is securely fastened
within 12 inches of termination.
Note: Separation is not required between low-voltage or limited-
energy cables, and raceways or cables containing power conductors.

Fig. 26-4 Horizontal Runs Must Be Secured
within 12 Inches of Termination

supported, Fig. 26-3.

Horizontal Runs. Flexible metal conduit run horizontally through framing members is considered supported if the raceway is securely fastened within 12 inches of termination, Fig. 26-4.

Note. Separation is not required between low-voltage or limited-energy cables and raceways or cables containing power conductors, Fig. 26-4.

System	Sections
CATV, MATV, CCTV	820-52(a)(2) Ex. 1
Class 2 and 3	725-54(a)(3) Ex. 1
Fire Alarm	760-54(a)(3) Ex. 1
Intrinsically Safe	504-30(a)(2) Ex. 1
Network-Powered	830-58(a)(2) Ex. 1
Radio and Television	810-18(b) Ex. 1
Telecommunications	800-52(a)(2) Ex. 1

350-20 Fittings

Only fittings listed for flexible metal conduit can be used with this raceway. This means you cannot use armored cable, metal-clad cable, or nonmetallic-sheathed cable fittings with flexible metal conduit, unless those fittings are also listed for use with flexible metal conduit.

Angle connector fittings cannot be installed in concealed locations.

350-22 **Trimming**

The cut ends of flex must be trimmed to remove the rough edges, but this is not necessary where fittings that thread into the raceway are used.

SUMMARY

☐ Flexible metal conduit is commonly called "Greenfield" or "Flex." ☐ Flexible metal conduit is permitted in wet locations if the conductors are TW, THW, XHHW, or THWN. ☐ Flexible metal conduit of any length is permitted in dropped ceiling spaces used for environmental air. ☐ There is no limit on the length of flex, except when used for service raceways, fixture whips, or installed in ducts and plenums. ☐ Flexible metal conduit is not approved as an equipment grounding conductor, except for 15 and 20 ampere rated circuits. ☐ Angle connector fittings are not permitted in concealed locations. ☐ Article 350 does not contain any requirements on the minimum bending radius of the raceway, but this Article does limit the number of bends to 360°. ☐ The number of conductors permitted in 3/8 inch flex is determined according to Table 350-12. ☐ Only listed fittings can be used with flexible metal conduit.

REVIEW QUESTIONS

1. Explain the uses permitted for flex.

2. Give a summary of the support requirements of flex.

3. When is a separate equipment grounding conductor required for flex?

Unit 27

Article 351

Liquidtight Flexible Metal Conduit and Liquidtight Flexible Nonmetallic Conduit

OBJECTIVES

After studying this unit, the student should be able to understand:
- why you would use liquidtight flexible conduit instead of flexible metal conduit.
- the requirements for installing liquidtight conduit outdoors.
- when liquidtight flexible conduit can be installed directly buried in the earth or concrete.

PART A. LIQUIDTIGHT FLEXIBLE METAL CONDUIT

351-2 Definition

Liquidtight flexible metal conduit is commonly called Sealtight® (a registered trade name), or simply "liquidtight." Liquidtight flexible metal conduit is of similar construction to flexible metal conduit, but has an outer liquidtight thermoplastic covering.

351-4 Use

(a) Permitted Use. Listed liquidtight flexible metal conduit can be installed either exposed or concealed at any of the following locations:

(1) Where flexibility or protection from liquids, vapors, or solids are required.

(2) Hazardous (classified) locations, such as:

- Class I, Division 2 Hazardous [501-4(b)]

- Class II, Division 1 [502-4(a)(2)]

- Class II, Division 2 [502-4(b)(2)]

- Class III, Division 1 [503-3(a)(2)].

(3) Directly buried in the earth if listed and marked for this purpose.

Note. When installed in other spaces used for environmental air such as a suspended ceiling, the raceway length must not exceed 6 feet [300-22(c) (1) Ex.].

There are no other *NEC* limits on the length of liquidtight metallic conduit, except when used for service raceways [230-43], or when used for fixture whips [410-67(c)].

(b) Not Permitted.   Liquidtight cannot be installed where exposed to physical damage.

351-5   Size

Liquidtight is manufactured in sizes from ½ inch to 4 inches.

Exception.   3/8 inch liquidtight can be used as permitted in Section 350-10(a).

351-6   Number of Conductors

(a) ½ Inch through 4 Inches.   When determining the number of conductors in liquidtight, the area of raceway fill must not exceed the percentages listed in Table 1 of Chapter 9, see Section 300-17.

(b) 3/8 Inch.   Use Table 350-12 to determine the number of conductors permitted in 3/8 inch liquidtight.

351-7   Fittings

Only fittings listed for liquidtight can be used and angle connectors cannot be installed in concealed locations.

351-8   Secured and Supports

Liquidtight used as a fixed raceway must be secured within 12 inches of every box, cabinet, or raceway termination. In addition liquidtight must be secured and supported at intervals not exceeding 4½ feet.

Exception No. 1: Fished.   Where liquidtight is fished, that portion of the raceway that cannot be reached is not required to be secured.

Exception No. 2: Flexibility Required.   Securing of the raceway is not required for lengths up to 3 feet from the last support to the raceway termination, where flexibility is required, such as for the connection to moving or vibrating equipment.

Exception No. 3: Fixture Whips.   Liquidtight conduit used as a 6-foot fixture whip as permitted in

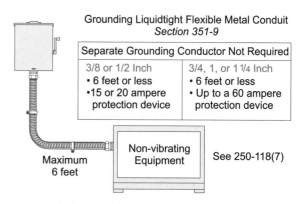

Note: An equipment ground is always required for liquidtight flexible nonmetallic conduit.

Fig. 27-1 Liquidtight Flexible Metal Conduit Can Serve as an Equipment Grounding Conductor

Section 410-67(c) is not required to be secured or supported.

351-9   Grounding

According to Section 250-118(7), liquidtight is not considered an equipment grounding conductor, except for, Fig. 27-1:

15 and 20 Ampere Circuits.   A separate grounding conductor is not required for 15 and 20 ampere circuits when installed in 3/8 or ½ inch liquidtight. The length of the liquidtight does not exceed 6 feet and the fittings are listed for grounding [250-118(7)].

25 to 60 Ampere Circuits.   A separate grounding conductor is not required for circuits not over 60 ampere when installed in ¾, 1, or 1¼ inch liquidtight. The length of the listed liquidtight cannot exceed 6 feet and the fittings must be listed for grounding [250-118(7)].

Note.   All liquidtight fittings up to 1¼ inch are listed for grounding.

When flexibility is required, such as to prevent the transmission of vibrations, a separate equipment grounding conductor must be installed. Vibrating and moving equipment tends to loosen fittings and locknuts, which impairs the equipment grounding path.

351-10 Bends and Fittings

The maximum bends between pull points cannot exceed 360°. Liquidtight must be installed so that the conductors can be pulled or removed without damaging the conductor insulation.

PART B. LIQUIDTIGHT FLEXIBLE NON-METALLIC CONDUIT

Requirements for liquidtight flexible nonmetallic conduit are the same requirements as those for liquidtight flexible metallic conduit.

351-23 Use

(a) Permitted Uses. Listed liquid flexible nonmetallic conduit can be installed exposed or concealed for the following applications:

(1) Where flexibility is necessary for installation, operation or maintenance.

(2) Protection from liquids, vapors, or solids.

(3) Outdoors where listed and marked.

(4) Direct burial where listed and marked.

(5) Liquidtight flexible nonmetallic conduit of the "gray type" can be installed in lengths over 6 feet, if secured in accordance with Section 351-27.

Note. There are 3 types of liquidtight flexible nonmetallic conduit listed in Section 351-22. Type (1) is orange, Type (2) is gray, and Type (3) is black.

(6) When part of listed manufactured pre-wired liquidtight flexible nonmetallic assemblies.

(b) Not Permitted. Liquidtight flexible nonmetallic conduit cannot be installed:

(1) Exposed to physical damage.

(2) The ambient temperature or operating temperature of the conductors exceed the temperature rating of the raceway.

(3) Longer than 6 feet, except liquidtight flexible nonmetallic conduit, as defined in Section 351-23(a)(5).

In addition, liquidtight flexible nonmetallic conduit is not permitted for:

• Ducts or plenums [300-22(b)]

• Areas used for environmental air [300-22(c)]

• Hazardous locations such as Class I (gas or vapor) [501-4], Class II (dust) [502-4(a)], Class III (fiber) [503-3(a)(2)]

• Pool and spa equipment [680-25].

351-27 Secured and Support

When liquidtight flexible nonmetallic conduit is installed as a fixed raceway, it shall be secured at intervals not exceeding 3 feet, and within 12 inches of termination.

SUMMARY

☐ Liquidtight flexible metal conduit is similar to flexible metal conduit, but has an outer liquidtight thermoplastic covering. ☐ Liquidtight flexible conduit can be installed exposed or concealed. ☐ There is no limit on the length of liquidtight, except for service raceways, fixture whips, or where installed in ducts, plenums, or other spaces used for environmental air. ☐ Liquidtight conduit must be secured (strapped) within 12 inches of every box, cabinet, or conduit termination, and at intervals not exceeding 4½ feet. ☐ Extreme cold may cause liquidtight nonmetallic conduit to become brittle and subject to physical damage. ☐ When an equipment grounding conductor is required, it must be installed according to Section 250-102. ☐ When liquidtight flexible nonmetallic conduit is installed as a fixed raceway, it shall be secured at intervals not exceeding 3 feet and within 12 inches on each side of every termination.

REVIEW QUESTIONS

1.   Why would you use liquidtight instead of flexible metal conduit?

2.   Give a brief description of the installation of liquidtight metal and nonmetallic conduit installed in ducts, plenums, and other spaces used for environmental air.

3.   What are some of the requirements when installing liquidtight outdoors?

4.   Can liquidtight or liquidtight flexible nonmetallic conduit be installed directly buried in the earth or in concrete?

5.   How many No. 12 THHN conductors are permitted in 3/8 liquidtight?

6.   What are the support requirements for liquidtight?

7.   When can liquidtight be used as a grounding conductor?

8.   What are the differences in the *NEC* requirements for liquidtight metal and nonmetallic conduits?

Unit 28

Article 352

Surface Raceways—Metallic or Nonmetallic

OBJECTIVES

After studying this unit, the student should be able to understand:
- the main reasons for using surface raceways.
- whether 277 or 480 volt lighting circuits can be installed in surface raceways.
- the NEC installation requirements for surface raceways, including uses, conductors, splices, fittings, and grounding.
- the main differences between metallic and nonmetallic surface raceways.

Surface raceways called Wiremold® and Walker Duct® (registered trade names), are available in different shapes and sizes, and can be mounted on walls, ceilings, or floors. Surface raceways generally have removable covers, which eliminates the need for wire pulling, Fig. 28-1.

Enclosures (boxes) that are approved for switches, receptacles, lighting fixtures, and other devices are identified by the markings on their packaging (cartons). These markings identify the type of surface raceway with which the enclosure can be used. Surface raceways are particularly helpful for additions to an existing location, and floor surface raceways are often used in large open areas such as offices.

Some surface raceways have two or more separate compartments, which permit the separation of cables and conductors as required in Section 352-6.

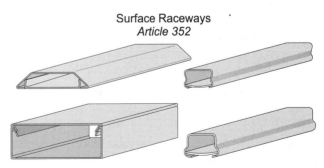

Surface Raceways
Article 352

Surface raceways are available in different shapes and sizes, and can be mounted on walls, ceilings, and floors.

Fig. 28-1 Types of Surface Raceways

261

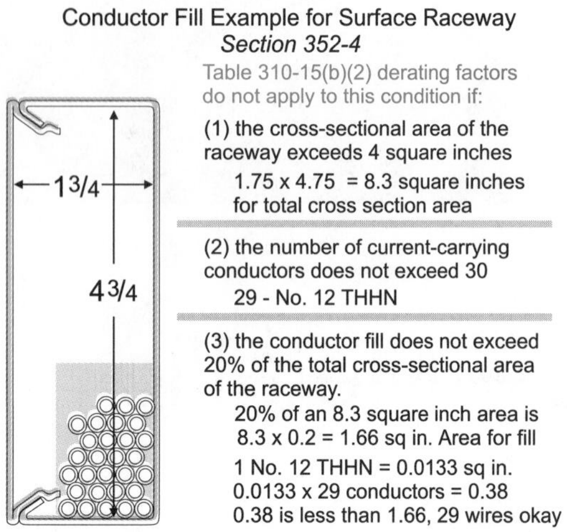

Conductor Fill Example for Surface Raceway
Section 352-4

Table 310-15(b)(2) derating factors do not apply to this condition if:

(1) the cross-sectional area of the raceway exceeds 4 square inches

1.75 x 4.75 = 8.3 square inches for total cross section area

(2) the number of current-carrying conductors does not exceed 30

29 - No. 12 THHN

(3) the conductor fill does not exceed 20% of the total cross-sectional area of the raceway.

20% of an 8.3 square inch area is
8.3 x 0.2 = 1.66 sq in. Area for fill

1 No. 12 THHN = 0.0133 sq in.
0.0133 x 29 conductors = 0.38
0.38 is less than 1.66, 29 wires okay

Fig. 28-2 Install No More than 30 Conductors at Any Cross-Sectional Area

PART A. SURFACE METAL RACEWAYS

352-1   Uses

(a) **Uses Permitted.** The use of surface metal raceways shall be permitted:

(1) In dry locations,

(2) In any hazardous (classified) location as permitted in Section 501-4(b) Exception, or

(3) Under raised floors, as permitted in Section 645-5(d)(2).

(b) **Uses Not Permitted.** Surface metal raceways shall not be used:

(1) Subject to severe physical damage,

(2) Where the voltage is 300 volt or more between conductors unless the metal has a thickness of at least 0.040 inch,

(3) Subject to corrosive vapors,

(4) In hoistways, or

(5) Concealed.

Note. Surface raceways can be installed above suspended ceiling because this area is considered to be an exposed location [Article 100].

352-3   Size of Conductors

The size, number, and type of conductors permitted in a surface raceway is marked on the raceway packaging.

Note. Because partial packages are often purchased, you may not always get this information.

352-4   Number of Conductors

The number of conductors permitted in a surface raceway shall not be more than the number for which the raceway has been listed, Fig. 28-2.

Ampacity. Conductor ampacity derating factors of Table 310-15(b)(2) do not apply to conductors in metal surface raceways when:

• The raceway exceeds 4 square inches in cross-sectional area

• The number of current carrying conductors do not exceed 30

• The conductor fill does not exceed 20 percent of the total cross-sectional area of the raceway area

352-5   Through Walls and Floors

Unbroken lengths of surface metal raceways can pass through dry walls, partitions, and floors.

352-6   Separate Compartments

Combination raceways have separate compartments within a single raceway with one compartment used for power and lighting and the other for low-voltage and limited-energy cables.

The interior finish of each separate compartment shall be of different and contrasting colors, and the relative position of these compartments must be maintained.

The following *Code* Sections prohibit the mixing of low-voltage and limited-energy conductors with power, Class 1, or nonpower-limited fire alarm circuit conductors:

System	Section
CATV	820-10(f)(1)
Control and Signaling	725-54(a)(1)
Fire Alarm	760-54(a)(1)
Intrinsically Safe	504-30(a)(2)
Instrument Tray Cable	727-5
Network Broadband	830-58(c)
Radio and Television	810-18(c)
Sound Systems	640-9(c)
Telecommunication	800-52(a)(1)

352-7 Splices and Taps

Splices and taps shall be accessible and shall not fill the raceway more than 75 percent of its cross-sectional area, including conductors that pass through the splice or tap point [300-15(a)].

352-9 Grounding

Surface raceway fittings must be mechanically and electrically joined together in a manner that does not subject the conductors to abrasion. Surface raceways that allow a transition to another wiring method such as knockouts for connecting conduits must have a means for the termination of an equipment grounding conductor.

PART B. SURFACE NONMETALLIC RACEWAYS

For all practical purposes, nonmetallic surface raceways must comply with the same requirements as metallic surface raceways.

352-21 Description

Surface nonmetallic raceways is manufactured as a flame-retardant plastic, resistant to moisture and chemical atmospheres. The raceway must be resistant to crushing and distortion under conditions likely to be encountered.

352-22 Use

(a) Uses Permitted. Nonmetallic surface raceways shall be installed in dry locations.

(b) Uses Not Permitted. Nonmetallic surface raceway shall not be installed:

(1) Concealed,

(2) Subject to physical damage,

(3) Where the voltage between conductors exceeds 300 volt,

(4) In hoistways,

(5) In hazardous (classified) location except as permitted in Section 501-4(b) Exception, or

(6) Where the ambient temperature or the operating temperature of the conductors exceeds the temperature rating of the raceway.

> **Note.** Article 352 contains no requirement on raceway supports, so installers should follow manufacturer's instructions.

PART C. STRUT-TYPE CHANNEL RACEWAY

352-41 Uses Permitted

Strut-type channel raceway can be installed concealed, exposed, in damp locations, locations subject to corrosive vapors when protected by finish judged suitable for the condition.

352-42 Uses Not Permitted

Strut-type channel raceway is not permitted in hazardous (classified) locations.

352-45 Number of Conductors Permitted

The number of conductors permitted in strut-type channel raceway shall not exceed the percentage fill values listed in Table 352-45. In addition, raceways containing internal joints shall not exceed 25 percent fill, and raceways with external joints can be filled up to 40 percent.

SUMMARY

☐ Surface raceways can be mounted on walls, ceilings, or floors. ☐ Surface raceways can be installed exposed only in dry locations. ☐ The size, number, and type of conductors permitted in a surface raceway will be marked on the raceway packaging. ☐ Surface raceways that allow a transition to another wiring method must have a means for the termination of an equipment grounding conductor. ☐ Nonmetallic surface raceways must comply with the same requirements as metallic surface raceways. ☐ Article 352 contains no requirement on supports, but be sure to read the instructions included with the surface raceway. ☐ Article 352 contains the requirements for strut-type channel raceway systems.

REVIEW QUESTIONS

1. What are the main reasons for using surface raceways?

2. Can 277 or 480 volt lighting circuits be installed in surface raceways?

3. Briefly describe the *NEC* installation requirements for surface raceways, such as uses, conductors, splices, fittings, and grounding.

4. What are the main differences between metallic and nonmetallic surface raceways?

Unit 29

Article 353
Multioutlet Assembly

OBJECTIVE

After studying this unit, the student should be able to understand:
* the similarities between surface raceways and multioutlet assemblies.

In the 1996 Code a multioutlet assembly was defined as a surface or flush raceway, but testing laboratories list freestanding raceways (power poles) which are intended to extend from the floor through a suspended ceiling in an office area. This product is labeled by the testing laboratories as a multioutlet assembly, which conflicts with the definition contained in the 1996 NEC. The new definition identifies a multioutlet assembly as a type of surface, flush, "or a freestanding (power pole) raceway" designed to hold conductors and receptacles, assembled in the field or at the factory, Fig. 29-1.

353-2 Uses

(a) Permitted. Multioutlet assembly shall only be permitted in dry locations.

(b) Not Permitted. Multioutlet assembly shall not be installed:

(1) Concealed,

(2) Subject to severe physical damage,

(3) Where the voltage is 300 volt or more between conductors unless the metal has a thickness of at least 0.040 inch,

(4) Subject to corrosive vapors,

(5) In hoistways,

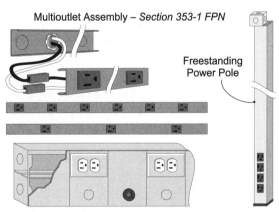

Multioutlet Assembly – *Section 353-1 FPN*

Freestanding Power Pole

Multioutlet Assembly: (Article 100 Definition) A type of surface, flush, or freestanding (power pole) raceway designed to hold conductors and receptacles, assembled in the field or at the factory.

Fig. 29-1 Multioutlet Assembly Types

(6) In any hazardous (classified) locations except as permitted in Section 501-4(b).

353-3 Through Partitions

Metal multioutlet assemblies can pass through a dry partition, provided no receptacle is concealed in the wall, and the cover of the exposed portion of the system can be removed.

Note. The feeder and service load of multioutlet assemblies shall be according with the requirements in Section 220-3(b)(8).

SUMMARY

☐ Multioutlet assemblies resemble metallic surface raceways with integrally installed receptacles. ☐ Multioutlet assemblies can be installed exposed only in dry locations where not subject to severe physical damage. ☐ Multioutlet assemblies cannot be installed where subject to corrosive vapors or in hazardous (classified) locations. ☐ This wiring method is not a raceway. ☐ Multioutlet assemblies can be installed through a dry partition.

REVIEW QUESTION

1. What are the similar requirements between surface raceways and multioutlet assemblies?

Unit 30

Article 362

Wireways—Metallic or Nonmetallic

OBJECTIVES

After studying this unit, the student should be able to understand:
- the main reasons for using wireways.
- the NEC installation requirements for wireways, such as permitted uses, conductors, splices, fittings, and grounding.

PART A. METALLIC WIREWAYS

362-1 Definition

Wireways are troughs that are used to form an exposed raceway system. Wireways are designed to permit extensions and taps. The conductors are laid in after the raceway has been installed as a complete system, Fig. 30-1.

Note. Many people call this wiring method an auxiliary gutter, which it is not.

362-2 Uses

(a) Permitted. Wireways can be installed:

(1) Exposed,

(2) Concealed as permitted for sound systems Article 640,

(3) Hazardous (classified) locations as permitted in Sections 501-4(b), 502-4(b), or 504-20.

(b) Not Permitted. Wireways cannot be installed where subject to severe physical damage or corrosive vapors.

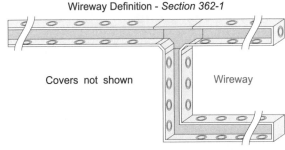

Wireway Definition - *Section 362-1*

Covers not shown   Wireway

Wireways are sheet metal or nonmetallic [362-14] troughs with hinged or removable covers for housing and protecting electrical wires and cables and in which conductors are laid in place after the wireway has been installed as a complete system.

Fig. 30-1 Wireway Definition

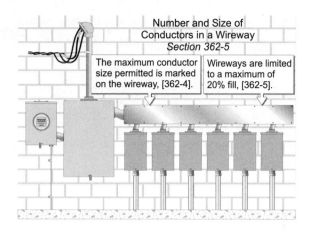

Fig. 30-2 Number and Size of Conductors
on a Wireway

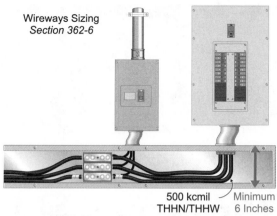

Dimensions corresponding to Table 373-6(a) shall apply
where insulated conductors are deflected within a wireway.

Fig. 30-3 Wireways Sized Based on the
Bending Dimensions Listed in Table 373-6(a)

362-4   Conductor - Maximum Size

The maximum size conductor permitted in a
wireway shall not be larger than that for which the
wireway is designed, Fig. 30-2.

362-5   Conductors - Maximum Number

The maximum number of conductors permitted
in a wireway is limited to 20 percent of the cross-
sectional area of the wireway, Fig. 30-2.

Ampacity. When more than 30 current-carrying conductors are installed in any cross-sectional
area of the wireway, the conductor allowable
ampacity must be reduced according to adjustment
factor listed in Table 310-15(b)(2).

Note. Signaling and motor control conductors are not considered current-carrying.

362-6   Wireway Sizing

Where conductors are bent within a wireway,
the wireway must be sized to meet the bending
radius requirements contained in Section 373-6(a)
based on one conductor per terminal.

Example: A wireway shall provide 6 inches
for conductor bending [Table 373-6(a)] if the wireway contains 500 kcmil conductors, Fig. 30-3.

So that conductor insulation will not be damaged when conductors are pulled in or out of
wireways, the distance between the conductor

entries shall not be less than 6 times the trade diameter of the raceway or cable connector.

Example: A wireway with 500 kcmil conductors pulled in and out must have the 3 inch raceways
spaced not less 18 inches apart, Fig. 30-4.

Note. This rule is not very effective in protecting conductor insulation when using the
wireway as a *pull box*. If this were a junction or pull box a minimum distance of 18
inches (6 × 3 inches) would be required
from the cable or raceway entry to the
"opposite wall" [370-28(a)].

362-7   Splices and Taps

Splices and taps shall be accessible and shall
not fill the wireway more than 75 percent of its

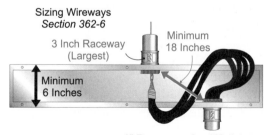

Where the same conductors enter and then exit a wireway
through a raceway, the distance between those raceway
entries shall not be less than 6 times the larger raceway.

Fig. 30-4 Distance between Raceways

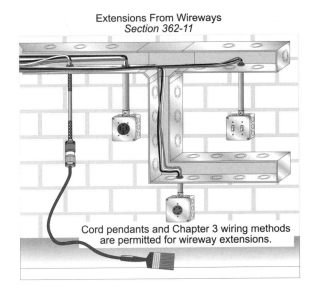

Extensions From Wireways
Section 362-11

Cord pendants and Chapter 3 wiring methods
are permitted for wireway extensions.

Fig. 30-5 Extensions from Wireways

362-8 Supports

Wireways shall be supported where run horizontally at each end and at intervals not to exceed 5 feet.

362-11 Extensions from Wireways

Wiring methods connected to the wireway are called "extensions." Extensions are permitted with any Chapter 3 wiring method or cord pendant. Extensions containing an equipment grounding conductor must have the grounding conductor terminate on to an approved terminal [250-8] onto a clean conductive surface [250-118], Fig. 30-5.

cross-sectional area, including conductors that pass through the splice or tap point.

PART B. NONMETALLIC WIREWAYS

Installation rules for nonmetallic wireway rules are the same as those for metal wireways.

SUMMARY

☐ Wireways are troughs used to form an exposed raceway system. ☐ Wireways can be installed exposed only where not subject to physical damage. ☐ Installation rules for nonmetallic wireway rules are the same as those for metal wireways.

REVIEW QUESTIONS

1. What are the main reasons for using wireways?

2. Briefly describe the *NEC* installation requirements for wireways, such as permitted uses, conductors, splices, fittings, and grounding.

Unit 31

Article 370

Outlet, Device, Pull and Junction Boxes, Conduit Bodies, and Fittings

OBJECTIVES

After studying this unit, the student should be able to understand:
- when round boxes are prohibited.
- when nonmetallic boxes can be used with metal raceways or cables.
- what must be done when equipment grounding conductors enter a metallic box.
- what the minimum cubic inch capacity is required for conduit bodies.
- when nonmetallic-sheathed cable is not required to be secured to a box.
- the two methods permitted for surface extensions.
- when boxes can be supported by raceways only.
- when a raceway can be used to support a conduit body with a fixture.
- how to size outlet boxes and conduit bodies.
- how to size pull and junction boxes.

PART A. **SCOPE AND GENERAL**

370-1 **Scope**

Article 370 contains installation requirements for outlet boxes, conduit bodies, pull and junction boxes.

370-3 **Nonmetallic Boxes**

Nonmetallic boxes can be used only with non-metallic cables and raceways.

Exception No. 1: Bonding Jumpers. Metal raceways and cables can be used with nonmetallic boxes if the metal raceways or cables are bonded together inside the box.

Exception No. 2: Integral Bonding Means. Listed nonmetallic boxes that contain an integral bonding means that bonds all threaded entries together can be used with metal raceways and cables.

370-4 **Metal Boxes**

Metal boxes must be grounded by an equipment grounding conductor, as listed in Section 250-118. To ensure the proper grounding of a metal box, all metal raceway fittings must be mechani-

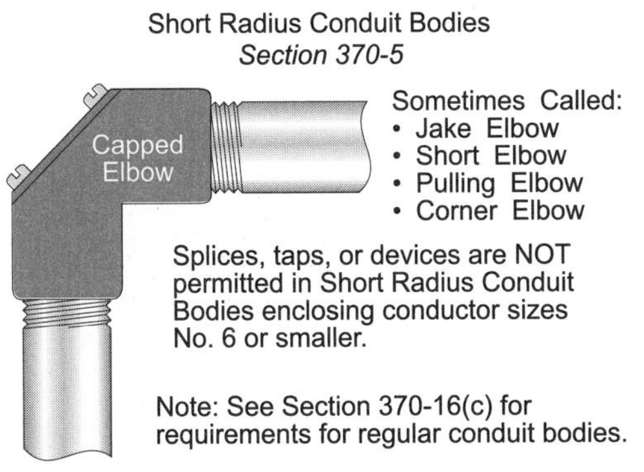

Short Radius Conduit Bodies
Section 370-5

Capped Elbow

Sometimes Called:
• Jake Elbow
• Short Elbow
• Pulling Elbow
• Corner Elbow

Splices, taps, or devices are NOT permitted in Short Radius Conduit Bodies enclosing conductor sizes No. 6 or smaller.

Note: See Section 370-16(c) for requirements for regular conduit bodies.

Fig. 31-1 Splices Not Permitted in Short Radius Conduit Bodies

cally as well as electrically continuous [300-10, 300-12]. Equipment grounding conductors must be spliced together within the box and a connection is required from the grounding conductors to the meal box [250-148].

370-5   Short Radius Conduit Bodies

Conduit bodies such as capped elbows, handy ells, and service-entrance elbows for conductors No. 6 or smaller shall not contain splices, taps, or devices, Fig 31-1.

PART B.   INSTALLATION

370-15   Damp, Wet, or Hazardous (Classified) Locations

(a) Damp or Wet Locations. Boxes and conduit bodies installed in damp or wet locations must prevent moisture from entering or accumulating in the enclosures and must be listed for use in wet locations.

Note. Raceways installed in wet locations must be raintight and arranged to drain [225-22] and must use compression fittings. Enclosures for switches [380-4], fixtures [410-4(a)], and receptacles [410-57] must be listed for use in wet locations, Fig. 31-2.

CAUTION: Some inspectors permit set screw connectors (not raintight) in wet locations if the location of the fitting is such that moisture will not enter the enclosure such as at the bottom of the enclosure, Fig. 31-2.

(b) Hazardous Locations. Boxes installed in hazardous (classified) locations must comply with the requirements contained in Sections 501-4, 502-4, and 503-3(a)(1), Fig. 31-3.

370-16   Number of Conductors in Boxes and Conduit Bodies

Boxes and conduit bodies shall be of sufficient size to provide free space for all conductors. Boxes and conduit bodies containing conductors No. 4 must be installed in accordance with the requirements contained in Sections 300-4(f) and 370-28.

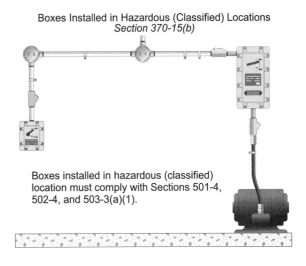

Boxes Installed in Hazardous (Classified) Locations
Section 370-15(b)

Boxes installed in hazardous (classified) location must comply with Sections 501-4, 502-4, and 503-3(a)(1).

Fig. 31-3 Boxes in Hazardous (Classified) Locations Must Comply with Chapter 5 Requirements

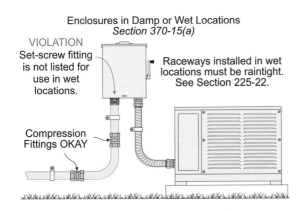

Enclosures in Damp or Wet Locations
Section 370-15(a)

VIOLATION
Set-screw fitting is not listed for use in wet locations.

Raceways installed in wet locations must be raintight. See Section 225-22.

Compression Fittings OKAY

Fig. 31-2 Enclosures in Damp or Wet Locations

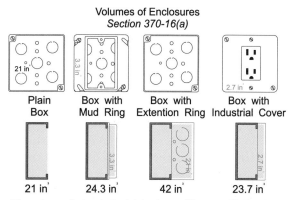

Volumes of Enclosures
Section 370-16(a)

Plain Box	Box with Mud Ring	Box with Extention Ring	Box with Industrial Cover
21 in³	24.3 in³	42 in³	23.7 in³

The volume of a box is the total volume of its assembled parts, including plaster rings, industrial (domed) covers, and extension rings. Extension rings are not required to be marked.

Fig. 31-4 Volume of a Box Is the Total Volume of Its Assembled Parts

(a) **Volume of Enclosure.** The volume of a box is the total volume of its assembled parts, including *plaster rings*, industrial (domed) covers, and extension rings. The total volume includes only those fittings that are marked with their volume in cubic inches, Fig. 31-4.

(1) **Standard Boxes.** The volume capacity of standard electrical boxes and *extension rings* are not required to be marked with their cubic inch capacity if their size is listed in Table 370-16(a).

(2) **Other Boxes.** Boxes not listed in Table 370-16(a) and nonmetallic boxes shall be marked by the manufacturer with their cubic inch capacity.

(b) **Box Fill Calculations.** The total conductor volume is determined by adding the sum of the volumes in (1) through (5) that follows.

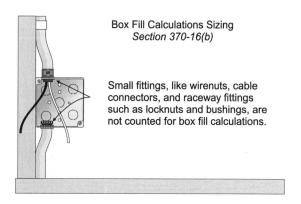

Box Fill Calculations Sizing
Section 370-16(b)

Small fittings, like wirenuts, cable connectors, and raceway fittings such as locknuts and bushings, are not counted for box fill calculations.

Fig. 31-5 Box Fill Calculations Sizing Does Not Include Small Fitting

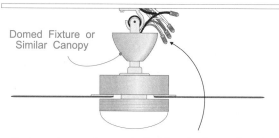

Conductors Not Counted for Box Fill
Section 370-16(b)(1) Exception

Domed Fixture or Similar Canopy

Equipment grounding conductor(s) and not more than 4 fixture wires sizes No. 18 or 16 can be omitted from the calculation where they enter a box from a domed canopy.

Fig. 31-6 Fixture Wires from a Canopy Not Counted for Box Fill

Note. *Wirenuts*, cable connectors, and raceway fittings such as locknuts or bushing are not counted, Fig. 31-5.

(1) **Conductors.** Each conductor running through the box without splice and each conductor that terminates in the box is considered as one conductor volume. Conductors such as pigtails that originate and terminate within the box, are not counted.

Exception: Conductors Not Counted. Equipment grounding conductor(s) and not more than 4 fixture wires (smaller than No. 14) can be omitted from the calculations where they enter the box from a domed fixture or similar canopy, such as a ceiling paddle fan canopy, Fig. 31-6.

Note. This exception permits pancake (½ inch deep) boxes to be used with lighting fixtures or paddle fans that have a domed canopy.

(2) **Cable Clamps.** Where 1 or more internal cable clamps are installed, the volume allowance shall be 1 conductor. Based on the largest conductor entering the box, Fig. 31-7.

(3) **Fixture Stud and Hickey.** Where 1 or more fixture studs or fixture *hickeys* are installed, the volume allowance shall be 1 conductor for each type. Based on the largest conductor entering the box, Fig. 31-8.

(4) **Device Strap (Yoke).** For each *yoke* or strap containing 1 or more devices, the volume

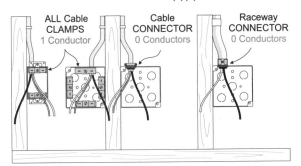

Box Fill - Cable Clamps and Connectors
Section 370-16(b)(2)

ALL Cable CLAMPS
1 Conductor

Cable CONNECTOR
0 Conductors

Raceway CONNECTOR
0 Conductors

One or more cable Clamps count as a total of one conductor (largest entering box).

One or more cable or raceway Connectors do not count for box fill calculations.

Fig. 31-7 Cable Clamps Count As One Conductor, Cable Connectors Do Not Count for Box Fill

allowance shall be 2 conductors. Based on the largest conductor that terminates on the strap or yoke, Fig. 31-9.

(5) Grounding Conductor. Where 1 or more equipment grounding conductors or bonding jumpers enter a box [250-148], the volume allowance shall be 1 conductor. Based on the largest grounding conductor entering the box, Fig 31-10.

An additional volume of 1 conductor shall apply for the second set of grounding conductor(s) installed for an isolating ground receptacle used to reduce electrical noise in accordance with Section 250-146(d).

Question. How many No. 14 THHN conductors could be pulled through a 4 inch square × 2-1/8 inch deep box, with a plaster ring of 3.6 cubic

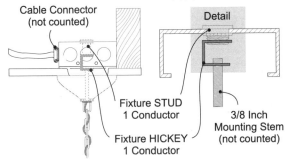

Box Fill - Conductor Equivalents for Fixture Stud and Hickey
Section 370-16(b)(3)

Cable Connector (not counted)

Detail

Fixture STUD
1 Conductor

Fixture HICKEY
1 Conductor

3/8 Inch Mounting Stem (not counted)

Fig. 31-8 Fixture Stud and Hickey Each Count As One Conductor

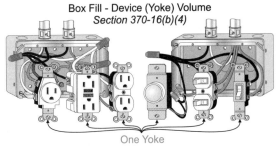

Box Fill - Device (Yoke) Volume
Section 370-16(b)(4)

One Yoke

Each individual yoke (strap) counts as a two conductor volume, based on the largest conductor connected on that device.

Fig. 31-9 Each Yoke Containing Devices or Equipment Count As Two Conductors

inches? The total volume of the box is 30.3 cubic inches + 3.6 cubic inches = 33.9 cubic inches according to Table 370-16(a) and it contains:

2 duplex receptacles
5 No. 12 THHN
2 No. 12 bare grounding conductors.

Answer. 5 conductors
Step 1. Conductor Volume
No. 12 THHN
5 conductors × 2.25 = 11.25 cubic inch

Devices, 2 straps × 2 conductors
4 conductors × 2.25 = 9 cubic inch

Grounding Conductor = 2.25 cubic inch

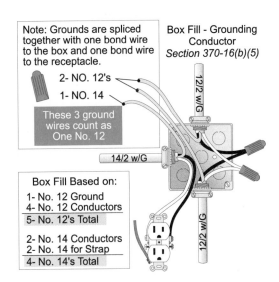

Note: Grounds are spliced together with one bond wire to the box and one bond wire to the receptacle.

Box Fill - Grounding Conductor
Section 370-16(b)(5)

2- NO. 12's
1- NO. 14

These 3 ground wires count as One No. 12

12/2 w/G

14/2 w/G

12/2 w/G

Box Fill Based on:
1- No. 12 Ground
4- No. 12 Conductors
5- No. 12's Total

2- No. 14 Conductors
2- No. 14 for Strap
4- No. 14's Total

Fig. 31-10 One or More Grounding Conductors Count As One Conductor

Step 2.

Conductors	11.25 cubic inch
Yokes	9.00 cubic inch
Grounding Conductor	2.25 cubic inch
	22.50 cubic inch

Step 3. Box volume
30.3 + 3.6 = 33.90 cubic inch
Spare Space   33.90 − 22.50 = 11.4 cubic inch

Step 4. No. 14 permitted in spare space
11.4 cu/in/2.00 cu/in = 5 conductors

(c) Conduit Bodies.

(2) Splices and Taps. Conductors No. 6 and smaller can be spliced within a conduit body that is marked with its cubic inch capacity [300-15(a)]. The number of conductors permitted in the conduit body shall be according to the requirements contained in Section 370-16(b).

Note. This subsection does not apply to short radius conduit bodies [370-5].

Question. How many No. 12 THHN conductors can be spliced in an LB conduit body that has a marked volume capacity of 15 cubic inches, Fig. 31-11?

Answer. Yes
15 cubic inches/2.25 = 6 conductors

370-17 Conductors Entering Boxes or Conduit Bodies

(b) Metal Boxes or Conduit Bodies. Raceways and cables must be mechanically fastened to metal boxes or conduit bodies by fittings designed for the wiring method [300-10, 300-12, 300-15(a)].

(c) Nonmetallic Boxes. Raceways and cables must be securely fastened to nonmetallic boxes or conduit bodies by fittings designed for the wiring method. When nonmetallic-sheathed cable is installed, it must extend at least ¼ inch into the nonmetallic box.

Note. Two nonmetallic-sheathed cables can be installed in a single cable connector, because most have been listed for this purpose.

Exception: Single Gang Boxes. Nonmetallic-sheathed cables installed in a 2¼ inch × 4 inch device box are not required to be secured to the box, but the cable must be securely fastened within 8 inches of the box.

370-18 Unused Openings

Each unused box and conduit body opening must be closed with a fitting identified for this purpose [110-12(a)], Fig. 31-12.

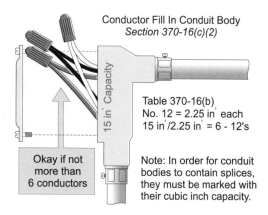

Conductor Fill In Conduit Body
Section 370-16(c)(2)

15 in³ Capacity

Table 370-16(b)
No. 12 = 2.25 in³ each
15 in³/2.25 in³ = 6 - 12's

Okay if not more than 6 conductors

Note: In order for conduit bodies to contain splices, they must be marked with their cubic inch capacity.

Fig. 31-11 Splices Permitted in Conduit Bodies

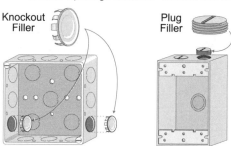

Unused Openings In Boxes - *Section 370-18*

Knockout Filler

Plug Filler

Unused cable or raceway openings in boxes and conduit bodies must be closed with a fitting identified for this purpose. Also see Sections 110-12(a) and 373-4.

Fig. 31-12 Unused Openings in Boxes Must Be Properly Closed or Sealed

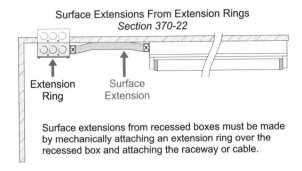

Surface Extensions From Extension Rings
Section 370-22

Extension Ring

Surface Extension

Surface extensions from recessed boxes must be made by mechanically attaching an extension ring over the recessed box and attaching the raceway or cable.

Fig. 31-13 Surface Extensions Must Be Made from Extension Ring

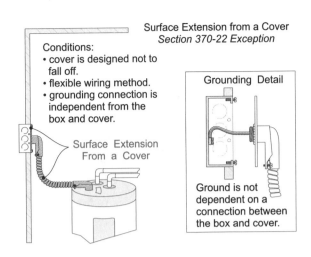

Surface Extension from a Cover
Section 370-22 Exception

Conditions:
• cover is designed not to fall off.
• flexible wiring method.
• grounding connection is independent from the box and cover.

Surface Extension From a Cover

Grounding Detail

Ground is not dependent on a connection between the box and cover.

Fig. 31-14 Surface Extension Permitted from a Cover if Flexible Wiring Method Used

370-20 Boxes Installed in Walls or Ceilings

Boxes installed in noncombustible materials such as drywall, concrete, or tile must have the front edge of the box or plaster ring set back no more than ¼ inch from the finished surface.

Boxes installed in combustible materials such as wood paneling must have the front edge of the box or plaster ring flush with or extend out from the finished surface.

Note. Plaster (mud) rings come in many depths to meet the above requirements.

370-21 Gaps or Spaces around Boxes

Gaps or spaces more than 1/8 inch around recessed boxes in plaster, drywall, or plasterboard surfaces must be filled.

370-22 Surface Extensions

Surface extensions from recessed boxes must be made by mechanically attaching an extension ring over the recessed box and attaching the raceway or cable to the extension ring, Fig. 31-13.

Note. A metal extension ring attached to a grounded recessed metal box is considered grounded.

Exception: Extensions from Covers. Surface extensions can be made from a cover if the cover is designed so it is unlikely to fall off or be removed if

its securing means becomes loose. The wiring method must include an equipment grounding conductor which is not dependent on the cover/box connection, Fig. 31-14.

Note. Only a flexible wiring method can be used for extensions attached to covers.

370-23 Supports of Boxes and Conduit Bodies

Boxes must be securely supported by any one of the methods listed in (a) through (h) below:

(a) Fastened to the Surface. Boxes can be fastened to any surface that provides adequate support.

(b) Attached to Structural Members. Boxes can be attached to the building structural members or from the earth's surface by the use of a suitable brace.

(c) Secured to Finished Surface. Boxes can be rigidly secured such as drywall or plaster walls or ceilings by clamps, anchors, or fittings identified for the use.

(d) Suspended Ceilings. An enclosure mounted to structural or supporting elements of a suspended ceiling shall be not more than 100 cubic inches in size and shall be securely fastened in place in one of the following ways:

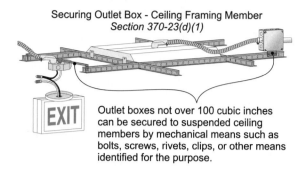

Securing Outlet Box - Ceiling Framing Member
Section 370-23(d)(1)

EXIT

Outlet boxes not over 100 cubic inches
can be secured to suspended ceiling
members by mechanical means such as
bolts, screws, rivets, clips, or other means
identified for the purpose.

Fig. 31-15 Outlet Boxes Permitted to Be Secured
to Suspended Ceiling Framing Members

(1) Ceiling Framing Members. Outlet boxes can be secured to suspended ceiling framing members by mechanical means such as bolts, screws, rivets, clips, or other means identified for the suspended ceiling framing member(s), Fig. 31-15.

Note. Section 410-16(c) permits lighting fixtures to be supported to ceiling framing members.

(2) Independent Support Wires. Outlet boxes can be secured with fittings identified for the purpose to independent support wires, which are taut and secured at both ends in accordance with Section 300-11(a), Fig. 31-16.

Note. See Section 300-11(a) on the use of independent support wires for raceways and cables.

(e) Raceways for Support—Boxes without Devices or Fixtures. Boxes that do not contain

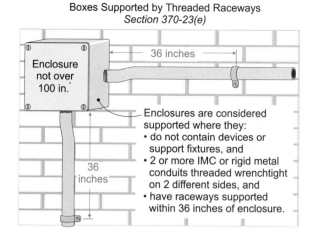

Boxes Supported by Threaded Raceways
Section 370-23(e)

36 inches

Enclosure
not over
100 in.

36
inches

Enclosures are considered
supported where they:
• do not contain devices or
  support fixtures, and
• 2 or more IMC or rigid metal
  conduits threaded wrenchtight
  on 2 different sides, and
• have raceways supported
  within 36 inches of enclosure.

Fig. 31-17 Boxes Can Be Supported by Threaded
Rigid Metal Conduit or IMC, 36 Inch Rule

devices or lighting fixtures are considered supported where 2 or more intermediate or rigid metal conduits are threaded wrenchtight on 2 or more different sides of the box, and the raceways are supported within 36 inches of the box, Fig. 31-17.

Exception: Conduit Bodies. Conduit bodies are considered supported when attached to rigid metal, rigid nonmetallic or intermediate metal conduit, or electrical metallic tubing, Fig. 31-18.

(f) Raceways for Support—Boxes with Devices or Fixtures. Boxes that contain devices or lighting fixtures are considered supported when 2 or more intermediate or rigid metal conduits raceways are threaded wrenchtight on 1 or more sides of the box, and the raceways are supported within 18 inches of the box, Fig. 31-19.

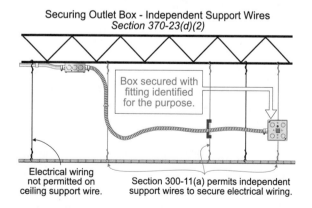

Securing Outlet Box - Independent Support Wires
Section 370-23(d)(2)

Box secured with
fitting identified
for the purpose.

Electrical wiring
not permitted on
ceiling support wire.

Section 300-11(a) permits independent
support wires to secure electrical wiring.

Fig. 31-16 Outlet Boxes Permitted to Be Secured
to Independent Support Wires

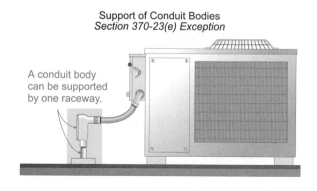

Support of Conduit Bodies
Section 370-23(e) Exception

A conduit body
can be supported
by one raceway.

Fig. 31-18 Conduit Body Can Be Supported by
Only One Conduit

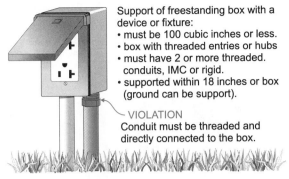

Raceway for Support of Device and Fixture Boxes
Section 370-23(f)

Support of freestanding box with a device or fixture:
• must be 100 cubic inches or less.
• box with threaded entries or hubs
• must have 2 or more threaded. conduits, IMC or rigid.
• supported within 18 inches or box (ground can be support).

VIOLATION
Conduit must be threaded and directly connected to the box.

Fig. 31-19 Boxes Can Be Supported by Threaded Rigid Metal Conduit or IMC, 18 Inch Rule

(g) Concrete or Masonry. Boxes embedded in concrete or masonry are considered supported.

(h) Pendant Cords. Boxes can be supported from a multiconductor pendant cord or cable; provided the conductors are protected from strain in accordance with the requirements contained in Section 400-10. This practice is very common in commercial and industrial applications, Fig. 31-20.

370-25 Covers and Canopies

When the installation is complete, a faceplate, cover, or fixture canopy must cover all outlet boxes openings [410-12].

(a) Nonmetallic or Metallic. Nonmetallic and metallic cover plates can be used with nonmetallic

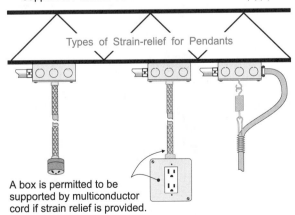

Support of Pendant Cord Boxes - *Section 370-23(h)(1)*

Types of Strain-relief for Pendants

A box is permitted to be supported by multiconductor cord if strain relief is provided.

Fig. 31-20 Boxes Can Be Supported by Pendant Cords

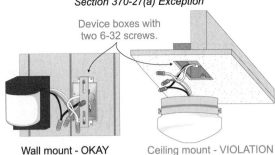

Device Box Supports Small Lighting Fixture
Section 370-27(a) Exception

Device boxes with two 6-32 screws.

Wall mount - OKAY   Ceiling mount - VIOLATION

Wall-mounted lighting fixtures up to 6 pounds can be mounted to device boxes with No. 6 screws.

Fig. 31-21 Device Box Can Support Small Lighting Fixture

or metallic outlet boxes. However, the metallic cover plates must be grounded according to Section 250-110.

FPN: It is generally accepted that metal cover plates are considered effectively grounded when mechanically and electrically attached to a grounded metal box or metal yoke [410-56(d)].

370-27 Outlet Boxes

(a) Boxes for Lighting Fixtures. Boxes and fittings for lighting fixtures must be designed to support a fixture of up to 50 pounds except as permitted in Section 410-16(a).

Boxes and fittings designed for the support of devices such as switches or receptacles cannot be used for the support of fixtures.

Exception. Wall mounted fixture weighing not more than 6 pounds and not exceeding 16 inches in any dimension can be supported to a device box with two 6/32 screws, Fig. 31-21.

(b) Floor Boxes. Floor boxes must be specifically listed for the purpose.

(c) Boxes for Paddle Fans. Outlet boxes cannot be used for the sole support for paddle fans.

Exception. Boxes specifically listed for the paddle fans can be used to support paddle fans not over 70 pounds [422-18(b) Ex.]. It is very important

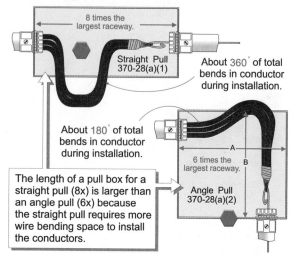

Fig. 31-22 Straight Pull Sized Larger than Angle Pull

that the outlet box be rigidly secured according to the requirements of Section 370-23.

CAUTION: Boxes for lighting fixtures and devices cannot be used for the support of paddle fans [370-27(c)].

370-28 Pull, Junction Boxes, and Conduit Bodies for Conductors No. 4 and Larger

Boxes and conduit bodies for conductors No. 4 and larger must be sized so that the conductor insulation will not be damaged.

(a) Minimum Size. When transposing cable size into raceway size in (1) and (2) below, the minimum trade size raceway required for the number and size of conductors in the cable shall be used.

(1) Straight Pulls. The minimum distance from where the conductors enter to the opposite wall must not be less than 8 times the trade size of the largest raceway, Fig. 31-22.

(2) Angle and U Pulls

Angle Pulls. The distance from the raceway entry to the opposite wall must not be less than 6 times the trade diameter of the largest raceway, plus

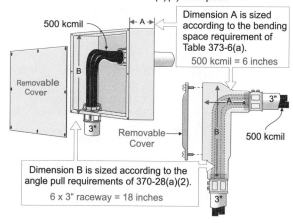

Fig. 31-23 Sizing Depth and Conduit Bodies Is Based on the Conductors Size and Table 373-6(a)

the sum of the diameters of the remaining raceways on the same wall and row, Fig. 31-22.

Exception: Opposite a Removable Cover. When conductors enter an enclosure with a removable cover such as a conduit body or wireway, the distance from where the conductors enter to the removable cover shall not be less than the bending distance as listed in Table 373-6(a) for one wire per terminal, Fig. 31-23.

Table 373-6(a) Minimum Size of Auxiliary Gutter

Largest conductor required to be bent (AWG or Circular Mils)	Minimum size auxiliary gutter
No. 4 through No. 2/0	Less than 4 inches
No. 3/0 and 4/0	4
No. 250 kcmil	4½
No. 300 and No. 350 kcmil	5
No. 400 and No. 500 kcmil	6

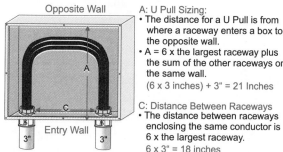

Fig. 31-24 U Pulls Sizing

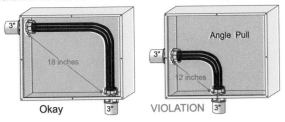

Distance Between Raceways Containing the Same Conductor
Section 370-28(a)(2)

The distance between raceways containing the same conductor shall not be less than 6 times the diameter of the larger raceway.

Minimum distance between raceways with the same conductor is:

6 inches x 3 inches = 18 Inches Minimum Distance

Fig. 31-25 Distance between Raceways Containing the Same Conductor

U Pulls. U pulls are when the conductor enters and leaves from the same wall. The distance from where the raceways enter to the opposite wall must not be less than 6 times the trade diameter of the largest raceway, plus the sum of the diameters of the remaining raceways on the same wall and row, Fig. 31-24.

Note. Each row of raceways is calculated individually, and the row that has the largest distance shall be used for that wall.

Distance between Raceways. The distance between raceways enclosing the same conductor shall not be less than 6 times the trade diameter of the largest raceway. The distance is measured from edge to edge as the crow flies, not from the center of the raceways, Fig. 31-25.

370-29 Wiring to Be Accessible

Wiring within enclosures must be **accessible** without having to remove any part of the building structure.

Exception: Boxes Listed for Underground Use. Enclosure designed and listed for underground installation can be buried where covered by gravel or other noncorrosive soil if the location is effectively identified and accessible for excavation.

SUMMARY

☐ A conduit body is a fitting for the installation of conductors and has a removable cover, such as an LB or LR. ☐ Round boxes that contain knockouts on the back are frequently used for mounting fixtures. ☐ Nonmetallic boxes are generally permitted only for nonmetallic cables and raceways. ☐ Metal boxes must be grounded, according to Article 250. ☐ Boxes and conduit bodies installed in damp or wet locations must prevent moisture from entering or accumulating in the enclosures. ☐ Raceways and cables must be securely (mechanically) fastened to the boxes and conduit bodies. ☐ Boxes installed in noncombustible materials must have the front edge of the box or plaster ring set back no more than ¼ inch from the finished surface. ☐ Boxes installed in combustible materials must have the front edge of the box or plaster ring flush with or extended out from the finished surface.

☐ Gaps or spaces more than 1/8 inch around recessed boxes in plaster, drywall, or plasterboard surfaces must be filled. ☐ Ceiling support wires that secure the suspended ceiling systems to the building structure cannot be used to support boxes. ☐ Boxes that contain devices or lighting fixtures can be supported when two raceways are threaded wrenchtight on 1 or 2 sides of the box. ☐ Boxes for lighting fixtures must be designed to support a fixture weighing up to 50 pounds. ☐ Floor boxes must be specifically listed for their use. ☐ Paddle fans cannot be supported by lighting fixture or device outlet boxes, and the fan must be supported to the building structure or a box listed for the purpose. ☐ Straight Pulls. The minimum distance from where conductors No. 4 and larger enter to the opposite wall must not be less than 8 times the trade size of the largest raceway. ☐ Angle and U-Pulls: The minimum distance from where conductors No. 4 and larger enter to the opposite wall must not be less than 6 times the trade diameter of the largest raceway, plus the sum of the diameters of the remaining raceways on the same wall and row. ☐ Wiring within boxes must be accessible.

REVIEW QUESTIONS

1. When are round boxes prohibited?

2. When can nonmetallic boxes be used with metal raceways or cables?

3. When equipment grounding conductors enter a metallic box, what must be done?

4. An 18 cubic inch nonmetallic box contains a 3-way switch with one 12/2 and one 12/3 nonmetallic-sheathed cable. Is this permitted?

5. What is the minimum cubic inch capacity required for 6 No. 10 conductors spliced in a conduit body?

6. When is nonmetallic-sheathed cable not required to be secured to a box?

7. Explain the two methods permitted for surface extensions.

8. What can suspended ceiling support wires be used to support?

9. When can boxes be supported by raceways only?

10. Can EMT or PVC be used to support conduit bodies?

11. When can a raceway be used to support a conduit body with a fixture?

12. When can a 2¼ inches × 4 inches device box be used to support lighting fixtures?

13. Why can't a fixture box be used to support a paddle fan?

14. A box has two 2 inch raceways entering on the left side and two 2 inch raceways entering from the bottom. What size pull box is required?

Unit 32

Article 373
Cabinets and Cutout Boxes

OBJECTIVES

After studying this unit the student should be able to understand:
- the protection requirements for conductors when installed in raceways.
- when a cabinet or cutout box can be used for splices and taps.

373-1 Scope

Article 373 covers the installation and construction specification of cabinets, cutout boxes, and meter socket enclosures, Fig. 32-1.

373-2 Damp, Wet, or Hazardous (Classified) Locations

(a) Damp and Wet Locations. Cabinets and cutout boxes must prevent moisture or water from entering or accumulating within the enclosure and must be listed as weatherproof.

When surface mounted, the enclosure must be mounted with at least a ¼ inch air space between the enclosure and the mounting surface.

(b) Hazardous Locations. Enclosures installed in hazardous locations must comply with the requirements contained in Articles 500 through 517.

373-3 Installed in Walls

Noncombustible Material. Cabinets, cutout boxes, and meter enclosures installed in noncombustible materials must have the front edge of the enclosure set back no more than ¼ inch from the finished surface.

Combustible Material. Cabinets, cutout boxes, and meter enclosures installed in combustible material such as wood, must have the front edge of the enclosure flush with, or extended out from, the finished surface.

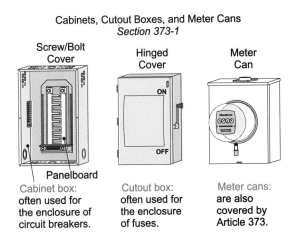

Cabinets, Cutout Boxes, and Meter Cans
Section 373-1

Screw/Bolt Cover   Hinged Cover   Meter Can

Panelboard
Cabinet box: often used for the enclosure of circuit breakers.

Cutout box: often used for the enclosure of fuses.

Meter cans: are also covered by Article 373.

Fig. 32-1 Cabinets, Cutout Boxes, and Meter Cans

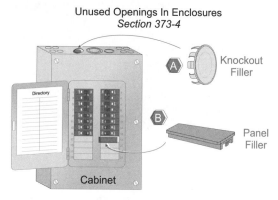

Unused openings in enclosures must be closed with a fitting that gives protection equivalent to that of the wall of the enclosure. Also see 110-12(a) and 370-18.

Fig. 32-2 Unused Openings in Enclosures Must Be Properly Closed or Sealed

373-4 Unused Openings

Unused openings in enclosures must be closed with a protection fitting equivalent to that of the wall of the enclosure [110-12(a) and 370-18], Fig. 32-2.

373-5 Cables

(c) Cables. Where cable is used, each cable shall be secured to the cabinet or cutout box.

Exception.   Cables are not required to be secured if the cables enter a nonflexible raceway not less than 18 inches, or more than 10 feet long that enters the top of a surface-mounted enclosure, if, Fig. 32-3:

The following are the conditions of the Exception

(a) Each cable is fastened within 12 inches from the end of the raceway

(b) The raceway does not penetrate a structural ceiling

(c) Fittings are provided on each end of the raceway to protect the cables from abrasion

(d) The raceway is sealed or plugged and fastened in place

(e) Nonmetallic-sheathed cable extends at least ¼ inch into the panelboard

(f) Raceway is properly secured

(g) Conductor fill is limited to 60 percent of the raceway cross-sectional area in accordance with Table 1 of Chapter 9.

373-8 Used for Raceway and Splices

Cabinet, cutout boxes, and meter socket enclosures can be used for conductors feeding through where the conductors do not fill the wiring space at any cross-section to more than 40 percent of the cross-sectional area of the space, Fig. 32-4.

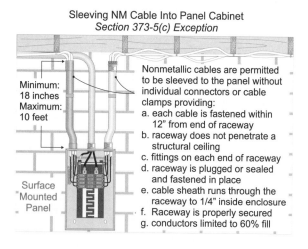

Sleeving NM Cable Into Panel Cabinet
Section 373-5(c) Exception

Minimum: 18 inches
Maximum: 10 feet

Surface Mounted Panel

Nonmetallic cables are permitted to be sleeved to the panel without individual connectors or cable clamps providing:
a. each cable is fastened within 12" from end of raceway
b. raceway does not penetrate a structural ceiling
c. fittings on each end of raceway
d. raceway is plugged or sealed and fastened in place
e. cable sheath runs through the raceway to 1/4" inside enclosure
f. Raceway is properly secured
g. conductors limited to 60% fill

Fig. 32-3 Sleeving NM Cable into Panel Cabinet

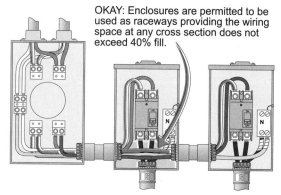

Cabinet and Cutout Box as Raceway
Section 373-8

OKAY: Enclosures are permitted to be used as raceways providing the wiring space at any cross section does not exceed 40% fill.

Note: Section 230-7 prohibits other conductors in the same raceway with service conductors. It does not prohibit service conductors with other conductors in cabinets and cutout boxes.

Fig. 32-4 Cabinets and Cutout Boxes Can Be Used as a Raceway

Note. Section 230-7 prohibits the mixing of service conductors with branch or feeder conductors in raceways or cables, but not enclosures.

Splices and taps are permitted in cabinets, cutout boxes, or meter socket enclosures if the splice or tap does not fill the wiring space at any cross-section to more than 75 percent, Fig. 32-5.

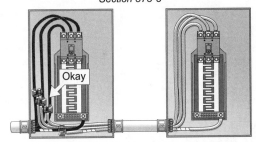

Splice and Tap in Cabinet and Cutout Box
Section 373-8

Splices and taps are permitted in cabinets, cutout boxes, and meter socket enclosures if the splice or tap does not fill more than 75% of the wiring space.

Fig. 32-5 Splices and Taps Permitted in Cabinets or Cutout Boxes

SUMMARY

Article 373 covers the installation and construction specifications of cabinets, cutout boxes, and meter socket enclosures. □ Cabinets are used to enclose panelboards, and cutout boxes are often used to enclose switches. □ Cabinet and cutout boxes must be installed to prevent moisture or water from entering or accumulating within the enclosure. □ Boxes installed in noncombustible materials must have the front edge of the enclosure set back no more than ¼ inch from the finished surface. □ Boxes installed in combustible materials must have the front edge of the enclosure flush with, or extended out from, the finished surface. □ Unused openings in enclosures must be closed with a protection fitting equivalent to that of the wall of the enclosure. □ Splices are permitted in panelboard and disconnect enclosures.

REVIEW QUESTION

1. When can a cabinet or cutout box be used for splices and taps?

Unit 33

Article 374
Auxiliary Gutters

OBJECTIVE

After studying this unit, the student should be able to understand:
- the differences between auxiliary gutters and wireways.

Auxiliary gutters and wireways are constructed the same and look similar. The principal difference between an auxiliary gutter and a wireway is not in their physical characteristics, but in their use. Wireways are used as a raceway system and can be run in unlimited lengths; but auxiliary gutters are for the purpose of supplemental wiring space and are limited to 30 feet lengths.

374-1 Uses

Auxiliary gutters supplement the wiring space of meter centers, distribution centers, switchboards, and similar arrangements. Auxiliary gutters can enclose busbars as well as conductors.

Auxiliary gutters cannot enclose switches, fuses, circuit breakers, appliances, or other similar equipment.

374-2 Length

An auxiliary gutter shall not extend greater than 30 feet from the equipment it supplements.

374-3 Supports

(a) Sheet Metal Auxiliary Gutters. Sheet metal auxiliary gutters shall be supported throughout their entire length at intervals not exceeding 5 feet.

(b) Nonmetallic Auxiliary Gutters. Nonmetallic auxiliary gutters shall be supported at intervals not to exceed 3 feet and at each end or joint, unless listed for other support intervals. In no case shall the distance between supports exceed 10 feet.

374-4 Covers

Covers shall be securely fastened to auxiliary gutter.

374-5 Number of Conductors

Auxiliary gutters shall not contain more than 30 current-carrying conductors at any cross section.

The number of conductors permitted in an auxiliary gutter is limited to 20 percent of the cross-sectional area.

374-6 Ampacity

Conductor. When more than 30 current-carrying conductors are installed in any cross-sectional area of the auxiliary gutter, Table 310-15(b)(2) derating factors must apply.

Busbar. The ampacity per square inch of the cross-sectional area of a *busbar* shall be 1,000 ampere for copper.

Question. What is the ampacity of a busbar 4 inches × ½ inch?

Answer. 2,000 ampere
Area = 4 inches × ½ inch = 2 square inch
2 square inches × 1,000 ampere = 2,000 ampere

374-8 Splices and Taps

(a) Cross-Sectional Area. Splices and taps must be accessible and cannot fill the auxiliary gutter to more than 75 percent of its cross-sectional area.

(c) Identification. All tap conductors must be identified at the gutter location to indicate the circuit or equipment they supply.

(d) Overcurrent Protection. Tap conductors are required to comply with the tap rules in Section 240-21.

374-9 Gutter Sizing

(a) Electrical and Mechanical Continuity. Gutters shall be constructed and installed to maintain electrical and mechanical continuity.

(d) Sizing of Auxiliary Gutters. Auxiliary gutters must be sized to allow the bending of conductors as listed in Table 373-6(a).

(e) Use. Auxiliary gutters installed in wet locations shall be suitable for the location.

SUMMARY

☐ Auxiliary gutters supplement the wiring space of meter centers, distribution centers, switchboards, and similar arrangements, and can enclose busbars and conductors. ☐ An auxiliary gutter is not a raceway but an enclosure. ☐ The maximum length of an auxiliary gutter is 30 feet, and it must be supported at intervals not exceeding 5 feet. ☐ The number of conductors permitted in an auxiliary gutter is limited to 20 percent of the cross-sectional area of the enclosure. ☐ When more than 30 conductors are installed in any cross-sectional space, the conductors' ampacity must be reduced. ☐ Splices and taps must be accessible and cannot fill the auxiliary gutter to more than 75 percent of its cross-sectional area. ☐ Tap conductors must be identified and installed according to the tap rules in Section 240-21. ☐ Auxiliary gutters must be sized to accommodate the bending radius of conductors according to Table 373-6(a), and cannot be used as a pull box.

REVIEW QUESTION

1. What are the differences between auxiliary gutters and wireways?

Unit 34

Article 380
Switches

OBJECTIVES

After studying this unit, the student should be able to understand:
- the conductor identification requirements for single-pole, 3- and 4-way switching.
- when the disconnect (switch) can be located above 6 feet 7 inches.
- what precautions must be taken when installing 277 and 480 volt switches.
- the mounting requirements of snap switches.
- when a circuit breaker can be used as a switch.
- when metal faceplates or switch straps must be grounded.

PART A. **INSTALLATION**

380-1 Scope

The requirements of Article 380 apply to all types of switches, such as snap (toggle) switches, *knife switches*, circuit breakers used as switches, and automatic switches such as time clocks. This Article also applies to switches used for disconnecting means.

> **Note.** The high-leg must terminate on the "B" phase for switchboards and panelboards [384-3(f)], but this rule does not apply to switches, Fig. 34-1.

380-2 Switch Connections

(a) Three-Way and Four-Way Switches. All three-way and four-way switching must be done with the ungrounded, Fig. 34-2.

> **Note.** White or gray conductor within a cable assembly can be used for single-pole, three-way or four-way switch loops if it is

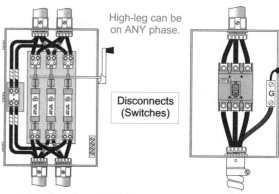

High-Leg Termination In Switch Enclosures (Disconnects)

High-leg can be on ANY phase.

Disconnects (Switches)

Note: There is no NEC Section requiring the high-leg conductor to terminate on the "B" (center) phase in switch enclosures. See Section 384-3(d) for panelboards.

Fig. 34-1 High-Leg Termination in Switch Enclosures (Disconnects)

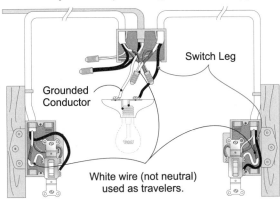

3-Way And 4-Way Switching - *Section 380-2(a)*

Switch Leg

Grounded
Conductor

White wire (not neutral)
used as travelers.

All switching must be done on the "hot" conductors. The grounded
conductor cannot be switched. The white wire is okay for travelers
but must be permanently re-identified [200-7(c)(2)].

Fig. 34-2 Grounded (Neutral) Conductor Not
Permitted to Be Used for 3 Way or 4 Way Switching

permanently re-identified to indicate its use as an ungrounded conductor at each location where the conductor is visible and accessible [200-7(c)(2)], Fig. 34-2.

When metal conduit or metal-clad cable is used for three-way and four-way switching, all conductors of the circuit must be grouped together in the same raceway or cable to prevent inductive heating [300-3(b), 300-20(a)].

Exception: Travelers and Switch Legs. Travelers and switch legs (switch loop conductors) are not required to be routed or grouped with the grounded (neutral) conductor.

(b) Switching Grounded (Neutral) Conductors. All switching must be done with the ungrounded conductor only, Fig. 34-3.

Exception No. 1: Grounded Conductor. Switching devices that open the grounded (neutral) conductor simultaneously with the ungrounded conductors can be used.

380-3   Switch Enclosures

(b) Used as a Raceway. Switch or circuit breaker enclosures can be used as a junction box for conductors feeding through, conductor splices or taps if the conductors do not fill the wiring space at any cross-section to more than 40 percent, and the splices and taps do not fill the wiring space at

Grounded Conductor Not Used For Switching
Section 380-2(b)

All switching must be done on the "hot" conductors.
The grounded conductor cannot be switched.

A

White colored conductor
(not neutral) feeding a switch.

Switch
Feed

Okay
The grounded (white)
conductor is not used
for switching.

See 200-7(c)(2)

Switch
Leg

B

Switched Grounded
(Neutral) Conductor

VIOLATION
The grounded (white) conductor
used for switching is not permitted.

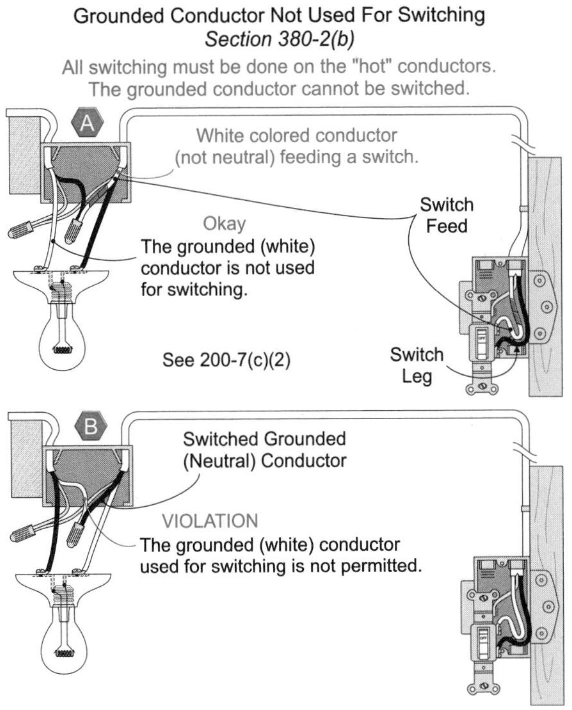

Fig. 34-3 Grounded (Neutral) Conductor Not
Permitted to Be Used for Switching

any cross-section to more than 75 percent [373-8], Fig. 34-4.

380-4   Wet Locations

Switches and circuit breakers installed in wet locations shall be installed in a weatherproof enclosure. The enclosure must be installed so that

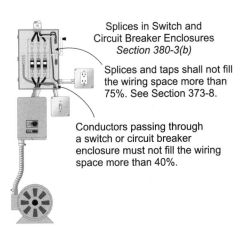

Splices in Switch and
Circuit Breaker Enclosures
Section 380-3(b)

Splices and taps shall not fill
the wiring space more than
75%. See Section 373-8.

Conductors passing through
a switch or circuit breaker
enclosure must not fill the wiring
space more than 40%.

Fig. 34-4 Switch and Circuit Breaker Enclosures
Can Be Used as a Raceway or Contain Splices

Switches in Wet Locations
Section 380-4

VIOLATION

Okay

Bathtub,
Hydromassage,
and Shower

Switches are not permitted within wet locations of bathtubs and
showers but are permitted to be located adjacent to this area.

**Fig. 34-5 Switches Not Permitted in Bathtub
or Shower Space**

least ¼ inch airspace is provided between the enclosure and the wall or other supporting surface [373-2(a)].

Bathtub and Shower Space. Switches must not be installed within a tub or shower space unless the switch and its assembly have been listed for this purpose by a qualified testing laboratory, Fig. 34-5.

Note. Switches must be located at least 5 feet from pools, and outdoor spas and hot tubs [680-6(c), 680-40], Fig. 34-6.

380-6 Position of Knife Switches

(a) Single-Throw Knife Switch. Single-throw knife switches must be installed so that gravity will not tend to close them.

380-7 Indicating

Switches, motor circuit switches, and circuit breakers must show whether they are in the "on" or "off" position. When the switch is operated vertically, it must be installed so the "up" position is the "on" position [240-81].

Exception: Double-throw switches such as three-way and four-way switches are not required to show whether they are in the "on" or "off" position.

380-8 Accessibility and Grouping

(a) Location. All switches and circuit breakers used as switches must be capable of being operated at no more than 6 feet 7 inches (2 meters) above a **readily accessible** location. The height is measured from the floor, or platform, to the center of the grip of the handle (in the highest position) of the switch or circuit breaker, Fig. 34-7.

Note. There is no minimum height for switches, however mobile and manufactured homes required the service

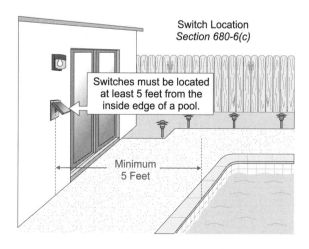

Switch Location
Section 680-6(c)

Switches must be located
at least 5 feet from the
inside edge of a pool.

Minimum
5 Feet

**Fig. 34-6 Switches Must Be at Least 5 Feet from
Outdoor Pools, Spa, and Hot Tubs**

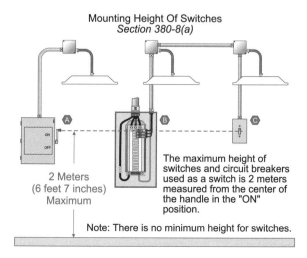

Mounting Height Of Switches
Section 380-8(a)

ON
OFF

A   B   C

2 Meters
(6 feet 7 inches)
Maximum

The maximum height of
switches and circuit breakers
used as a switch is 2 meters
measured from the center of
the handle in the "ON"
position.

Note: There is no minimum height for switches.

**Fig. 34-7 Switches Must Not Be Mounted above
6 feet, 7 Inches**

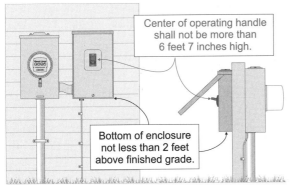

Height of Mobile and Manufactured Home
Outside Disconnecting Means – *Section 550-23(f)*

Center of operating handle
shall not be more than
6 feet 7 inches high.

Bottom of enclosure
not less than 2 feet
above finished grade.

Fig. 34-8 Disconnecting Means Must Be Mounted
No Less than 2 Feet from Finished Grade

disconnecting means to be mounted a minimum of 2 feet from the finish grade
[550-23(f)], Fig. 34-8.

Exception No. 2: Equipment Disconnect.
Switches or circuit breakers used as a switch can be
mounted higher than 6 feet 7 inches if located next
to the equipment it supplies and it is accessible by
portable means, Fig. 34-9.

Note. This exception applies to appliances
[422-31(b), 422-32, 422-33], electric space
heating equipment [424-19(a), 424-19(b)],
and electric duct heating equipment [424-
19(a), 424-65].

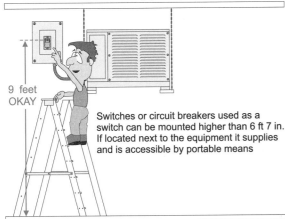

Switches Located Adjacent to Equipment
Section 380-8(a) Exception 2

9 feet
OKAY

Switches or circuit breakers used as a
switch can be mounted higher than 6 ft 7 in.
If located next to the equipment it supplies
and is accessible by portable means

Fig. 34-9 Switches Can Be Located above 6 Feet,
7 Inches if Adjacent to Equipment It Controls

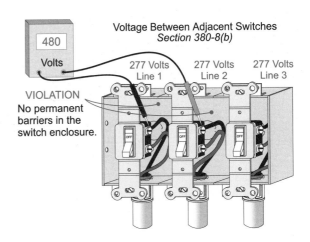

Voltage Between Adjacent Switches
Section 380-8(b)

480
Volts

277 Volts
Line 1

277 Volts
Line 2

277 Volts
Line 3

VIOLATION
No permanent
barriers in the
switch enclosure.

Fig. 34-10 Voltage between Adjacent Switches Not
to Exceed 300 Volts

(b) **Voltage Between Adjacent Switches.**
Switches must be arranged so that the voltage
between adjacent switches does not exceed 300
volt, Fig. 34-10.

Note. The rule does not cover voltage
between 277 volt switches and 125 volt
receptacles.

380-9 Snap Switch Faceplates

(a) **Mounting.** Faceplates shall be installed so
they completely cover the wall opening and seat
against the wall surface.

(b) **Grounding.** All snap switches (including
dimmers) must be designed so that they can ground
a metal faceplate whether or not a metal faceplate is
installed, Fig. 34-11.

Note. This means that switches with plastic
yokes are no longer permitted.

(1) Metal Box. A metal *yoke* secured with
metal screws to a metal box is generally considered
grounded.

(2) Nonmetallic Box. A metal yoke installed on
to a nonmetallic box must have an equipment
grounding conductor terminate to the switch yoke.

Exception. Existing snap switches installed
where no grounding means exist in the outlet box
can be replaced without grounding the switch yoke.

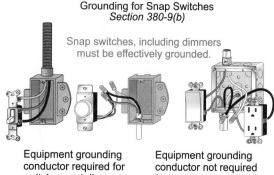

Fig. 34-11 Snap Switches and Dimmers Must Be Grounded So That Covers Can Be Grounded

Note. A snap switch is just a fancy name for the typical toggle switch that is used everyday.

380-10   Mounting Snap Switches

(a) Mounting of Snap Switches to Boxes. Snap switches installed in recessed boxes must have the yoke seated against the finished wall surface. Because drywall installers are often very aggressive when cutting out the opening for electrical outlet boxes, this rule is difficult to comply with.

Note. The maximum distance the box can be recessed from the finish surface is ¼ inch [370-20], and all gaps larger than 1/8 inch around the box must be filled [370-21].

Boxes that are flush or surface mounted must have the yoke seated against the box or plaster ring.

380-11   Circuit Breakers Used As Switches

A hand-operable circuit breaker can be used as a switch and must clearly show when it is in the "on" (closed) or "off" (open) position [240-81, 380-7].

Note. Circuit breakers used to switch fluorescent electric discharge lighting must be rated for switching duty "SWD" [240-83(d)].

380-12   Grounding

Metal Enclosures.   Metal enclosures and faceplates for switches or circuit breakers must be grounded according to Sections 250-42 and 250-148.

Nonmetallic Enclosures.   Nonmetallic enclosures containing switches or circuit breakers can be used with metal raceways and cables if the metal raceways or connectors are grounded in accordance with the exceptions to Section 370-3.

380-14   Rating and Use of Snap Switches

(a) AC General Use Snap Switch. Alternating current general use snap switches can control:

(1) Resistive and inductive loads, including electric-discharge lamps, not exceeding the ampere rating of the switch at the voltage involved.

(2) Tungsten-filament lamp loads not exceeding the ampere rating of the switch at 120 volt.

(3) Motor loads, 2 horsepower or less, that do not exceed 80 percent of the ampere rating of the switch [430-109(c)].

(c) CO/ALR Snap Switches. Snap switches and receptacles [410-56(b)] rated 15 or 20 ampere must be marked CO/ALR when connected to aluminum wire.

Note. Aluminum conductors are not permitted on screwless (push-in) terminals of snap switches or receptacles.

SUMMARY

☐ AC snap switches are rated for motor and inductive loads. ☐ Single-pole switches make or break the connection of one conductor. ☐ The grounded (neutral) conductor cannot be used as a switch leg or for travelers between 3- and 4-way switches. ☐ Switches and circuit breakers cannot switch the grounded (neutral) conductor. ☐ Single-throw knife switches must be installed so that gravity will not tend to close them. ☐ Switches or circuit breakers operated vertically must be installed so the "up" position is the "on" position. ☐ The maximum voltage between adjacent switches is 300 volt. ☐ All switches and circuit breakers used as switches must be mounted no higher than 6 feet 7 inches above a readily accessible location. ☐ A hand-operable circuit breaker can be used as a switch and must clearly indicate when it is in the "on" or "off" position. ☐ Enclosures for switches must be grounded and nonmetallic enclosures are permitted with metal raceways and cables. ☐ Circuit breakers used to switch fluorescent electric discharge lighting must be rated SWD. ☐ Snap switches rated 15 and 20 ampere must be marked CO/ALR when connected to aluminum wire.

REVIEW QUESTIONS

1. What are the conductor identification requirements for single-pole, 3- and 4-way switching?

2. When can a disconnect (switch) be located above 6 feet 7 inches?

3. When installing 277 and 480 volt switches, what precautions must be taken?

4. Provide a brief description of the mounting requirements of snap switches.

5. When can a circuit breaker be used as a switch?

6. When are metal faceplates or switch straps (yokes) required to be grounded?

Unit 35

Article 384
Switchboards and Panelboards

OBJECTIVES

After studying this unit, the student should be able to understand:
- the requirements for installing the high-leg conductor in panelboards.
- working space and dedicated space.
- the rules that apply to lighting and appliance branch circuit panelboards.
- the purpose of back-feeding circuit breakers and the Code rules that apply to this type of installation.
- the *NEC* requirements for grounding and grounded terminals in panelboards.

384-1 Scope

Article 384 covers the specific requirement for switchboards, panelboards, and distribution boards that control light and power circuits. For the purposes of this book, we will cover only the requirement of panelboards. In the trade, the nickname term for panelboard is "guts," Fig. 35-1.

Note. The installation requirements for motor control centers enclosures must comply with Article 384, but motor control center functions are covered in Article 430.

384-3 Arrangement of Busbars and Conductors

(e) High-Leg Marking. Panelboards supplied by a 4-wire delta, 3-phase system, where the midpoint of one phase winding is grounded (high-leg), must have that phase busbar or conductor marked orange, or identified with other effective means [215-8, 230-56], Fig. 35-2.

(f) Phase Arrangement Panelboards. Panelboards supplied by a 4-wire delta, 3-phase system, where the midpoint of one phase winding is grounded (high-leg) must have the high-leg conductor connected to the "B" (center) phase, Fig. 35-2.

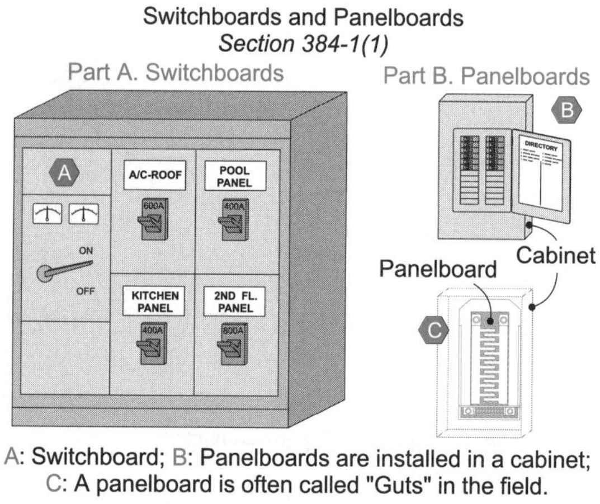

Switchboards and Panelboards
Section 384-1(1)
Part A. Switchboards   Part B. Panelboards

A: Switchboard; B: Panelboards are installed in a cabinet;
C: A panelboard is often called "Guts" in the field.

Fig. 35-1 Switchboards and Panelboards

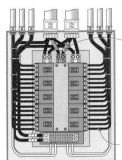

High-leg Identification
in Panelboards
Section 384-3(e)

The high-leg phase conductor is
required to be identified by an orange
color or other effective means.

The high-leg conductor must terminate
to the B (center) phase [384-3(f)].

Note: There are no NEC requirements
for identifying the high-leg conductor
on branch circuits.

Fig. 35-2 High-Leg Conductor Must Be Identified with the Color Orange, and Terminate onto B Phase

> **WARNING:**
> The *ANSI* standard for meter equipment requires the high-leg conductor to be installed to the right or to the C phase. For meter equipment, the high-leg conductor must terminate on the C phase.

384-4 Working Clearance and Dedicated Space

See Section 110-26(f) for the requirements for dedicated equipment space.

PART C. PANELBOARDS

384-13 Circuit Identification

All panelboard circuit breakers shall be legibly marked to identify their purpose [110-22]. The identification shall be posted on the panelboard, Fig. 35-3.

384-14 Lighting and Appliance Branch Circuit Panelboard

This *Code* Section contains the definition of a lighting and appliance branch circuit panelboard and a power panelboard.

(a) Lighting and Appliance Panelboard. A lighting and appliance branch-circuit panelboard is one having more than 10 percent of its overcurrent devices protecting "lighting and appliance circuits."

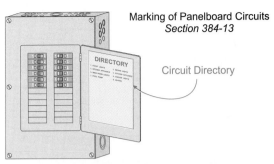

Marking of Panelboard Circuits
Section 384-13

DIRECTORY

Circuit Directory

Panelboard circuits must be legibly marked to identify their purpose. The identification shall be posted on the panelboard. Also see Section 110-22.

Fig. 35-3 All Panelboard Circuit Must Be Legibly Marked to Identify Their Purpose

A lighting and appliance branch circuit is a branch circuit that has a connection to the neutral of the panelboard and that has overcurrent protection of 30 amperes or less, Fig. 35-4.

384-15 Number of Overcurrent Devices

The maximum number of overcurrent devices that can be installed in a lighting and appliance branch circuit panelboard is 42 (not counting the main breaker). When counting the number of devices, a 2-pole circuit breaker counts as 2 overcurrent devices and a 3-pole circuit breaker counts as 3 overcurrent devices, Fig. 35-4.

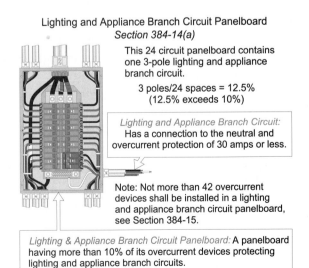

Lighting and Appliance Branch Circuit Panelboard
Section 384-14(a)

This 24 circuit panelboard contains one 3-pole lighting and appliance branch circuit.

3 poles/24 spaces = 12.5%
(12.5% exceeds 10%)

Lighting and Appliance Branch Circuit:
Has a connection to the neutral and overcurrent protection of 30 amps or less.

Note: Not more than 42 overcurrent devices shall be installed in a lighting and appliance branch circuit panelboard, see Section 384-15.

Lighting & Appliance Branch Circuit Panelboard: A panelboard having more than 10% of its overcurrent devices protecting lighting and appliance branch circuits.

Fig. 35-4 Definition—Lighting and Appliance Branch Circuit Panelboard

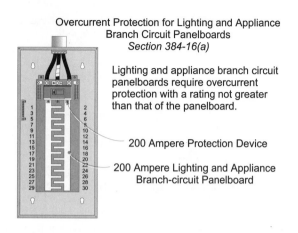

Overcurrent Protection for Lighting and Appliance
Branch Circuit Panelboards
Section 384-16(a)

Lighting and appliance branch circuit
panelboards require overcurrent
protection with a rating not greater
than that of the panelboard.

200 Ampere Protection Device

200 Ampere Lighting and Appliance
Branch-circuit Panelboard

Fig. 35-5 Overcurrent Protection Required for
Lighting and Appliance Branch Circuit Panelboard

384-16 Overcurrent Protection of Panelboard

(a) Lighting and Appliance Branch Circuit Panelboard. Each lighting and appliance branch circuit panelboard shall have overcurrent protected, Fig. 35-5.

Exception No. 1: Feeder Protection. The panelboard is considered protected by the feeder protection device if the feeder protection device rating does not exceed the rating of the panelboard, Fig. 35-6.

CAUTION: When tap conductors [240-21] supply a lighting and appliance branch circuit panelboard, overcurrent protection is required ahead of the panelboard. Lighting and appliance branch circuit panelboards on the secondary of a transformer also require overcurrent protection [384-16(e)].

(d) Maximum Continuous Load on Overcurrent Device. The maximum continuous load on an overcurrent protection device shall not exceed 80 percent of the overcurrent device rating.

Question. What is the maximum continuous load permitted on a 100 ampere breaker?

Answer.   80 ampere
100 ampere × 0.80 = 80 ampere

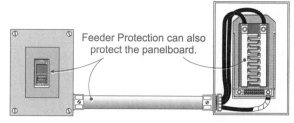

Feeder Protection Protecting Panelboard
Section 384-16(a) Exception 1

Feeder Protection can also
protect the panelboard.

A feeder protection device can protect a panelboard if the rating
of the device does not exceed the rating of the panelboard.

Fig. 35-6 Feeder Overcurrent Protection Can Be
Used for Panelboard Overcurrent Protection

Exception. When the overcurrent protection device and its assembly are listed for continuous duty, the overcurrent protection device can be sized to 100 percent of the continuous load.

Note. There are no circuit breakers rated less than 225 ampere listed for continuous load operation.

(e) Panelboards Supplied through a Transformer. When a lighting and appliance branch circuit panelboard is supplied from a transformer, the panelboard protection device must be located on the *secondary* side of the transformer, Fig. 35-7.

Exception. Panelboards located on the secondary side of 2-wire transformers can be protected by the *primary* protection device if the primary protection device is sized according to Section

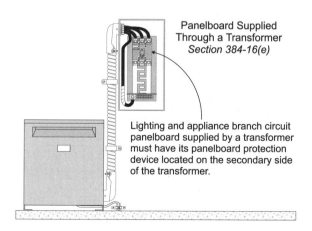

Panelboard Supplied
Through a Transformer
Section 384-16(e)

Lighting and appliance branch circuit
panelboard supplied by a transformer
must have its panelboard protection
device located on the secondary side
of the transformer.

Fig. 35-7 Panelboard on Secondary Side of a
Transformer Is Not Protected by Primary Protection

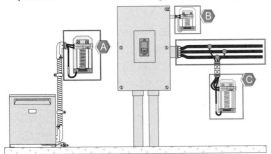

Back-fed Devices - *Section 384-16(g)*

Back-fed circuit breakers must be fastened to the panelboard with fasteners designed for that purpose.

A: Back-fed breaker protecting a panelboard on the load side of a transformer.
B: Back-fed breaker protecting a panelboard tapped from a disconnect.
C: Back-fed breaker protecting a panelboard tapped from feeder conductors.

Fig. 35-8 Back-Fed Circuit Breakers Must Be Secured to Panelboard

Grounding in Panelboards
Section 384-20
A grounding terminal is bonded to the enclosure and isolated from the neutral terminal.

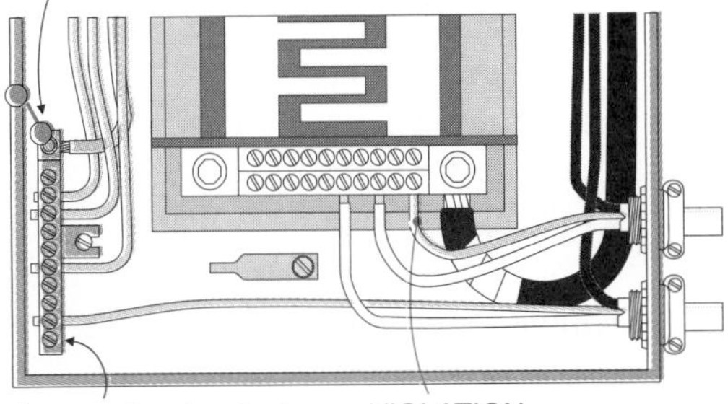

A grounding terminal bar is required for all grounding conductors.

VIOLATION
Grounding conductors and neutral conductors must be kept separate.

Fig. 35-9 Grounding Terminal in a Panelboard Must Be Isolated from Grounded (Neutral) Terminal

450-3(b) and it does not exceed the panelboard rating by multiplying the primary-to-secondary voltage ratio. Secondary conductors can also be protected by the primary protection device for 2-wire transformers [240-3(f)].

(g) Back-Fed Devices. Plug-in circuit breakers or plug-in main lug assemblies *back-fed* with field installed ungrounded supply conductors must be secured in place by an additional fastener. The purpose of the fastener is to prevent the overcurrent protection device from being accidentally released from the panelboard, thereby exposing energized parts, Fig. 35-8.

Note. Circuit breakers are often back-fed to provide overcurrent protection for lighting and appliance panelboards [384-16(a)]. According to UL, circuit breakers marked "LINE" and "LOAD" cannot be back-fed.

384-20 Grounding of Panelboards

Metal panelboard cabinets and frames must be grounded to an equipment grounding conductor [250-118].
Service equipment panelboards [250-24(b), 384-3(c)], separate building disconnect panelboards

[250-32(a)], and panelboards on the secondary side of transformers [250-30(a)(2)] can be grounded to the grounded (neutral) conductor in accordance with Section 250-142.

Grounding Terminal Bar. Equipment grounding conductors in panelboards must terminate to a grounding terminal bar. The grounding terminal bar must be bonded to the panelboard cabinet and it must be insulated from the grounded (neutral) terminal bar, Fig. 35-9.

Note. Only one conductor is permitted under a terminal unless the terminal is listed otherwise. This means that you cannot twist all the ground wires together and install them under one lug [110-14(a)].

Exception: Isolated Ground Circuits. Isolated equipment grounding conductors installed for the reduction of electrical noise [250-146(d)] can pass through the panelboard without terminating on to the equipment grounding terminal of the panelboard.

Neutral-to-Ground Connection. Grounded (neutral) conductors must not terminate on the grounding terminal bar, except as permitted for services, separately derived systems, or separate building disconnects in accordance with Section 250-142.

SUMMARY

Article 384 covers the specific requirements for switchboards, panelboards, and distribution boards that control light and power circuits. ☐ Panelboards supplied by a 4-wire delta, 3-phase system shall have the high-leg phase busbar or conductor marked orange or identified with other effective means, and connected to the B (center) phase. ☐ All panelboard circuits shall be legibly marked to identify the purpose of the circuit; this identification shall be on the panelboard directory. ☐ A lighting and appliance branch-circuit panelboard is one having more than 10 percent of its overcurrent devices protecting lighting and appliances. ☐ A lighting and appliance branch circuit is a branch circuit that has a connection to the neutral of the panelboard and that has overcurrent protection of 30 amperes or less. ☐ A power panelboard is one having 10 percent or fewer of its overcurrent devices protection lighting and appliance branch circuits. ☐ The maximum number of overcurrent devices that can be installed in a lighting and appliance branch circuit panelboard is 42, not counting the main. ☐ Lighting and appliance branch circuit panelboards must be individually protected by an overcurrent protection device. ☐ Plug-in type breakers that are back-fed shall be secured to the panelboard with fasteners designed for this purpose. ☐ Panelboards must have a grounding terminal designed for the purpose and it must be secured to the panelboard cabinet. ☐ The grounded (neutral) terminal shall not have grounding electrode conductors connected to it.

REVIEW QUESTIONS

1. What are the requirements for installing the high-leg conductor in panelboards?

2. Give a summary of the rules that apply to lighting and appliance branch circuit panelboards.

3. Explain the purpose of back-feeding circuit breakers and any *Code* rules that apply to this type of installation.

4. Explain the *NEC* requirements for grounding and grounded terminals in panelboards.

CHAPTER 4
EQUIPMENT

Scope of Chapter 4 Articles

Chapter 1 of the *Code* covers the general rules for electrical installations; Chapter 2 deals with the specific rules that apply to wiring and protection of services, feeders, and branch circuits; Chapter 3 contains the wiring method rules for conductors, cables, raceways, and enclosures; and Chapter 4 contains the rules for wiring of specific equipment, such as fixtures, appliances, motors, air-conditioners, and transformers.

This is the last chapter of the book, and you have worked very hard to get to this point. Thank you for your determination in sticking with me. The *NEC* contains 5 more chapters, but these deal with special occupancies, equipment, and conditions, such as health care facilities, motion picture studios, agricultural buildings, signs, elevators, X-ray equipment, and so on. These subjects are beyond the scope of this book, which deals with general *Code* rules applying to most occupancies.

ARTICLE 400 FLEXIBLE CORDS AND CABLES

This Article applies to flexible cords and cables, which are not considered a wiring method but can be used for temporary wiring [305-4].

ARTICLE 402 FIXTURE WIRES

This Article covers the general requirements for wires used for lighting fixtures, Class 1 remote-control circuits, and nonpower-limited fire alarm systems.

ARTICLE 410 LIGHTING FIXTURES AND RECEPTACLE INSTALLATIONS

Article 410 contains the installation requirements for fixtures, lampholders, and receptacles.

ARTICLE 422 APPLIANCES

The scope of Article 422 includes appliances and room air-conditioners that are fastened in place, permanently connected, or cord- and plug-connected in any occupancy. This Article does not apply to electric space-heating equipment, motors, or air-conditioning equipment.

ARTICLE 424 FIXED ELECTRIC SPACE HEATING EQUIPMENT

This Article covers fixed electric equipment used for space heating, including heating cable, unit heaters, boilers, central systems, or other types of approved fixed electric space heating equipment. This Article does not apply to air-conditioning equipment.

ARTICLE 430 MOTORS

This Article covers the specific rules for branch circuit and feeder conductor sizing and protection, overload protection (heaters), control circuit conductors, motor controllers, and disconnecting means.

ARTICLE 440 AIR-CONDITIONING AND REFRIGERATION

This Article applies to electric-driven air-conditioning and refrigeration equipment that has a hermetic refrigerant motor compressor.

ARTICLE 450 TRANSFORMERS

Article 450 covers requirements for transformers and reactors.

Unit 36

Article 400
Flexible Cords and Cables

OBJECTIVES

After studying this unit, the student should be able to understand:
- the most common uses for flexible cords.
- when an attachment plug may be used as a disconnect for appliances.
- some of the common violations when flexible cords are used.

400-1 Scope

This Article applies to flexible cords and cables.

Flexible cords and cables cannot be used as a wiring method, but they can be used for temporary wiring [305-4].

400-3 Suitability

Flexible cords, cables, and their fittings must be approved for the use and must be suitable for the location. For example, when using cords in wet locations, the cord and the fittings must be approved for the wet location.

400-4 Types

Flexible cords and cables must comply with the installation requirements contained in Table 400-4.

Note. Data processing cables are not required to comply with the requirements of this Article, but they must comply with the requirements contained in Article 645.

400-5 Ampacity of Flexible Cords and Cables

Tables 400-5(A) and 400-5(B) list the allowable ampacity for copper conductors in cords and cables. Conductor bundling and ambient temperature factors also apply; see Notes to Tables 400-5(A) and (B).

Note. The following tables give a summary of flexible cord ampacity and overcurrent protection as required in the various Sections of the *NEC*.

Table 400-5(a) Allowable ampacity and protection for flexible service cords used with listed appliances, portable lamps and fixtures [410-14 and 410-30].

Size (AWG)	Three-Conductor Cables		Two-Conductor Cables	
	Ampacity	Maximum protection 240-4 Ex 1	Ampacity	Maximum protection 240-4 Ex 1
18	7 Amperes	20 Amperes	10 Amperes	20 Amperes
16	10 Amperes	20 Amperes	13 Amperes	20 Amperes
14	15 Amperes	30 Amperes	18 Amperes	30 Amperes
12	20 Amperes	50 Amperes	25 Amperes	25 Amperes
10	25 Amperes	50 Amperes	30 Amperes	50 Amperes
8	35 Amperes	50 Amperes	40 Amperes	50 Amperes
6	45 Amperes	50 Amperes	55 Amperes	55 Amperes

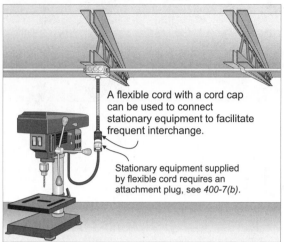

Flexible Cords For Stationary Equipment
Sections 400-7(a)(6)

A flexible cord with a cord cap can be used to connect stationary equipment to facilitate frequent interchange.

Stationary equipment supplied by flexible cord requires an attachment plug, see *400-7(b)*.

Fig. 36-1 Flexible Cords Permitted for Stationary Equipment

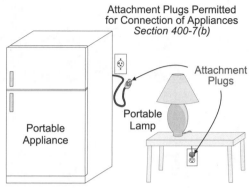

Attachment Plugs Permitted for Connection of Appliances
Section 400-7(b)

Attachment Plugs

Portable Appliance

Portable Lamp

Flexible cords and cables used as permitted by Section 400-7(a)(3, 6, and 8) are required to have an attachment plug and be connected to a receptacle.

Fig. 36-2 Attachment Plugs Permitted for Connection of Portable Appliances

Table 400-5(b)   Allowable ampacity and protection of flexible service cords used as pendant wiring for power and receptacles [210-50(a) and 370-23(g)].

	Three-Conductor Cables		Two-Conductor Cables	
Size (AWG)	Ampacity	Maximum protection 240-4 Ex 1	Ampacity	Maximum protection 240-4 Ex 1
14	15 Amperes	15 Amperes	18 Amperes	20 Amperes
12	20 Amperes	20 Amperes	25 Amperes	25 Amperes
10	25 Amperes	30 Amperes	30 Amperes	30 Amperes
8	35 Amperes	40 Amperes	40 Amperes	40 Amperes
6	45 Amperes	50 Amperes	55 Amperes	50 Amperes

400-7 Uses Permitted

(a) Uses.   Cords and cables must be run exposed and can be used for, Fig. 36-1:

Equipment	Section
Appliances	422-16
Data Processing Cables	645-5
Lighting Fixtures	410-14, 410-30
Pendant Receptacles	210-50(a)
Pendant Boxes	370-13(h)
Lamps and Appliances	400-7(b)
Prevent Noise or Vibration	422-16(a)
Stationary Equipment	422-16(a) 400-7(b)
Temporary Wiring	305-4(b)(c)
Cooking Appliances	422-32(b)

(b) Attachment Plugs. **Attachment plugs** are required for cords used for the connection of portable lamps, portable appliances, stationary equipment, or appliances listed for cord connection, Fig. 36-2.

Note. Attachment plugs can serve as the disconnect for stationary appliances [422-32] and room air-conditioners [440-63].

400-8 Uses Not Permitted

Unless specifically permitted in Section 400-7, flexible cords and cables shall not be used:

(1) As a substitute for the fixed wiring of a structure.

(2) Where run through holes in walls, structural ceilings, suspended/dropped ceilings, or floors, Fig. 36-3.

Note. A cord run through a cabinet for an appliance is generally not considered a violation of this Section, Fig. 36-4.

(3) Where run through doorways, windows, or similar openings.

(4) Where attached to building surfaces.

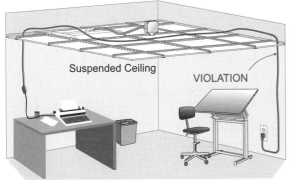

Flexible Cords and Cables - Not Through Ceilings
Section 400-8(2)

Suspended Ceiling

VIOLATION

Cords and cables cannot be run through holes in walls,
structural ceilings, suspended ceilings, or dropped ceilings,
floors, or through doorways, windows, or similar openings.

Fig. 36-3 Flexible Cords and Cables Not Permitted
through Ceilings

(5) Where concealed behind building walls, structural ceilings, suspended/dropped ceilings, or floors.

(6) Where installed in raceways, except as otherwise permitted in this *Code*.

400-10 Pull at Joints and Terminals

Flexible cords shall be installed so that tension will not be transmitted to the conductor terminals. Tension to conductor terminals can be prevented by knotting the cord, winding with tape, or using fittings that are designed for the purpose, such as a strain-relief fitting, Fig. 36-5.

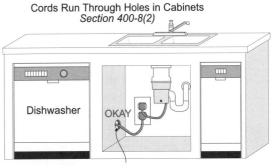

Cords Run Through Holes in Cabinets
Section 400-8(2)

Dishwasher   OKAY

A hole in a cabinet for an appliance cord does
not violate the intent of Section 400-8(2).

Fig. 36-4 Flexible Cord Not Permitted to Be Run
through Holes in Cabinets

400-13 Overcurrent Protection

Cords must be protected against overcurrent according to Section 240-4.

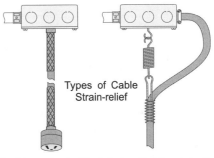

No Tension at Terminals - *Section 400-10*

Types of Cable
Strain-relief

Flexible cords shall installed so that tension will
not be transmitted to the conductor terminals.

Fig. 36-5 Tension Not Permitted to Be
Transmitted to the Terminals

SUMMARY

☐ Attachment plugs are required when flexible cords are used for the connection of portable lamps or appliances, the connection of stationary equipment, or for appliances listed for cord connection. ☐ Attachment plugs can be used as a disconnect for some appliances. ☐ Cords and cables cannot be concealed, be used as a replacement for fixed wiring, attached to the building, installed in raceways, run through holes in walls, ceilings, or floors, or run through doorways, windows, or similar openings. ☐ Flexible cords shall be installed so that tension will not be transmitted to the conductor terminals. ☐ Cords must have overcurrent protection according to Section 240-4.

REVIEW QUESTIONS

1. What are the most common uses for flexible cords?

2. When can an attachment plug be used as a disconnect for appliances?

3. Give a brief summary of some common violations when cords are used.

4. A pendant receptacle rated 20 ampere, 120 volt, as permitted in Section 210-50(a), requires what size conductor and overcurrent protection device?

Unit 37

Article 402
Fixture Wires

OBJECTIVES

After studying this unit, the student should be able to understand:
- the uses permitted and not permitted for fixture wires.
- the smallest size fixture tap permitted for recessed lighting fixtures and what size overcurrent protection device is required.

402-1 Scope

This Article covers the general requirements for fixture wires used for fixtures, Class 1 remote-control circuits, and nonpower-limited fire alarm systems.

402-3 Types

Fixture wires must comply with the installation requirements contained in Table 402-3.

> FPN: Take care when installing any conductor in temperatures below 14°F because the insulation may stiffen. Thermoplastic insulation can be deformed at locations of pressure, such as the conductor bending contact point.

402-5 Allowable Ampacity of Fixture Wires

Table 402-5 gives a summary of fixture wire ampacity and overcurrent protection, as required in the various Sections of the *NEC*.

Table 402-5 Allowable Ampacity and Protection for Fixture Wires [402-5].

Size (AWG)	Ampacity	Maximum Overcurrent (240-4)
18	6 Amperes	20 Amperes up to 50 ft.
16	8 Amperes	20 Amperes up to 100 ft.
14	17 Amperes	30 Amperes
12	23 Amperes	50 Amperes
10	28 Amperes	50 Amperes

Temperature Rating. Fixture wires must not be installed in any way that the conductors' insulation temperature rating is exceeded. Factors that must be considered include ambient temperature, conductor bundling, and mixing of conductors with different insulation ratings. See my comments in Sections 310-10 and 310-15.

402-6 Minimum Size

The smallest size fixture wire permitted in the *NEC* is No. 18.

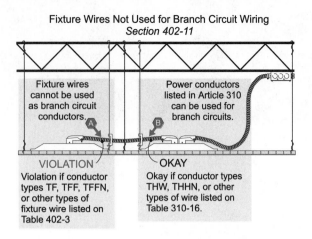

Fixture Wires Not Used for Branch Circuit Wiring
Section 402-11

Fixture wires cannot be used as branch circuit conductors. **A**

Power conductors listed in Article 310 can be used for branch circuits. **B**

VIOLATION
Violation if conductor types TF, TFF, TFFN, or other types of fixture wire listed on Table 402-3

OKAY
Okay if conductor types THW, THHN, or other types of wire listed on Table 310-16.

Fig. 37-1 Fixture Wires Can Not Be Used for Branch Circuit Wiring

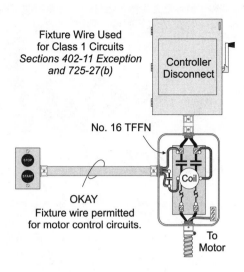

Fixture Wire Used for Class 1 Circuits
Sections 402-11 Exception and 725-27(b)

Controller Disconnect

No. 16 TFFN

STOP
START

Coil

OKAY
Fixture wire permitted for motor control circuits.

To Motor

Fig. 37-2 Fixture Wires Can Be Used for Class 1 Control Wiring

402-7   Number of Conductors in Conduit or Tubing

The number of fixture wires permitted in a raceway shall not exceed the percentage fill specified in Table 1, Chapter 9.

Note. See Appendix C for raceway fill tables for fixture wires.

402-8   Grounded Conductor

Grounded conductors for lighting fixtures must be connected to the screw shell of the lampholder [200-10(c), 410-23] and this conductor must be identified according to the requirements contained in Section 400-22.

402-10   Uses Permitted

Fixture wires can be used for the connection of lighting fixtures to the branch circuit wiring, Fig. 37-1.

402-11   Uses Not Permitted

Fixture wires cannot be used for branch circuits, Class 2 or Class 3 control or signaling circuits [725-52], or power-limited fire protective signaling circuits [760-52].

Exception. Fixture wires can be used for Class 1 control or signaling circuits [725-27(b)], Class 3 circuits [725-71(g)], and nonpower-limited fire alarm circuits [760-27(b)], Fig. 37-2.

402-12   Overcurrent Protection

Fixture wires must be protected against overcurrent according to the requirements contained in Section 240-4. The overcurrent protection requirements of Section 240-4 have been incorporated in Table 402-5 in this book.

Note. Fixture wire used for motor control circuits taps must have overcurrent protection, according to Section 430-72(a). Class 1 motor control circuits must have overcurrent protection according to Section 725-23.

SUMMARY

☐ Take care when installing fixture wires in temperatures below 14°F because the insulation may stiffen. ☐ Fixture wires shall not be installed in any way that the temperature rating of the conductor will be exceeded. ☐ The ampacity of fixture wires is listed in Table 402-5. ☐ The number of fixture wires in a raceway is limited to the percent fill of Table 1 of Chapter 9. ☐ The grounded (neutral) conductor must be identified by means of stripes or by means described in Section 400-22(a) through (e). ☐ Fixture wires can be used for wiring of lighting fixtures, Class 1 remote-control, and nonpower-limited fire alarm circuits. ☐ Fixture wires cannot be used for branch circuits, Class 2, Class 3, or power-limited fire alarm circuits. ☐ Fixture wire shall have overcurrent protection according to Section 240-4.

REVIEW QUESTION

1. What are uses permitted and not permitted for fixture wires?

Unit 38

Article 410
Lighting Fixtures, Lampholders, Lamps, and Receptacles

OBJECTIVES

After studying this unit, the student should be able to understand:
- the requirements for installing lighting fixtures around pools and spas.
- the requirements for installing lighting fixtures in commercial cooking hoods.
- the requirements for installing lighting fixtures in the bathtub zone.
- the requirements for installing lighting fixtures in clothes closets.
- the requirements for installing electric-discharge lighting fixtures not exceeding 1,000 volt.
- the NEC requirements for installing lighting fixtures on metal poles.
- the use of suspended ceilings for the support of fixtures.
- when cords can be used for the permanent wiring of fixtures.
- the receptacle requirements of Article 410.
- the installation requirements for recessed lighting fixtures when combustible materials and thermal insulation are present.
- the installation requirements for recessed lighting fixtures with tap conductors.
- the installation requirements for track lighting.

PART A. GENERAL

410-1 Scope

Article 410 contains the requirements for fixtures, lampholders, receptacles, and receptacle covers, Fig. 38-1.

Note. Because of the many types and applications of lighting fixtures, manufacturer instructions are very important and helpful for proper installation. Underwriters Laboratories (UL) produces a pamphlet called the "Fixture Marking Guide," which provides information for properly installing common types of incandescent, fluorescent, and high intensity discharge (HID) fixtures.

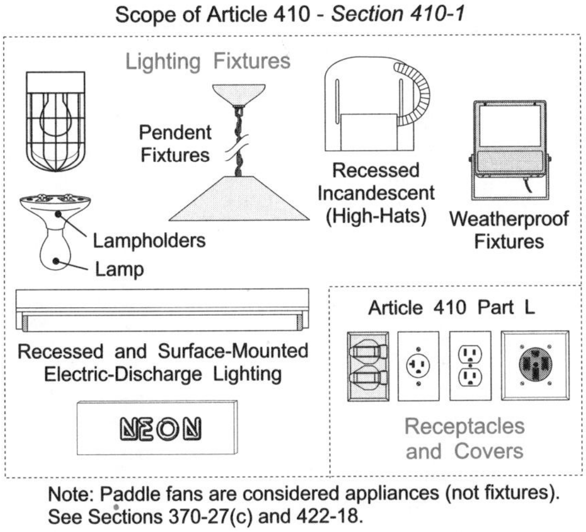

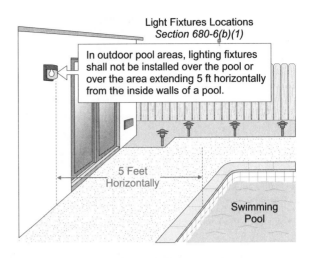

Fig. 38-3 Light Fixtures Must Not Be Closer than 5 Feet from Outdoor Pools, Spas, or Hot Tubs

Lighting fixtures marked "Suitable for Wet Locations" can be installed in wet, damp, or dry locations (according to testing laboratory publications).

Note. Lighting fixtures installed in and around permanently installed pools and outdoor spas must be installed in accordance with Section 680-6(b)(1) and 680-20, Fig. 38-3.

(b) Corrosive Locations. Fixtures installed in corrosive locations must be approved for the location.

(c) Commercial Hood Fixtures. Fixtures installed in commercial cooking hoods must be identified for use in commercial hoods. The wiring methods that supply these fixture(s) cannot be exposed within the cooking hood [300-22(a)], Fig. 38-4.

Note. Cooking grease and oils can cause grounds, shorts, and fires. Fixtures installed in cooking hoods must be specially constructed and listed for this purpose. Standard gasketed lighting fixtures are not permitted because they permit the accumulation of grease and oil deposits that can result in a fire because of high temperatures on the glass globe.

(d) Above Bathtubs. No part of cord-connected lighting fixtures, hanging lighting fixtures, lighting track, *pendants* or suspended ceiling paddle

---

Fig. 38-1 Article 410 Covers Fixtures, Lampholders as Well as Receptacles

CAUTION: Paddle fans are listed as an appliance, not as a lighting fixture, and must be installed according to Sections 370-27(c) and 422-18.

PART B. FIXTURE LOCATIONS

410-4 Specific Locations

(a) Installed in Wet and Damp Locations. Lighting fixtures installed in wet or damp locations must be installed in a manner that will prevent water from accumulating in any part of the fixture.

Fixtures marked "Suitable for Dry Locations Only" must be installed in dry locations only. Fixtures marked "Suitable for Damp Locations" can be installed in damp or dry locations, Fig. 38-2.

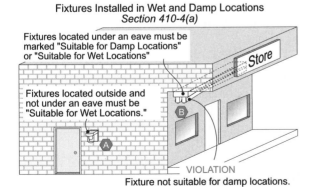

Fig. 38-2 Fixtures Installed in Wet or Damp Locations Must Be Suitable for the Location

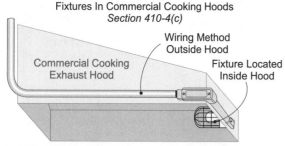

Fixtures In Commercial Cooking Hoods
Section 410-4(c)

Wiring Method
Outside Hood

Commercial Cooking
Exhaust Hood

Fixture Located
Inside Hood

Lighting Requirements:
• Fixture must be rated for use in commercial cooking hoods.
• Fixture must be constructed so grease and oil do not get on the lamp or in the wiring compartment.
• Fixture must be resistant or protected against corrosion.
• Wiring method must not be exposed inside the cooking hood.

Fig. 38-4 Fixtures Must Be Listed for the Location, When Installed in Commercial Cooking Hoods

fans shall be located within a zone measured 3 feet horizontally and 8 feet vertically from the top of the bathtub rim or shower stall threshold, Fig. 38-5.

Note. This Section specifies the types of lighting fixtures that are not permitted in the bathtub and shower zone area. It would be better if it specifically identified what is permitted because the way the rule is written, wall mounted lighting fixtures could be installed!

GFCI protection is not required for lighting fixtures installed within the bathtub or shower zone. However, lighting fixtures mounted over indoor spas and hot tubs require GFCI protection if the fix-

Fixtures in Bathtub or Shower Area - *Section 410-4(d)*
Exhaust fan and/or light is okay.

Paddle Fan

VIOLATION

8 feet

3 feet

Shower
Stall
Zone

Shower Stall Threshold

No parts of cord-connected fixtures, hanging fixtures, lighting track, pendants, or ceiling paddle fans shall be located within the bathtub/shower zone.

Fig. 38-5 Ceiling Paddle Fan Not Permitted in the Bathtub or Shower Zone

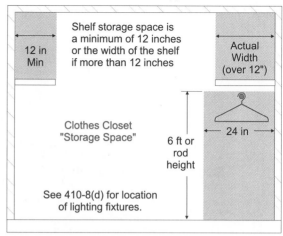

Definition of Storage Space in Clothes Closets
Section 410-8(a)

12 in
Min

Shelf storage space is a minimum of 12 inches or the width of the shelf if more than 12 inches

Actual
Width
(over 12")

Clothes Closet
"Storage Space"

6 ft or
rod
height

24 in

See 410-8(d) for location of lighting fixtures.

Fig. 38-6 Definition of Storage Space in Clothes Closets

ture is located less than 7 feet, 6 inches, above the maximum water level [680-41(b)(1)].

410-8 Clothes Closets

The *Code* does not require lighting fixture to be installed in a clothes closet, but if a lighting fixture is installed, we must prevent the lamps from igniting combustible material.

(a) Definition of Storage Space. Storage space shall be defined as a volume extending from the closet floor vertically to a height of 6 feet or the highest clothes-hanging rod, parallel to the walls at a horizontal distance of 24 inches from the sides and back of the closet walls, and continuing vertically to the closet ceiling a horizontal distance of 12 inches or the width of the shelf, whichever is greater, Fig. 38-6.

(b) Fixture Types Permitted in Clothes Closets. Fixtures that prevent hot particles from coming in contact with combustible materials such as totally enclosed incandescent and any fluorescent fixtures can be installed in clothes closets, Fig. 38-7.

(c) Fixture Types Not Permitted in Clothes Closets. Incandescent lighting fixtures that have open lamps and pendant type lighting fixtures cannot be installed in clothes closets, Fig. 38-8.

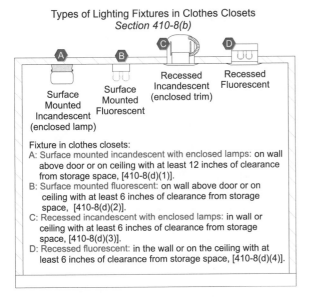

Fig. 38-7 Types of Lighting Fixtures Permitted in a Clothes Closet

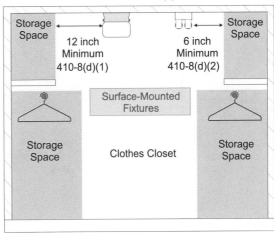

Fig. 38-9 Location of Surface Mounted Lighting Fixtures in a Clothes Closet

(d) Installation of Fixtures in Clothes Closets.

Surface Mounted. Totally enclosed incandescent lighting fixtures must maintain a minimum clearance of 12 inches from the storage space. Surface mounted fluorescent lighting fixtures must maintain a minimum clearance of 6 inches from the storage space. Fig. 38-9.

Recessed Mounted. Totally enclosed recessed incandescent lighting fixtures must maintain a minimum clearance of 6 inches from the storage area. Recessed fluorescent lighting fixtures must maintain a minimum clearance of 6 inches from the storage area, Fig. 38-10.

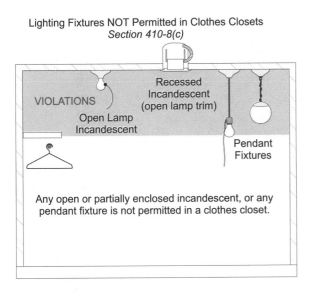

Fig. 38-8 Types of Lighting Fixtures Not Permitted in a Clothes Closet

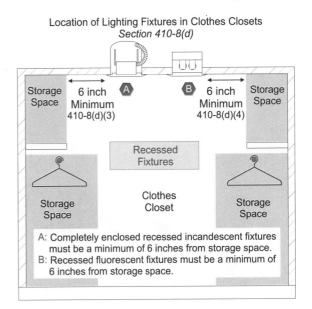

Fig. 38-10 Location of Recessed Lighting Fixtures in a Clothes Closet

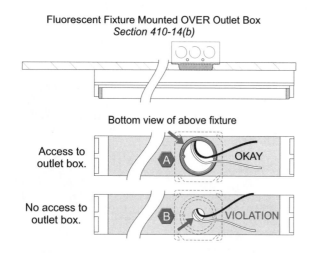

Fluorescent Fixture Mounted OVER Outlet Box
Section 410-14(b)

Bottom view of above fixture

Access to outlet box.   Ⓐ   OKAY

No access to outlet box.   Ⓑ   VIOLATION

Fig. 38-11 Fluorescent Fixture Mounted over Outlet Box, Must Permit Access to Wiring

PART C. FIXTURE OUTLET BOXES AND COVERS

410-12 Outlet Box to Be Covered

All lighting outlet boxes must be covered with either a fixture, canopy, lampholder, receptacle, or blank cover [370-25].

410-14 Connection of Electric Discharge Lighting Fixtures

(a) Fixture Supported Independent of the Outlet Box. Electric-discharge lighting fixtures supported independently of the outlet box must be

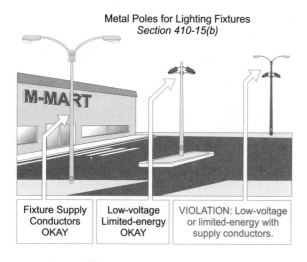

Metal Poles for Lighting Fixtures
Section 410-15(b)

M-MART

| Fixture Supply Conductors OKAY | Low-voltage Limited-energy OKAY | VIOLATION: Low-voltage or limited-energy with supply conductors. |

Fig. 38-12 Metal Poles Can Not Contain Low-Voltage or Limited-Energy Wiring

wired with a raceway, metal-clad cable, armored cable, or nonmetallic-sheathed cable.

Individual electric discharge lighting fixtures can be cord connected if the lighting fixtures require aiming [410-30(b)] or if the cord is visible for its entire length and is plugged into a receptacle [410-30(c).

(b) Access to Outlet Box. When an electric discharge fixture is mounted over an outlet box, the fixture must permit access to the outlet box, Fig. 38-11.

PART D. FIXTURE SUPPORTS

410-15 Metal Poles

(a) General Support Requirement. Fixtures, lampholders, and receptacles [410-56(f)] must be securely supported.

(b) Metal Poles. Metal poles can be used to support lighting fixtures and serve as a raceway enclosing lighting fixture supply conductors. Low-voltage and limited-energy system conductors and cables for cameras, speakers, etc. cannot be installed within the pole with lighting fixture power conductors according to the following *Code* Sections, Fig. 38-12:

System	Section
CATV	820-10(f)(1)
Control and Signaling	725-54(a)(1)
Fire Alarm	760-54(a)(1)
Radio and Television	810-18(c), 810-70
Telecommunications	800-52(a)(1)

(1) Handhole. The metal pole must have an accessible 2 inch × 4 inch handhole with a raintight cover. The handhole must provide access to the supply conductors within the pole.

Exception No. 1: Poles 8 Feet or Less. The handhole can be omitted on metal poles 8 feet or less in height above finish grade if the supply conductors continue to the fixture and the splices are accessible by removing the fixture.

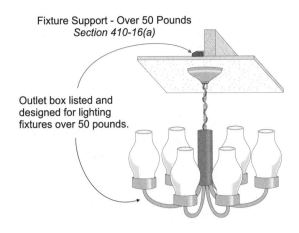

Fig. 38-13 Outlet Box Can Support Lighting Fixtures That Weighs More Than 50 Pounds

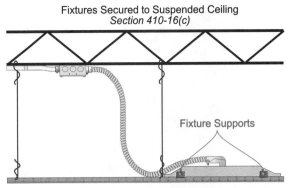

Fixtures must be securely fastened to the ceiling framing member by screws, bolts, rivets, or listed clips identified for the purpose.

Fig. 38-14 Lighting Fixtures Must Be Secured to Suspended Ceiling Required

Exception No. 2: Poles 20 feet or Less. The handhole can be omitted on metal poles 20 feet or less above finish grade if the pole is provided with a hinged base.

(2) Without Handhole. When the supply raceway or cable does not enter the pole, a threaded fitting or nipple must be welded, brazed, or tapped opposite the handhole opening.

(3) Grounding Terminal. An accessible grounding terminal must be installed inside the pole.

Exception: Poles 8 Feet or Less. The grounding terminal can be omitted on metal poles 8 feet or less in height above finish grade if the supply conductors continue to the fixture and the splices are accessible by removing the fixture.

(5) Grounding the Metal Pole. The metal pole must be grounded to an equipment grounding conductor [250-118] sized according to Table 250-122. If the supply raceway serves as the equipment grounding conductor, it must be bonded to the metal pole.

(6) Conductor Vertical Supports. Conductors run greater than 100 feet vertically must be supported according to Section 300-19.

410-16 Support

(a) To Outlet Boxes. Outlet boxes or fittings specifically designed and listed for the support of lighting fixtures [370-27] can support lighting fixtures that weigh up to 50 pounds. Fixtures that weigh more than 50 pounds must be supported independently of the outlet box or fitting unless the outlet box or fitting is designed and listed for the weight of the fixture, Fig. 38-13.

(c) To Suspended Ceiling Framing Members. The proper support of lighting fixtures to a suspended ceiling framing member is important for the protection of firefighters when they enter the room to put out a fire. Lighting fixtures attached to the suspended ceiling framing shall be secured to the framing member with screws, bolts, rivets, or clips listed and identified for use with the type of ceiling framing members and fixtures involved, Fig. 38-14.

Note. All wiring behind removable panels must be supported in accordance with Section 300-11(a). Outlet boxes can be secured to the ceiling grid framing members in accordance with Section 370-23(d).

CAUTION: Independent support wires for suspended ceiling fixtures are sometimes required by local or state Codes, such as those in California and Washington State (earthquakes). But they are not required by the *National Electrical Code.*

(h) Fixtures Supported to Trees. Trees can support lighting fixtures and other electrical

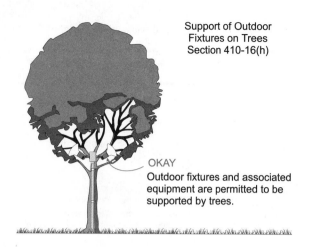

Support of Outdoor
Fixtures on Trees
Section 410-16(h)

OKAY
Outdoor fixtures and associated
equipment are permitted to be
supported by trees.

Fig. 38-15 Outdoor Lighting Fixtures Can be Supported onto Trees

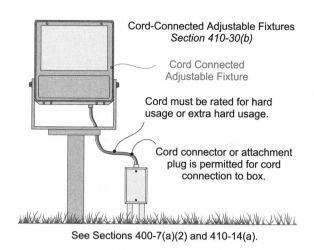

Cord-Connected Adjustable Fixtures
Section 410-30(b)

Cord Connected
Adjustable Fixture

Cord must be rated for hard
usage or extra hard usage.

Cord connector or attachment
plug is permitted for cord
connection to box.

See Sections 400-7(a)(2) and 410-14(a).

Fig. 38-16 Cords Can be Used for the Connection of Lighting Fixtures that Require Adjustment

equipment, but trees cannot be used for the support of overhead conductor spans except for temporary lighting [225-26], Fig. 38-15.

PART F. WIRING OF FIXTURES

410-23 Polarization of Fixtures

Fixtures shall have the grounded (neutral) connected to the screw shell of the lampholder [200-10(c)], and the grounded (neutral) conductor must be properly identified [200-6 and 402-8].

410-30 Cord-Connected Lighting Fixtures

(b) Adjustable Fixtures. Fixtures that require adjusting or aiming after installation shall not be required to be equipped with an attachment plug or cord connector provided the exposed cord is of the hard usage or extra-hard usage type and is not longer than that required for maximum adjustment. The cord shall not be subject to strain or physical damage, Fig. 38-16.

(c) Electric-Discharge Fixtures.

(1) Listed Fixture or Assembly. A single listed electric discharge fixture or a listed fixture assembly can be cord-connected if, Fig. 38-17:

(a) The fixture is mounted directly below the outlet box or busway, and

(b) The flexible cord is visible its entire length outside the fixture, not subject to stress or physical damage [400-10], and the receptacle outlet must not be rated less than 15 ampere [410-56(a)].

Note. The Code does not specify locking plugs or receptacles, but an involuntary disconnect of the circuit could occur without them.

410-31 Fixtures Used As Raceway

Lighting fixtures shall not be used as a raceway for circuit conductors other than the lighting fixture, Fig. 38-18.

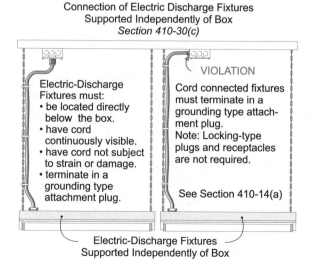

Connection of Electric Discharge Fixtures
Supported Independently of Box
Section 410-30(c)

Electric-Discharge
Fixtures must:
• be located directly
  below the box.
• have cord
  continuously visible.
• have cord not subject
  to strain or damage.
• terminate in a
  grounding type
  attachment plug.

VIOLATION

Cord connected fixtures
must terminate in a
grounding type attach-
ment plug.
Note: Locking-type
plugs and receptacles
are not required.

See Section 410-14(a)

Electric-Discharge Fixtures
Supported Independently of Box

Fig. 38-17 Electric Discharge Fixtures Can Be Cord and Plug Connected

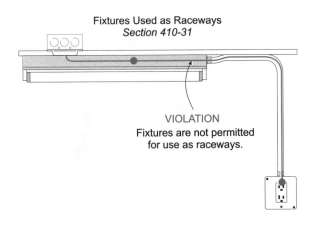

Fig. 38-18 Lighting Fixtures Can Not Be Used
as a Raceway for Other Conductors

Exception No. 1: Listed as a Raceway. Fixtures can be used as a raceway if the fixture is marked "Suitable for Use as Raceway." These lighting fixtures will be marked with the number of conductors, the size of the conductors, and the conductor insulation rating.

Note. This exception means that lighting fixtures marked suitable for use as a raceway can contain branch circuit wiring for other circuits. I have never seen such a fixture.

Exception No. 2: End-to-End Assembly. Fixtures designed for end-to-end assembly (fluorescent strips) or connected together by wiring method can carry through the fixture supply conductors, Fig. 38-19.

Exception No. 3: Night Lights. Fixtures designed for end-to-end assembly or connected together by wiring method can carry through an

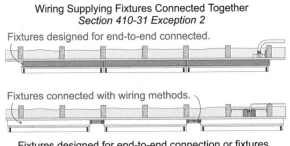

Fig. 38-19 Wiring Requirements for Fixtures
That Are Connected Together

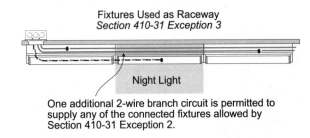

Fig. 38-20 Night Light Fixture Wiring Permitted

additional 2-wire circuit through the lighting fixtures to supply 1 or more fixtures, Fig. 38-20.

PART H.   LAMPHOLDERS

410-47   Receptacle Adapter

Lampholders are designed for lamps only, and a receptacle adapter cannot be installed in the lampholder, Fig. 38-21.

Note. Receptacle adapters are commonly installed for Christmas and decorative lighting in violation of this Section.

PART L.   RECEPTACLES

410-56   Receptacle Installation

(b) Receptacles Connected to Aluminum Conductors. Receptacles and snap switches [380-14(c)] rated 15 and 20 ampere connected to aluminum wire must be marked CO/ALR.

Note. Copper and copper-clad aluminum conductors can terminate at 15 and 20

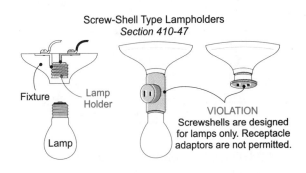

Fig. 38-21 Receptacle Adapters Are Not Permitted

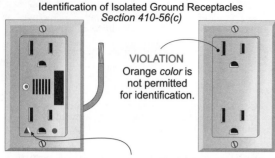

Identification of Isolated Ground Receptacles
Section 410-56(c)

VIOLATION
Orange *color* is
not permitted
for identification.

Isolated ground receptacles are identified by an orange
triangle located on the face of the receptacle.

**Fig. 38-22 Isolated Ground Receptacles Must Be
Identified by an Orange Color Triangle**

ampere receptacles not marked CO/ALR.
Push-in type terminals on receptacles and
switches are never suitable for aluminum or
stranded wire.

(c) Isolated Ground Receptacles. Isolated-
ground receptacles used for the reduction of
electric noise [250-146(d)] must have an orange triangle marking on the exposed receptacle surface,
Fig. 38-22.

Isolated ground receptacles installed in nonmetallic boxes must be covered with a nonmetallic
faceplate. A metal faceplate covering is not permitted for an isolated ground receptacle installed in
nonmetallic box because the cover cannot be
grounded. This is due to the fact that the metal
mounting yoke of the isolated ground receptacle is
not connected to the circuit equipment grounding
conductor, Fig. 38-23.

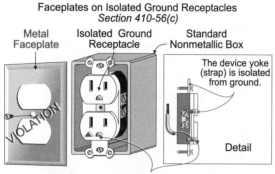

Faceplates on Isolated Ground Receptacles
Section 410-56(c)

Metal
Faceplate

Isolated Ground
Receptacle

Standard
Nonmetallic Box

The device yoke
(strap) is isolated
from ground.

VIOLATION

Detail

A metal faceplate cannot be installed on an isolated
ground receptacle in a nonmetallic box because the
faceplate cannot be grounded.

**Fig. 38-23 Faceplates on Isolated Ground
Receptacles Must Be Able to Be Grounded**

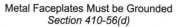
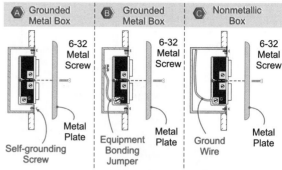

Metal Faceplates Must be Grounded
Section 410-56(d)

A Grounded Metal Box	B Grounded Metal Box	C Nonmetallic Box
6-32 Metal Screw	6-32 Metal Screw	6-32 Metal Screw
Metal Plate	Metal Plate	Metal Plate
Self-grounding Screw	Equipment Bonding Jumper	Ground Wire

Note: The 6-32 metal faceplate screw grounds
the metal faceplate to the strap and box.

**Fig. 38-24 Metal Receptacle Faceplates
Must Be Grounded**

Exception. An isolated ground receptacle can
be installed in a nonmetallic box with a metal faceplate, if the box contains a feature or accessory that
effectively grounds the metal faceplate.

(d) Grounding Metal Faceplate(s). Metal
faceplates for receptacles must be grounded. This
can be accomplished by securing the metal faceplate to a receptacle [517-13(a) Ex. 2], Fig. 38-24.

(e) Faceplate Installation. Faceplates for
receptacles must completely cover the outlet openings, and they must seat firmly against the mounting
surface.

(f) Receptacle Faceplate Requirements.
Before installing a receptacle faceplate, the box
must be properly mounted [370-20] and secured
[370-23].

(1) Recessed Box. The receptacle contained
in a recessed outlet box must be mounted so that the
yoke is held rigidly to the building surface.

(2) Flush or Surface Mounted Box. Receptacle installed in a flush or surface mounted box must
be mounted so that the yoke is held against the box
or box cover.

(3) Secured to Cover. Receptacles mounted
to and supported by a cover shall be secured by two
or more screws. However, devices or assemblies
listed and identified for this purpose can have the

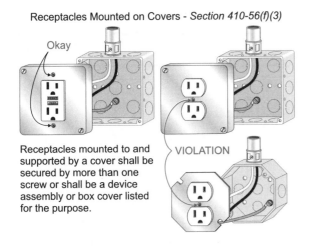

Receptacles Mounted on Covers - *Section 410-56(f)(3)*

Okay

Receptacles mounted to and supported by a cover shall be secured by more than one screw or shall be a device assembly or box cover listed for the purpose.

VIOLATION

Fig. 38-25 Receptacles Mounted on Covers Must Be Secured with Two Screws

receptacle installed to a cover by a single screw, Fig. 38-25.

410-57 Receptacles in Damp or Wet Locations

(a) Receptacles in Damp Locations. Receptacles installed outdoors protected from the weather or in other damp locations, must be installed in a **weatherproof** enclosure when the **attachment plug** cap is not inserted and the receptacle cover is closed.

Receptacles shall be considered to be protected from the weather where not subject to beating rain or water runoff, such as under roofed open porches, canopies, and marquees.

(b) Receptacles in Wet Locations.

(1) Fixed Equipment. A receptacle installed in a wet location shall have a cover that is listed as weatherproof while the attachment plug is inserted. This is common for water sprinkler controllers, landscape lighting, holiday lighting, or other loads that are intended to have an attachment plug inserted into the receptacle, Fig. 38-26.

(2) Portable Equipment. A receptacle installed in a wet location for portable equipment shall have a listed cover that is weatherproof when the attachment plug cap is inserted or removed, Fig. 38-26.

(c) Bathtub and Shower Space. There is no minimum distance that a receptacle must be from the bathtub or shower space, but a receptacle cannot be installed in the bathtub or shower space, Fig. 38-27.

Note. Receptacles must be a minimum of 5 feet away from an indoor spa or hot tub [680-41(a)(1)], Fig. 38-28.

PART M.   RECESSED FIXTURES

410-65 Thermally Protected

(c) Recessed Incandescent Fixtures. Recessed incandescent lighting fixtures must be listed as thermally protected.

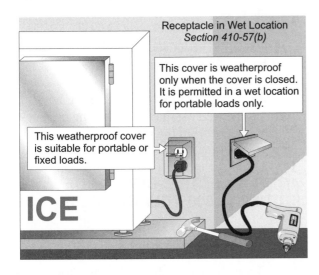

Receptacle in Wet Location
Section 410-57(b)

This cover is weatherproof only when the cover is closed. It is permitted in a wet location for portable loads only.

This weatherproof cover is suitable for portable or fixed loads.

ICE

Fig. 38-26 Receptacles Installed in Wet Location Must Be Protected from the Weather

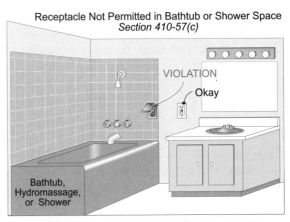

Receptacle Not Permitted in Bathtub or Shower Space
Section 410-57(c)

VIOLATION

Okay

Bathtub, Hydromassage, or Shower

There is not a requirement for how far a receptacle must be located away from bathtubs and showers.

Fig. 38-27 Receptacle Not Permitted in Bathtub or Shower Space

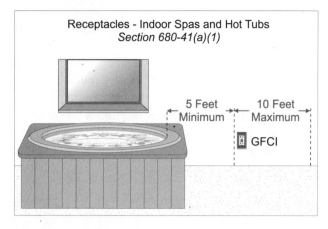

Fig. 38-28 Receptacles Must Be a Minimum of 5 Feet Away from an Indoor Spa or Hot Tub

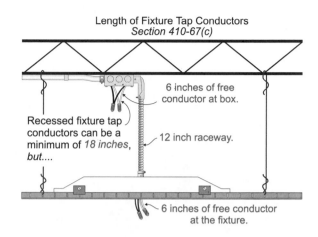

Fig. 38-29 Fixture Tap Conductors Can Not Be Less than 18 Inches or More than 6 Feet

Exception No. 2: IC Rated Fixtures. Fixtures designed and constructed to provide protection similar to thermally protected fixtures such as Type IC rated fixtures.

> **Note.** When higher wattage lamps or improper trims are installed, recessed fixture can overheat, activating the thermal protection device causing the fixture cycling off and on (blinking).

410-66   Recessed Fixture Clearances

(a) Clearances from Combustible Materials.

(1) Except at the points of support, recessed lighting fixtures must maintain a ½ inch air space from combustible materials.

(2) Recessed lighting fixtures listed for direct contact with insulation (Type IC) do not have to maintain a ½ inch air space from combustible materials.

(b) Clearances from Thermal Insulation. Thermal insulation is not permitted within 3 inches of a recessed fixture, wiring compartment, or ballast unless the fixture is listed for direct contact with insulation (Type IC).

> **Note.** Since it is beyond the control of the electrician where blown insulation ends up, it is advisable to use "Type IC" recessed fixtures.

410-67   Wiring

(c) Tap Conductors (Whips). Fixture tap conductors are permitted if the tap conductors comply with the following, Fig. 38-29.

Insulation. Tap conductors must have an insulation temperature rating suitable for the recessed fixture termination.

Outlet Box. The outlet box must be located at least 1 foot from the recessed fixture.

Wiring Methods. Tap conductors must be installed in a raceway, armored cable, or metal-clad cable.

Length. The length of the raceway or cable for the tap conductors is not less than 18 inches and no more than 6 feet.

Ampacity. The ampacity of the tap conductors cannot be less than 15 ampere [210-19(d) Ex. 1].

> **Note.** Section 300-14 requires at least 6 inches of free conductor where the conductors emerge from the raceway.

PART P. ELECTRIC-DISCHARGE LIGHTING (1,000 VOLT OR LESS)

410-73   General

(e) Thermal Protection. Ballasts for electric-discharge lighting installed indoors must have

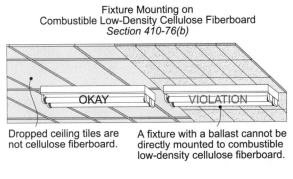

Fig. 38-30 Lighting Fixture Can Not Be Directly Mounting to Combustible Low-Density Fiberboard

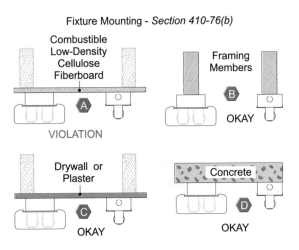

Fig. 38-31 Lighting Fixtures Can Be Directly Mounting to Wood, Drywall or Concrete Surfaces

protection from overheating (Class P).

Class P ballasts have an internal thermal protection device that opens at a predetermined temperature to prevent overheating of the ballast and related components. When the ballast cools off, it automatically resets (closes the circuit) and the light comes back on. When you see a fluorescent fixture cycling on and off, the ballast is overheating and needs to be replaced.

Note. Supplementary overcurrent protection such as a in-line fuse is not required by the *NEC* to meet the requirements of Section 410-73(e).

410-76 Fixture Mounting

(b) Surface Mounted Fixtures (with Ballast). Surface mounted lighting fixtures containing a ballast must have a minimum of 1½ inch clearance from combustible low density fiberboard unless the fixture is marked "Suitable for Surface Mounting on Combustible Low Density Cellulose Fiberboard," Fig. 38-30.

Note. Fluorescent lighting fixtures can be mounted directly to wood, plaster, concrete, or drywall, Fig. 38-31.

PART Q. ELECTRIC DISCHARGE LIGHTING MORE THAN 1,000 VOLTS

410-80 General

(b) Dwelling Occupancies. Electric-discharge lighting operating at over 1,000 volt such as

neon lighting is not permitted in or on a dwelling unit [600-32(i)].

PART R. TRACK LIGHTING

410-100 Definition

Track lighting is manufactured as an assembly to support readily removable fixtures.

410-101 Installation

(a) Track Lighting. Track lighting must be permanently installed and permanently connected to the branch circuit wiring. Lampholders for track lighting are designed for lamps only and a receptacle adapter is not permitted [410-47].

(b) Circuit Rating. Track lighting shall not be connected to a circuit that exceeds the rating of the track. This means that 15 ampere rated track cannot be connected to a 20 ampere circuit, nor could 20 ampere rated track be connected to a 30 ampere circuit.

(c) Uses. Track lighting shall not be installed at any of the following:

(1) Where likely to be subjected to physical damage.
(2) In wet or damp locations.
(3) Where subject to corrosive vapors.

(4) In storage battery rooms.

(5) In hazardous (classified) locations.

(6) Where concealed.

(7) Where extended through walls or partitions.

(8) Less than 5 feet above the finished floor except where protected from physical damage or track operating at less than 30 volt RMS open-circuit voltage.

(9) Within the zone measured 3 feet (914 mm) horizontally and 8 feet vertically from the top of the bathtub rim.

410-104 Fastening

Track lighting shall be securely mounted to support the weight of the fixtures. Individual track lengths of 4 feet or less require two supports, and continuous lengths must be supported every 4 feet.

SUMMARY

Part A. General □ Article 410 contains the requirements for fixtures, lampholders, and receptacles. □ Manufacturers' instructions are very important and helpful for proper installation of fixtures.

Part B. Fixture Locations □ Fixtures installed in wet or damp locations must be installed to prevent water from accumulating in any part of the fixture. □ Lighting fixtures installed in and around permanently installed pools and outdoor spas have special requirements. □ Fixtures mounted over indoor spas and hot tubs have special requirements. □ Fixtures installed in corrosive locations must be approved for the location. □ Fixtures installed in commercial cooking hoods must be identified for the purpose. □ Pendant, hanging type, paddle fans and track lighting fixtures cannot be installed in the bathtub zone. □ It is generally accepted that the bathroom area is not a damp or wet location. □ Fixtures near combustible material must be constructed, installed, or equipped to prevent the combustible materials from being exposed to temperatures over 90°C. □ Incandescent lighting fixtures that totally enclose the lamps and fluorescent fixtures can be installed in clothes closets. □ Storage area shall include the space above the shelf not less than 12 inches in depth extending to the ceiling. □ Totally-enclosed surface-mounted incandescent lighting fixtures in clothes closets must maintain a minimum clearance of 12 inches from the storage area. □ Surface-mounted fluorescent lighting fixtures and recessed incandescent lighting fixtures must maintain a minimum clearance of 6 inches from the storage area. □ The *Code* does not require lighting outlets in clothes closets.

Part C. Outlet Boxes □ When an electric-discharge fixture such as fluorescent or metal halide fixtures is mounted over an outlet box, the fixture must permit access to the box.

Part D. Fixture Supports □ The metal pole for lighting fixtures must have a 2 inch × 4 inch accessible handhole with a raintight cover and grounding terminal. □ The handhole and grounding terminal can be omitted on metal poles 8 feet or less above finish grade. □ The metal pole must be grounded to an equipment grounding conductor. □ Paddle fans are not fixtures, but are appliances and must be installed according to Sections 422-18 and 370-27(c). □ Fixtures can be supported by the suspended ceiling framing members. □ Trees can support lighting fixtures and other electrical equipment, but not overhead conductor spans.

Part E. Grounding □ Exposed metal parts of lighting fixtures shall be grounded according to Article 250.

Part F. Wiring of Fixtures □ Fixtures shall have the grounded (neutral) conductor connected to the screw shell of the lampholder. □ Fixtures cannot be used as a raceway for circuit conductors unless listed for this purpose.

Part L. Receptacles □ 15 and 20 ampere receptacles connected to aluminum conductors must be marked CO/ALR. □ Copper and copper-clad aluminum conductors can terminate to 15 and 20 ampere receptacles that are not marked CO/ALR. The push-in type terminals on 15 and 20 ampere receptacles and switches are never suitable for aluminum or stranded wire. □ Isolated ground receptacles used for the reduction of electric noise [250-146(d)], must have an orange triangle marking on the exposed receptacle surface. □ Metal receptacle faceplates must be grounded. □ Receptacles installed outdoors protected from the weather or other damp locations shall be installed in an enclosure that is weatherproof when no attachment plug is inserted. □ Receptacles installed in other wet locations shall be installed in an enclosure that is weatherproof.

Part M. Recessed Fixtures □ Recessed incandescent lighting fixtures must be listed as thermally protected. □ Recessed lighting fixtures must maintain a ½ inch space from combustible materials. □ Thermal insulation is not permitted within 3 inches of the recessed fixture, wiring compartment, ballast. □ Recessed lighting fixtures listed for direct contact with insulation (Type IC) can be encapsulated with insulation.

Part P. Electric-Discharge Not Over 1,000 volt □ Ballasts installed indoors are required to have internal thermal protection (Class P).

Part Q. Electric-Discharge Over 1,000 volt □ Dwelling units cannot have neon lighting with voltage exceeding 1,000 volt.

Part R. Track Lighting □ Track lighting shall be permanently installed and permanently connected to the branch circuit wiring.

REVIEW QUESTIONS

1. Explain the requirements for installing lighting fixtures around pools and spas.

2. Explain the requirements for installing lighting fixtures in commercial cooking hoods.

3. Explain the requirements for installing lighting fixtures in the bathtub zone.

4. Explain the requirements for installing lighting fixtures in a clothes closet.

5. Summarize the installation requirements for installing electric-discharge lighting fixtures not exceeding 1,000 volt.

6. When installing lighting fixtures on metal poles, what are the *NEC* requirements?

7. Provide a brief summary on the use of suspended ceilings for the support of fixtures.

8. When can cords be used for the permanent wiring of fixtures?

9. Give a brief summary of the receptacle requirements of Article 410.

10. Explain the installation requirements for recessed fixtures, combustible materials, and thermal insulation.

11. Explain the installation requirements for recessed lighting fixtures with tap conductors.

12. Give a brief summary of the installation requirements for track lighting.

Unit 39

Article 422

Appliances

OBJECTIVES

After studying this unit, the student should be able to understand:
- how conductors are sized for appliances.
- when flexible cords can be used to supply appliances.
- the rules for installing wall-mounted ovens and counter-mounted cooking units.
- the installation requirements of paddle fans.
- the rules for using circuit breakers or switches for disconnects.
- the rules for using attachment plugs and receptacles for disconnects.
- the rules for using unit switches for disconnects.
- the rules for overcurrent protection of appliances.

PART A. GENERAL

422-1 Scope

The scope of Article 422 includes appliances that are fastened in place, permanently connected, or cord- and plug-connected in any occupancy. This Article does not apply to electric space-heating equipment [Article 424], Fig. 39-1.

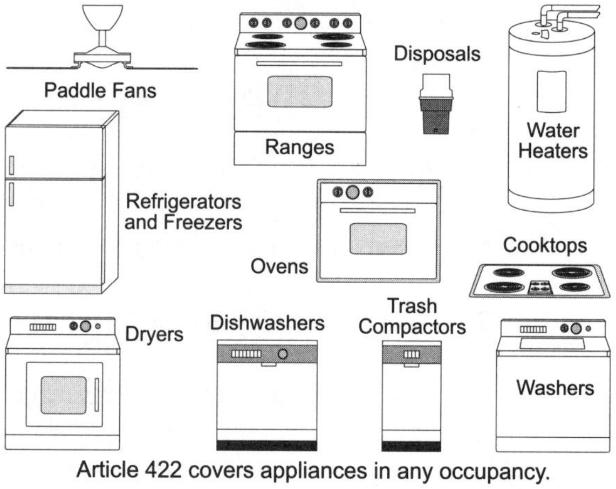

422-3 Other Articles

Appliances used in hazardous locations must comply with the requirements of Articles 500 through 517. Motor operated appliances must comply with Article 430, and hermetic refrigerant motor compressor(s) must comply with Article 440, except where specifically modified by this Article.

Fig. 39-1 Scope of Article 422—Appliances

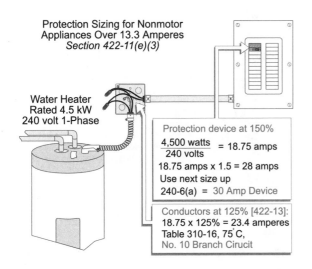

Protection Sizing for Nonmotor
Appliances Over 13.3 Amperes
Section 422-11(e)(3)

Water Heater
Rated 4.5 kW
240 volt 1-Phase

Protection device at 150%
$$\frac{4,500 \text{ watts}}{240 \text{ volts}} = 18.75 \text{ amps}$$
18.75 amps x 1.5 = 28 amps
Use next size up
240-6(a) = 30 Amp Device

Conductors at 125% [422-13]:
18.75 x 125% = 23.4 amperes
Table 310-16, 75°C,
No. 10 Branch Cirucit

Fig. 39-2 Overcurrent Protection Shall Not Exceed 150 Percent of Appliance Current Rating

PART B. BRANCH CIRCUIT REQUIREMENTS

422-10 Branch Circuit Sizing

(a) Individual Circuits. The ampacity of branch circuit conductors to an individual appliance shall not be less than required by the appliance marking or instructions [110-3(b)].

Appliances that are continuously loaded must have the conductor ampacity sized not less than 125 percent of the appliance marked rating [220-2(a)].

> **Note.** Overcurrent protection devices such as breakers must be sized no less than 125 percent of the continuous load [384-16(c)].

The branch circuit conductors to ranges, cooktops, ovens, and other household cooking appliances can be sized according to Table 220-19, Note 4.

422-11 Overcurrent Protection

Appliances must be protected against overcurrent by an overcurrent protection device and overcurrent devices for continuous loads must be sized no less than 125 percent of the appliance load [384-16(d)].

(a) Branch-Circuit. Branch circuit conductors must have overcurrent protection according to Sections 240-3.

If the protective device rating is marked on the appliance, the overcurrent device rating shall not exceed the protective device rating marked on the appliance.

(e) Nonmotor Appliances. The rating of the appliance overcurrent protection device shall not exceed the rating marked on the appliance. Appliances that are not marked with the overcurrent protection device rating must have an overcurrent protection device sized according to the following:

(1) Appliances rated 13.3 ampere or less. The overcurrent protection shall not exceed 20 ampere.

(2) Appliances rated over 13.3 ampere. The overcurrent protection device shall not exceed 150 percent of the appliance current rating.

(3) When 150 percent of the appliance rating does not correspond with the standard ampere ratings of overcurrent protection devices [240-6(a)], the next higher standard rating is permitted.

Question. What size overcurrent protection is permitted for a 4,500 watt water heater rated 240 volt, single-phase, Fig. 39-2?

> *Answer.*   30 ampere
> 4,500 watts/240 volt = 18.75 ampere
> 18.75 ampere × 1.5 = 28 ampere,
> The next size up, 30 ampere [240-6(a)]

422-12 Central Heating Equipment

Central heating, other than fixed electric space heating equipment, must be supplied by a separate individual branch circuit. This Section is intended to apply to gas, oil, coal, or other fossil fuel fired central heating equipment units.

Exception. Auxiliary equipment such as pumps, valves, humidifiers, and electrostatic air cleaners associated with the central heating equipment, can be connected to the central heater branch circuit.

> **Note.** Central heating units containing resistance heaters must be installed accord-

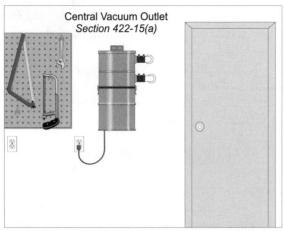

A separate circuit is not required for a central vacuum outlet but the ampere rating of the equipment must not exceed 50% of the ampere rating of the circuit, see Section 210-23(a).

Fig. 39-3 Central Vacuum Appliance Does Not Always Required Separate Circuit

ing to the requirements of Article 424—Electric Space Heating Equipment.

422-13 Water Heaters

Branch circuit conductors to fixed electric storage type water heaters having a capacity of no more than 120 gallons shall be sized at not less than 125 percent of the appliance rating, Fig. 39-2.

Note. See Section 422-11(e) for sizing the appliance overcurrent protection device.

422-15 Central Vacuum

(a) Circuit Loading. A separate circuit is not required for a central vacuum receptacle outlet if the load of the central vacuum does not exceed 50 percent of the ampere rating of the **general purpose circuit** [210-23(a)], Fig. 39-3.

The following explains how a central vacuum system can be supplied:

• Central vacuum rated up to 7.5 ampere, 15 ampere general purpose receptacle circuit.

• Central vacuum rated 7.6—10 ampere, 20 ampere general purpose receptacle circuit.

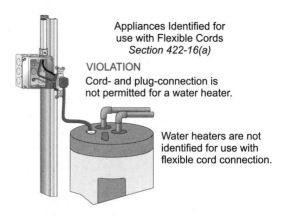

Fig. 39-4 Cords Not Permitted, Unless Appliance Is Specifically Identified for Use with Flexible Cords

• Central vacuum rated 10.1—12 ampere, individual 15 ampere circuit.

• Central vacuum rated 12.1—16 ampere, individual 20 ampere circuit.

422-16 Flexible Cords

(a) Uses Permitted. Flexible cords are permitted for appliances that are frequently interchanged or to prevent the transmission of noise and vibration. In addition flexible cords can be used to facilitate the removal of appliances fastened in place, where the fastening means and mechanical connections are specifically designed to permit ready removal.

CAUTION: Flexible cords are permitted only for appliances that are specifically identified for use with flexible cords [400-7(a)]. Flexible cords cannot be installed on water heaters, range exhaust hoods, forced air units (furnaces) and other fixed appliances unless the appliances are specifically identified to be used with flexible cords, Fig. 39-4.

(b) Specific Appliances.

(1) Waste Disposal. A waste disposal can be cord- and plug-connected, but the cord length must not be less than 18 inches or more than 36 inches.

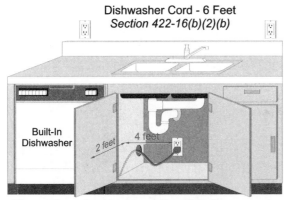

Dishwasher Cord - 6 Feet
Section 422-16(b)(2)(b)

Built-In
Dishwasher

2 feet   4 feet

The cord must not be longer than 4 feet measured
from the back of the appliance (4 ft behind plus 2 ft
under to appliance junction box).

COPYRIGHT 1998© Mike Holt Enterprises, Inc.

Fig. 39-5 Dishwasher and Trash Compactor Cord
Length Not to Exceed 6 Feet

The cord must be protected from physical damage
and the receptacle for the waste disposal must be
accessible.

(2) Dishwasher and Trash Compactor.
Dishwashers and trash compactors can be cord- and
plug-connected, but the cord length must not be less
than 36 inches or more than 48 inches measured
from the back of the appliance (6 feet). The cord
must be protected from physical damage and the
receptacle for the appliances must be accessible and
located in the space occupied by the appliances or
adjacent thereto, Fig. 39-5.

Note. Cords cannot be run through holes in
walls, ceilings, or floors [400-8], but most
inspectors permit a dishwasher cord to be
run through a cabinet, Fig. 39-5.

(3) Wall-Mounted Oven. Wall-mounted
ovens and counter-mounted cooking units can be
permanently connected or cord- and plug-connected.

Note. An accessible plug and receptacle is
permitted to serve as the disconnecting
means for cord- and plug-connected appliances [422-32].

422-18 Paddle Fans

(a) Paddle Fans Not Over 35 Pounds. Paddle fans that do not exceed 35 pounds (with or

without accessories) can be supported directly by
boxes that are specifically listed for the support of
paddle fans [370-27(c) Ex.]. These boxes must be
securely fastened in accordance with the requirements of Section 370-23.

(b) Paddle Fans over 35 Pounds. Paddle
fans that weigh more than 35 pounds (with or without accessories) must be supported independently of
the outlet box [370-27(c)].

Exception. Listed outlet boxes or outlet box
systems that are identified for the purpose can support suspended ceiling paddle fans of 70 pounds or
less, Fig. 39-6.

PART C. DISCONNECT

422-30 Disconnecting Means

Disconnecting means must be provided for
each appliance, according to Sections 422-31
through 422-35.

422-31 Permanently Connected Appliance Disconnect

(b) Circuit Breaker or Switch. A circuit
breaker or switch can serve as the disconnect for
permanently connected appliances rated over 300
VA, or 1/8 horsepower. The disconnect must be
located **within sight** from the appliance, Fig. 39-7.

If the circuit breaker or switch is not within
sight from the appliance, the circuit breaker or

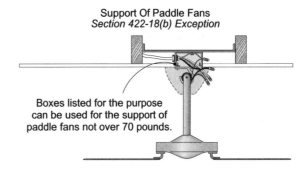

Support Of Paddle Fans
Section 422-18(b) Exception

Boxes listed for the purpose
can be used for the support of
paddle fans not over 70 pounds.

Fig. 39-6 Boxes Listed for the Purpose Can
Support Paddle Fans That Weigh Up to 70 Pounds

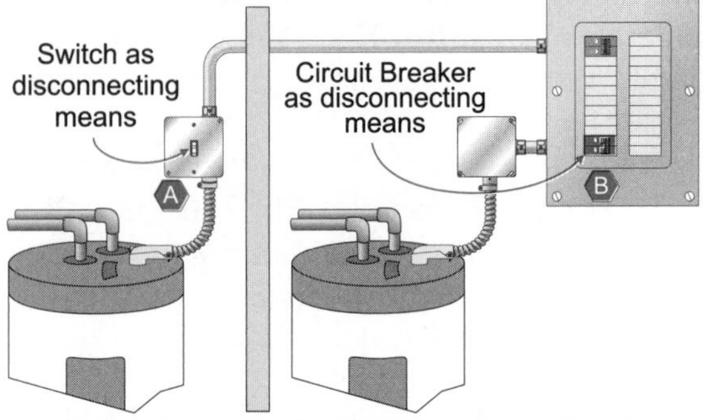

Appliance Disconnecting Means
Section 422-31(b)

Switch as disconnecting means

Circuit Breaker as disconnecting means

A disconnecting means for an appliance can be (A) a switch or (B) a circuit breaker within sight of the appliance. If the disconnect is not within sight of the equipment, it must be capable of being locked in the open position.

Fig. 39-7 Switch or Circuit Breaker If Properly Located Can Serve as Appliance Disconnect

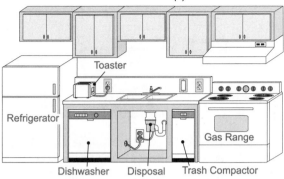

Disconnection of Cord-and Plug-Connected Appliances
Section 422-32(a)

Toaster

Refrigerator

Gas Range

Dishwasher   Disposal   Trash Compactor

For cord- and plug-connected appliances, the attachment plug and accessible receptacle can serve as the disconnecting means.

Fig. 39-8 Cord and Plug Can Serve as the Appliance Disconnect

switch must be capable of being locked in the open position, Fig. 39-7.

Note. Unit switches that have an OFF setting and that disconnect all ungrounded conductors such as a dishwasher can serve as the appliance disconnect [422-33].

422-32 Cord- and Plug-Connected Appliance Disconnect

(a) Cord- and Plug-Connected Appliances. An accessible plug and receptacle can serve as the disconnecting means for cord- and plug-connected appliances, Fig. 39-8.

(b) Cord- and Plug-Connected Ranges. The attachment plug and receptacle for cord- and plug-connected ranges can serve as the range disconnect. The **attachment plug** and receptacle must be located at the rear base of the range and it must be accessible from the front by removal of the bottom drawer.

422-33 Unit Switch As Disconnect

A unit switch that has an OFF setting and disconnects all ungrounded conductors can serve as the appliance disconnect, Fig. 39-9.

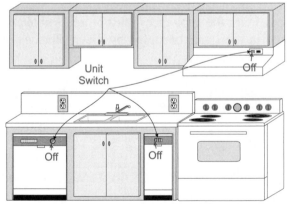

Unit Switch(es) as Disconnecting Means
Section 422-33

Unit Switch

Off

Off   Off

Unit switch(es) that has a marked "Off" position can serve as the appliance disconnecting means.

Fig. 39-9 Unit Switch Can Serve as the Appliance Disconnect

SUMMARY

Part A. General □ Article 422 does not apply to electric space heating, motors, and air-conditioning equipment.

Part B. Branch Circuit Requirements □ The conductor ampacity to an individual appliance shall not be less than required by the appliance marking or instruction. □ Appliances that are continuously loaded must have the conductor and overcurrent protection device sized not less than 125 percent of the appliance rating. □ Overcurrent protection for appliances must not exceed the protective device rating marked on the appliance. □ For appliances rated 13.3 ampere or less, the overcurrent protection shall not exceed 20 ampere. □ For appliances rated over 13.3 ampere, the overcurrent protection device shall not exceed 150 percent of the appliance current rating.

Part C. Installation of Appliances □ Central heating equipment other than fixed electric space-heating equipment must be supplied by an individual branch circuit. □ Disposals can be cord- and plug-connected with a maximum cord length of 36 inches. □ Dishwashers and trash compactors can be cord- and plug-connected with a maximum cord.length of 48 inches. □ Cords cannot be run through holes in walls, ceilings, or floors. □ Wall-mounted ovens and counter-mounted cooking units can be permanently connected or cord- and plug-connected. □ Listed paddle fans that do not exceed 35 pounds can be supported directly to boxes specifically listed for the support of paddle fans.

Part D. Disconnect and Protection □ Circuit breakers or switches can serve as the disconnect for permanently connected appliances rated over 300 VA. □ Unit switches that have an "Off" setting and that disconnect all ungrounded conductors can serve as the disconnect for the appliance. □ An accessible plug and receptacle can serve as the disconnecting means for cord- and plug-connected appliances and ranges.

REVIEW QUESTIONS

1. How are conductors sized for appliances?

2. Give a summary of the rules for overcurrent protection of appliances.

3. When can flexible cords be used for appliances?

4. Provide a summary of the rules for installing wall-mounted ovens and counter-mounted cooking units.

5. Explain the installation requirements for paddle fans.

6. What are the rules for using circuit breakers or switches for disconnects?

7. What are the rules for using attachment plugs and receptacles for disconnects?

8. What are the rules for using unit switches for disconnects?

Unit 40

Article 424

Fixed Electric Space Heating Equipment

OBJECTIVES

After studying this unit, the student should be able to understand:
- how to size the conductors and overcurrent protection device for a heating system.
- the requirements for disconnects.

PART A. GENERAL

424-1 Scope

Article 424 contains the installation requirements for fixed electric equipment used for space heating such as heating cable, unit heaters, boilers, central systems, or other types of fixed electric space heating equipment. This Article does not apply to air-conditioning equipment [Article 440].

Note. This book covers only the rules that apply to fixed electric space heating, not unit heaters, boilers, central systems, or other types of fixed electric space heating equipment.

424-3 Branch Circuits

(b) Conductors and Overcurrent Protection. The branch circuit conductor and overcurrent protection device for fixed electric space heating equipment shall not be smaller than 125 percent of the total load.

Note. Where the ampacity of the installed conductors does not correspond with the standard ampere rating of overcurrent protection devices [240-6(a)], the next larger overcurrent device can be used [240-3(b)].

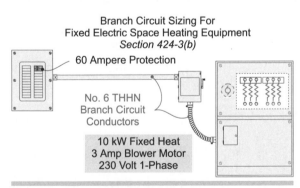

Branch Circuit Sizing For
Fixed Electric Space Heating Equipment
Section 424-3(b)

60 Ampere Protection

No. 6 THHN
Branch Circuit
Conductors

10 kW Fixed Heat
3 Amp Blower Motor
230 Volt 1-Phase

The overcurrent protection device and the branch circuit conductors are sized at 125% of the total load of the motors and the heaters.
Combined Load of Motors and Heaters

$I = \dfrac{Watts}{Volts} = \dfrac{10,000 \ watts}{230 \ volts} = 43 \ amps + 3 \ amps = 46 \ amps$

46 amps x 1.25 = 58 amps
Table 310-16, No. 6 rated 65 amps at 75ºC, 240-6(a) 60 amp protection

Fig. 40-1 Branch Circuit Sizing for Fixed Electric Space Heating Equipment

326

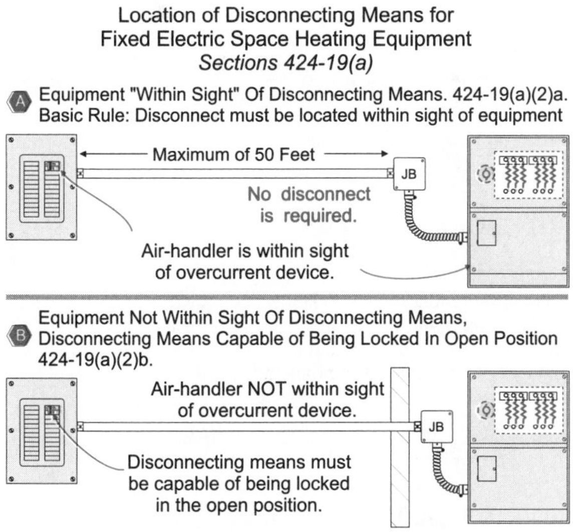

Location of Disconnecting Means for
Fixed Electric Space Heating Equipment
Sections 424-19(a)

Ⓐ Equipment "Within Sight" Of Disconnecting Means. 424-19(a)(2)a.
Basic Rule: Disconnect must be located within sight of equipment

Maximum of 50 Feet

No disconnect
is required.

Air-handler is within sight
of overcurrent device.

Ⓑ Equipment Not Within Sight Of Disconnecting Means,
Disconnecting Means Capable of Being Locked In Open Position
424-19(a)(2)b.

Air-handler NOT within sight
of overcurrent device.

Disconnecting means must
be capable of being locked
in the open position.

Fig. 40-2 Location of Disconnecting Means for
Fixed Electric Space Heating Equipment

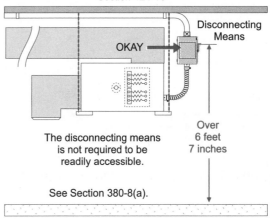

Disconnecting Means for Fixed Electric Space Heating
Section 424-19

Disconnecting
Means

OKAY

Over
6 feet
7 inches

The disconnecting means
is not required to be
readily accessible.

See Section 380-8(a).

Fig. 40-3 Disconnecting Means for Fixed Electric
Space Heating

Question. What size THHN conductor (75°C
terminals) and protection device is required for a 10
kW fixed electric space heating unit with a 3 ampere
motor, Fig. 40-1?

Answer.   No. 6, 60 ampere protection
Step 1. Total Load
I = 10,000 VA/230 volt = 43 ampere
43 ampere + 3 ampere = 46 ampere
46 ampere × 1.25 = 58 ampere

Step 2. Conductor [110-14(c), Table 310-16]
No. 6, rated 65 amperes, at 75°C.

Step 3. Protection [240-3(b) and 240-6(a)]
Protection not be less than 58 amperes.

Note. Fixed electric space heating with hermetic motor compressors (package units)
must also comply with Article 440. For conductor sizing, see Section 440-32 and for
overcurrent protection, see Section 440-22.

424-9 Baseboard Heaters

Factory-installed receptacle outlets provided as
a separately listed assembly on permanently
installed electric baseboard heaters can be used as
the required receptacle outlets for dwelling unit wall
space [210-52(a)].

PART C.   ELECTRIC SPACE HEATING EQUIPMENT

424-19 Disconnecting Means

The following summary contains the rules for
disconnects as required in Sections 424-19(a), (b),
and 424-65:

*General Rule (Heating Equipment with
Motors Over 1/8 Hp).* The disconnect must disconnect the heating equipment from all ungrounded
conductors.

Location of the Disconnects. The disconnect
must be located within sight of the heating equipment, Fig. 40-2.

If the disconnect is not within sight of the heating equipment, the disconnect must be capable of
being locked in the open position, Fig. 40-2.

Note. The disconnect is not required to
be readily accessible and it can be mounted
above 6 feet 7 inches [240-24(a)(4),
380-8(a) Ex 2], Fig. 40-3.

Unit Switch as Disconnect. Unit switches
that have an OFF position and that disconnect all
ungrounded conductors can serve as the heating
equipment disconnect [424-19(c)], Fig. 40-4.

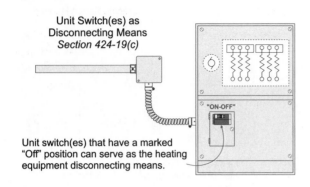

Unit Switch(es) as
Disconnecting Means
Section 424-19(c)

Unit switch(es) that have a marked
"Off" position can serve as the heating
equipment disconnecting means.

Fig. 40-4 Unit Switch(es) as Disconnecting Means

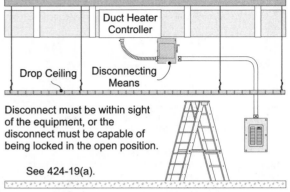

Duct Heater Control Equipment
Location of Disconnecting Means
Section 424-65

Duct Heater
Controller

Drop Ceiling   Disconnecting
Means

Disconnect must be within sight
of the equipment, or the
disconnect must be capable of
being locked in the open position.

See 424-19(a).

Fig. 40-5 Duct Heater Control Equipment Location
of Disconnecting Means

Note. Some air handlers have a circuit breaker (unit switch) built into the unit to serve as the equipment disconnect. Access to the unit switch requiring the use of tools is permitted because there is no requirement that the disconnect for electric heating equipment be readily accessible.

PART F. DUCT HEATERS

424-65 Disconnect for Electric Duct Heater Controller

The disconnect for an electric duct heater controller must be accessible (not readily accessible)

and it must be installed according to the requirements listed in Section 424-19 in this book, Fig. 40-5.

The disconnect for the duct heater can be located above or within the suspended ceiling [240-24(a)(4), 380-8(a) Ex 2].

SUMMARY

☐ The branch circuit conductor and overcurrent protection device for fixed electric space heating equipment shall not be smaller than 125 percent of the total load (heater plus motor). ☐ A disconnect must disconnect the controller and the heater. ☐ The disconnect must be located within sight of the heater. ☐ If the disconnect is not within sight of the heater, the disconnect must be capable of being locked in the open position. ☐ Unit switches with an OFF position that disconnect all ungrounded conductors can serve as the unit disconnect. ☐ The disconnect is not required to be readily accessible.

REVIEW QUESTIONS

1. Size the conductors and overcurrent protection device to a 10 kW central heater (3.7 ampere motor). The system voltage is 208 volt, single-phase.

2. Give a summary of the requirements for disconnects.

Unit 41

Article 430
Motors, Motor Circuits, and Controllers

OBJECTIVES

After studying this unit, the student should be able to understand:
- the requirements of motor controllers.
- the requirements of controller disconnects.
- the requirements of motor disconnects.

PART A. **GENERAL**

430-1 Scope

Article 430 contains the specific rules for conductor sizing, overcurrent protection, control circuit conductor, motor controllers, and disconnecting means, Fig. 41-1.

> FPN No. 1: Section 384-4 contains the installation requirements for motor control centers, and Article 440 contains the installation requirements for air-conditioning and refrigerating equipment.

430-2 Adjustable Speed Drive Systems

The branch circuit or feeder to adjustable speed drive equipment shall be based on the rated ampere input of the equipment.

> FPN: Electrical resonance may result from the interaction of nonsinusoidal currents from adjustable speed drives with power

factor correction capacitors. Capacitors placed too close to solid-state motor drives could overheat, blow fuses, and, worse yet, explode because of the interaction of nonsinusoidal (harmonic) currents. This topic is beyond the scope of this book. For more details call 1-888 NEC CODE.

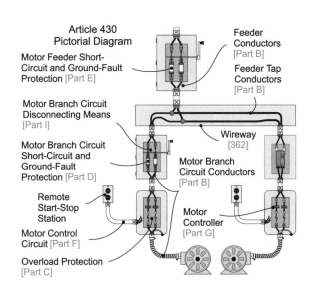

Fig. 41-1 Article 430 Pictorial Diagram

430-6   Conductor Sizing

(a) General Motor Applications. For general motor applications, current ratings shall be determined based on the following, Fig. 41-2:

(1) Table Full Load Current. The FLC values given in Table 430-147, 430-148, or -150, including notes, shall be used to determine the ampacity of conductors or ampere ratings of switches, branch circuit short circuit and ground-fault protection instead of the actual current rating marked on the motor nameplate [430-22(a), 430-24].

(2) Motor Nameplate Current Rating. Separate motor overload protection shall be based on the motor nameplate current rating [430-32(a)(1)].

430-9   Motor Controllers Terminal Requirements

(b) Copper Conductors.   Motor controllers and terminals of control circuit devices shall be connected with copper conductors unless identified otherwise.

(c) Torque Requirements.   Motor control conductors No. 14 and smaller must be *torqued* at a minimum of 7 lb-inches for screw-type pressure terminals, unless identified otherwise.

430-14   Location of Motors

(a) Ventilation and Maintenance.   When installing motors, ventilation and maintenance clearances must be considered.

430-17   The Highest Rated Motors

When selecting the conductors according to Sections 430-24 and 430-25, the highest rated motor shall be the motor with the highest rated motor full-load current.

Question. Which of the following motors has the highest rating?
(a) 10-hp, 3-phase, 208 volt,
(b) 5-hp, 1-phase, 208 volt, or
(c) 3-hp, 1-phase, 120 volt

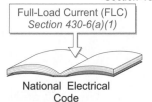

FLC Versus Nameplate Motor Currents
Section 430-6(a)

Full-Load Current (FLC)
Section 430-6(a)(1)

Nameplate (FLA)
Section 430-6(a)(2)

National Electrical Code

FLC comes from Tables 430-147 through 430-150 of the NEC and is used for sizing:
• conductors and disconnects
• short-circuit and ground-fault devices

FLA (full-load amperes) comes from the motor nameplate and is used for sizing motor overloads based on either Section 430-32 or 430-34.

Fig. 41-2 When to Use Motor FLC Versus Motor Nameplate Current Rating

Answer.   (c) 3 hp motor
(a) 10 hp = 30.8 ampere, [Table 430-150]
(b) 5 hp = 30.8 ampere, [Table 430-148]
(c) 3 hp = 34.0 ampere, [Table 430-148]

PART B.   CONDUCTOR SIZE

430-22   Single Motor Conductor Size

(a) Branch Circuit Conductor Size.   Motor branch circuit conductors to a single motor shall be sized no less than 125 percent of the motor current rating as listed in Table 430-147, 430-148, or -150, in accordance with Section 430-6(a).

Note. Many in the electrical industry improperly size motor branch circuit conductors to the motor nameplate current rating.

NOTICE:
Conductors are sized at 125 percent of motor full-load current [430-6(a)(1) and 430-22(a)]. The motor short-circuit and ground-fault protection device is sized from 150 percent up to 300 percent of motor full-load current. There is no relationship between the sizing of the conductor (125 percent) and the sizing of the short-circuit and ground-fault protection device (150 percent up to 300 percent). See Example D8 in Appendix D of the *NEC*.

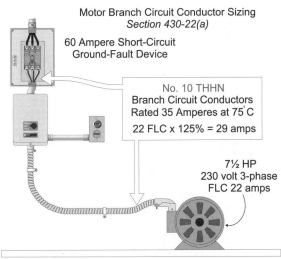

Motor Branch Circuit Conductor Sizing
Section 430-22(a)

60 Ampere Short-Circuit
Ground-Fault Device

No. 10 THHN
Branch Circuit Conductors
Rated 35 Amperes at 75°C
22 FLC x 125% = 29 amps

7½ HP
230 volt 3-phase
FLC 22 amps

A motor branch circuit conductor ampacity shall not be less
than 125 percent of the motor full-load current rating.

Fig. 41-3 Motor Branch Circuit Conductors Are Sized at 125 Percent of Motor Full Load Current

Question. What size branch circuit conductor is required for a 7½ hp 230 volt, 3-phase motor? Motor conductors are to be sized at 75°C, Fig. 41-3.

Answer. No. 10
Step 1. Branch Circuit Conductor
[430-22(a) and Table 430-150]
22 ampere × 1.25 = 29 ampere
No. 10, rated 35 ampere at 75°C, Table 310-16

Note. Branch circuit conductors must be protected against short-circuits and ground-faults according to Section 430-52. For an inverse time breaker:
22 ampere × 2.5 = 55 ampere
Next size up = 60 ampere

430-24 Feeder Conductor Size

Conductors that supply several motors shall be sized not less than 125 percent of the largest motor FLC as listed in Table 430-147, -148, or -150 plus the sum of the FLCs of the other motors (on the same phase).

Question. What size feeder conductor and circuit breaker short-circuit and ground-fault protection is required for two 7½ hp 230 volt, 3-phase motors? Motor conductors to be sized at 75°C ampacity, Fig. 41-4.

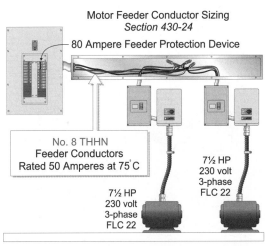

Motor Feeder Conductor Sizing
Section 430-24

80 Ampere Feeder Protection Device

No. 8 THHN
Feeder Conductors
Rated 50 Amperes at 75°C

7½ HP
230 volt
3-phase
FLC 22

7½ HP
230 volt
3-phase
FLC 22

Motor feeder conductors shall be sized not less than 125%
of the largest motor FLC plus the sum of the FLC's of the
other motors on the same phase.

Fig. 41-4 Motor Feeder Conductor Sizing

Answer. No. 8, 80 ampere circuit breaker
Step 1. Branch Circuit Conductor
[430-24, Table 430-150]
(22 ampere × 1.25) + 22 ampere = 49.5 ampere
No. 8 rated 50 ampere at 75°C, Table 310-16

Step 2. Branch Circuit Protection
[240-6(a), 430-52(c)(1) Ex. 1, Table 430-150]
Inverse Time Breaker
22 ampere × 2.5 = 55 ampere = 60 ampere

Note. Feeder circuit conductors must be protected against short-circuits and ground-faults according to Section 430-62. For an inverse time breaker:
60 ampere + 22 ampere = 82 ampere
Next size down, 80 ampere

430-26 Diversity and Balancing

Where motors operate intermittently or if all motors do not operate at one time, the authority having jurisdiction can grant permission for feeder conductors to have an ampacity less than specified in Section 430-24.

430-28 Motor Tap Conductors

Motor conductors tapped from a feeder must have an ampacity according to Section 430-22(a)

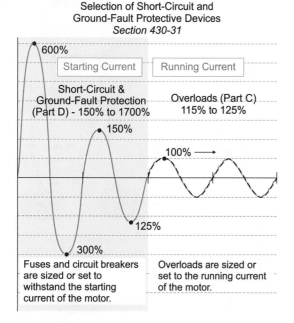

Fig. 41-5 Starting Current (Short-Circuit Protection) Versus Running Current (Overload Protection)

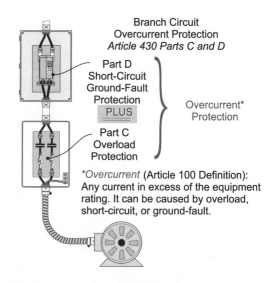

Fig. 41-6 Part C—Overload Protection
Part D—Short-Circuit and Ground-Fault Protection

Part C of Article 430 contains the requirements of overload protection, and Part D of Article 430 contains the requirements of short-circuit and ground-fault protection, Fig. 41-6.

and they must terminate in a branch circuit protective device sized in accordance with Section 430-52.

(1) 10 Foot Tap. Tap conductors not over 10 feet in length must have an ampacity of at least 1/10 the ampacity of the feeder overcurrent protection device and must be installed in a raceway.

(2) 25 Foot Tap. Tap conductors over 10 feet but not over 25 feet must have an ampacity of at least 1/3 the ampacity of the feeder conductor and must be installed in a raceway.

PART C. OVERLOAD PROTECTION

Motor circuit conductors must be protected against overcurrent (short-circuits, ground-faults, and overloads). Because of the special characteristics of induction motor loads (high inrush current), the overcurrent protection of motors is generally accomplished by having the overload protection (responsive to running current sized between 115 to 125%) separate from short-circuit and ground-fault protection (sized between 150-1700%), Fig. 41-5.

430-31 Overload Purpose

Overload protection devices called "heaters" are installed in motor *starters* and are intended to provide overload protection for the motor, the motor control equipment, and branch circuit conductors, Fig. 41-7.

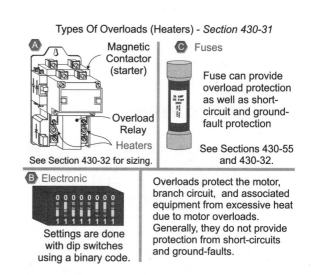

Fig. 41-7 Types of Overloads (Heaters)

Overload is the condition where equipment operates at overcurrent, which, if it persists for a sufficient length of time, can result in damage or dangerous overheating of the equipment. The overload protection device is not intended to protect against short-circuit or ground-fault currents.

Note. In reality, motor overload protection is accomplished by following the instruction on the inside of the starter cover.

430-32 Overload Sizing

Overload protection devices for continuous duty motors (with no integral thermal protector) must be sized according to (a) through (c) of this Section.

(a) Motors Rated More than One-Horsepower. Motors rated more than 1 horsepower without integral thermal protection must have an overload device sized to the motor nameplate current rating [430-6(a)(2)]. The sizing of the overload device is dependent on the following nameplate markings:

Service Factor. Motors with a nameplate service factor (S.F.) of 1.15 or 1.25 shall have the overload device sized no more than 125 percent of the motor nameplate current rating.

Note. Service factor of 1.15 means that the motor is designed to be loaded to 115 percent of its rated horsepower continuously.

Temperature Rise. Motors with a nameplate temperature rise rating not over 40°C shall have the overload device sized no more than 125 percent of motor nameplate current rating.

Note. Nameplate temperature rise not over 40°C means that the motor is designed to operate so as not to heat up over 40°C.

All Other Motors. All other motors shall have the overload device sized at no more than 115 percent of the motor nameplate ampere rating.

(b) Motors One Horsepower or Less (Nonautomatically Started). Permanently installed nonautomatically started motors rated one horsepower or less without integral thermal overload protection can use the branch circuit protection device (if within sight of the controller) and serve as the overload protection device.

(c) Motors One Horsepower or Less (Automatically Started). Motors rated one horsepower or less without integral thermal protection and automatically started must have the overload protection according to Section 430-32(a).

430-36 Fuses for Overcurrent Protection

If remote control is not necessary for the motor, considerable savings in installation cost can be achieved by using fuses (preferably dual-element) to protect the motor and the circuit conductors against overload, short-circuit and ground-fault current. However, a fuse must be installed in each ungrounded conductor [430-37 and 430-55], and the fuse must be sized to the overload requirements contained in Section 430-32.

430-37 Number of Overload Devices

An overload protection device must be installed in each ungrounded conductor of the motor [Table 430-37].

430-42 Cord- and Plug-Connected Motors

(c) Overload Device Not Required. Overload protection is not required for cord- and plug-connected motors rated not over 1 horsepower [430-42(a)]. Cord- and plug-connected motors rated 1 horsepower or higher must have overload protection that is an integral part of the motor [430-42(b)].

430-43 Automatic Restarting

A motor overload device that can restart a motor automatically after overload tripping shall not be installed if automatic restarting of the motor can result in injury to persons.

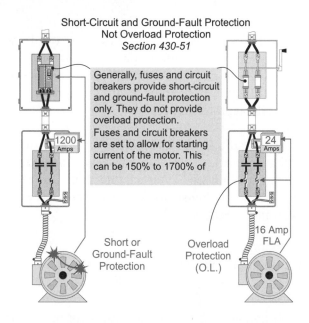

Fig. 41-8 Short-Circuit and Ground-Fault Protection Does Not Protect Against Overload

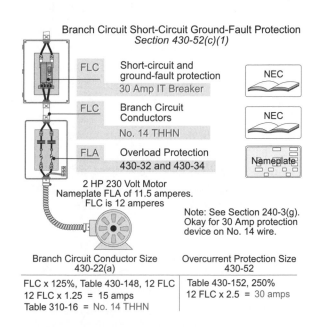

Fig. 41-9 Branch Circuit Short-Circuit Ground-Fault Protection

Note. This Section permits the automatic restarting of a motor after a power outage, but a motor shall not restart automatically after overload tripping if automatic restarting of the motor can result in injury to persons.

430-44 Overload Alarm

If orderly shutdown is necessary to reduce hazards to persons, the overload sensing device can be connected to a supervised alarm instead of causing the motor to shut down.

PART D. BRANCH CIRCUIT SHORT-CIRCUIT AND GROUND-FAULT PROTECTION

430-51 General

The branch circuit short-circuit and ground-fault protection device protects the motor, the motor control apparatus, and the conductors against overcurrents due to short-circuits or ground-faults, but not overload, Fig. 41-8.

The ground-fault protection referred to in this part is not the type required for personnel [210-8], feeders [215-9 and 240-13], services [230-95], or receptacles for temporary wiring [305-6(a)].

430-52 Branch Circuit Short-Circuit and Ground-Fault Protection

(c) Protection Size. The branch circuit short-circuit and ground-fault protection device is sized based on the type of motor (induction, synchronous, wound-rotor, etc.) and the type of protection device such as fuses or breakers.

(1) Table 430-152. Each motor branch circuit must be protected against short-circuit and ground-fault by a protection device sized no greater than the percentages listed in Table 430-152.

Question. What size conductor, overload, and short-circuit ground-fault protection (inverse time circuit breaker) would be required for a 2 hp, 230 volt motor that has a nameplate current rating of 11.5 amperes and a FLC rating of 12 amperes, Fig. 41-9?

Answer.   No. 14, 40 ampere breaker

Branch Circuit Short-Circuit Ground-Fault Protection
Section 430-52(c)(1) Exception 1

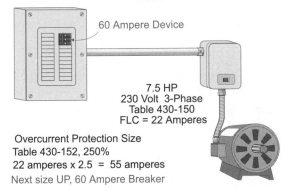

60 Ampere Device

7.5 HP
230 Volt 3-Phase
Table 430-150
FLC = 22 Amperes

Overcurrent Protection Size
Table 430-152, 250%
22 amperes x 2.5 = 55 amperes
Next size UP, 60 Ampere Breaker

Fig. 41-10 Motor Branch Circuit Short-Circuit and
Ground-Fault Protection at 250 Percent

Step 1. Conductor [Table 310-16]
12 ampere (FLC) × 1.25 = 16 ampere
No. 14 is rated 20 ampere

Step 2. Overload Protection
14 ampere (nameplate) × 1.15 = 16.1 ampere

Step 3. Short-Circuit, Ground-Fault Protection
12 ampere (FLC) × 2.50 = 30 ampere

Exception No. 1: Next Size Up Rule. Where the short-circuit and ground-fault values derived from Table 430-152 don't correspond with the standard rating of overcurrent protection devices as listed in Section 240-6(a), the next size larger protection device can be installed.

Question. What size conductor and inverse time circuit breaker are required for a 3-phase 7½ hp motor? Voltage rating of 230, service factor of 1.25, FLC 22 ampere [Table 430-150] and a nameplate rating of 20 ampere, Fig. 41-10?

Answer. No. 10, 60 ampere breaker
Step 1. Conductor [Table 310-16]
22 ampere (FLC) × 1.25 = 27.5 ampere,
No. 10 is rated 30 ampere [110-14(c)]

Step 2. Overload Protection [240-6(a)]
20 ampere (nameplate) × 1.25 = 25 ampere

Step 3. Short-Circuit, Ground-Fault Protection
22 ampere (FLC) × 2.50 = 55 ampere
Next size up, 60 ampere breaker [240-6(a)]

Note. I know this bothers many in the electrical industry to see a No. 10 conductor protected at 60 amperes. Remember, the branch circuit conductors are protected against overcurrent by 25 ampere overloads and the 60 ampere short-circuits and ground-fault protection device.

Exception No. 2: Motor Won't Start. Where the short-circuit and ground-fault protection device sized according to Table 430-152 as modified by Exception No. 1 is not sufficient to start the motor, then (a) through (c) can be applied.

(a) The rating of a nontime-delay fuse (not exceeding 600 ampere) or a time-delay Class CC fuse can be sized up to 400 percent of the full-load current.

(b) The rating of a time-delay (dual-element) fuse can be sized up to 225 percent of the full-load current.

(c) The rating of an inverse time circuit breaker can be sized up to:

(1) 400 percent for FLCs of 100 ampere or less
or

(2) 300 percent for FLCs greater than 100 ampere.

430-55 Single Protective Device

A motor can be protected against overcurrent (overload, short-circuit and ground-fault protection) by a single protection device size (best to use a dual-element fuse) sized in accordance with Section 430-32.

Question. If a dual-element fuse is used for overload protection, what size dual-element fuse is required for a 10 hp, 208 volt, 3-phase motor, service factor 1.15, motor nameplate current rating of 28 ampere [430-6(a)(2), 430-32(a)(1)], Fig. 41-11?

Answer. 35 ampere
28 ampere × 1.25 = 35 ampere [240-6(a)]

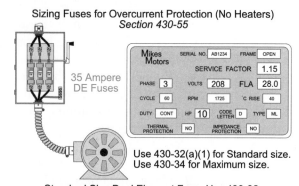

Sizing Fuses for Overcurrent Protection (No Heaters)
Section 430-55

35 Ampere
DE Fuses

Use 430-32(a)(1) for Standard size.
Use 430-34 for Maximum size.

Standard Size Dual Element Fuse, Use 430-32.
The nameplate value is used to size overloads.
Service Factor of 1.15, use 125%
28 nameplate amperes x 1.25 = 35 amperes
240-6(a) = 35 Ampere Dual Element Fuses

Fig. 41-11 Overcurrent Protection by a Single Protection Device (fuses)

PART E. FEEDER SHORT-CIRCUIT AND GROUND-FAULT PROTECTION

430-62 Feeder Protection

(a) Protection Size. Feeder conductors sized according to Section 430-24 must be protected against short-circuits and ground-faults, but not against overload. The feeder short-circuit and ground-fault protection device must not be greater than the rating of the largest branch circuit short-cir-

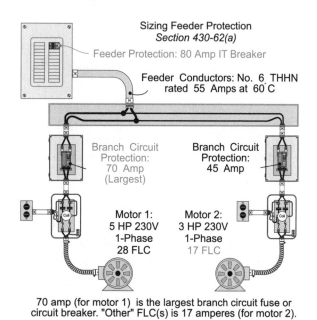

Sizing Feeder Protection
Section 430-62(a)

Feeder Protection: 80 Amp IT Breaker

Feeder Conductors: No. 6 THHN
rated 55 Amps at 60°C

Branch Circuit
Protection:
70 Amp
(Largest)

Branch Circuit
Protection:
45 Amp

Motor 1:
5 HP 230V
1-Phase
28 FLC

Motor 2:
3 HP 230V
1-Phase
17 FLC

70 amp (for motor 1) is the largest branch circuit fuse or
circuit breaker. "Other" FLC(s) is 17 amperes (for motor 2).

Fig. 41-12 Sizing Feeder Protection

cuit and ground-fault protection device [430-52] of any motor of the group, plus the sum of full-load currents of the other motors of the group (on same phase).

Question. What size THHN feeder conductor and inverse time circuit breaker is required for a 5 hp, 230 volt single-phase motor, and a 3 hp, 230 volt single-phase motor, Fig. 41-12.

Answer.   No. 6, 80 ampere breaker
Step 1. Feeder Conductors [430-24]
(28 ampere × 1.25) + 17 ampere = 52 ampere
No. 6 based on 75°C terminals.

Step 2. Branch Circuit Protection
5 hp, 28 ampere × 2.5 = 70 ampere
3 hp, 17 ampere × 2.5 = 43 ampere, 45 ampere

Step 3. Feeder Protection [430-62].
70 ampere + 17 ampere = 87 ampere
Next size down, 80 ampere [240-6(a)].

430-63 Feeder Protection for Motors and Other Loads

Feeder conductors that supply motor and other loads [430-24] must be protected against short-circuits and ground-faults, but not overloads. The protection device must not be sized greater than the largest branch circuit short-circuit and ground-fault protection device for any motor of the group, plus the sum of full-load currents of the other motors, plus the other loads as determined in Articles 210 or 220.

PART F. MOTOR CONTROL CIRCUITS

430-71 Definition of Motor Control Circuit

A motor control circuit carries the electrical signals directing the performance of the controller. See Section 430-81(a) for the definition of a controller, Fig. 41-13.

Note. Most motor control circuit conductors are Class 1 [725-21] and must be

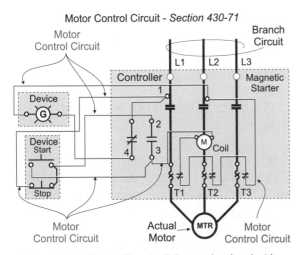

Motor Control Circuit - Section 430-71

Motor Control Circuit: The circuit that carries the electric signals that directs the performance of the controller, but does not carry the main power current.

Fig. 41-13 Definition—Motor Control Circuit

installed in a Chapter 3 wiring method [725-25].

430-72 Overcurrent Protection for Control Circuits

(a) Control Conductors Not Tapped. Motor control conductors not tapped from the branch circuit protection device are considered to be Class 1 remote control conductors, and they must have overcurrent protection in accordance with Section 725-23.

Table 430-72(b) Column A Overcurrent protection for motor control conductors tapped from the load side of the motor branch circuit protection device.

Conductor Size (AWG)	Conductor Ampacity	Overcurrent Protection Size Table 430-72(b) Column A
No. 18	6 Amperes (402-5)	7 Amperes
No. 16	8 Amperes (402-5)	10 Amperes
No. 14 and larger	*	According to Section 240-3

*Ampacities as listed in Table 310-16

(b) Control Conductors Tapped. Motor control circuit conductors tapped from the motor branch circuit protection device must have overcurrent protection in accordance with Table 430-72(b) Column A. These tapped conductors are not considered branch circuit conductors and they can be

protected by either a supplemental overcurrent protection device [240-10] or by the use of a branch circuit protection device.

Exception No. 2: Two-wire Transformers. The conductors on the secondary side of a 2-wire transformer are considered protected by the *primary* protection device, if the primary protection rating protects the secondary conductors according to the secondary-to-primary voltage ratio of Section 430-72(c).

(c) Control Circuit Transformers Protection. Transformers for motor control circuit conductors must have overcurrent protection on the primary side.

Exception: Hazard Created. If a hazard is created by the opening of the overcurrent protection device, the control transformer protection can be omitted.

Note. Many control transformers have very little iron, resulting in very high inrush currents. One-time and dual-element fuses at times will not accommodate the high inrush current of motor control transformers, therefore manufacturers have designed fuses for this purpose. Save yourself some aggravation; use the proper fuses for control transformer protection.

(2) Transformers 2–9 Ampere. Transformers whose primary current rating is 2 ampere or more, but less than 9 ampere must have the primary overcurrent protection device set no more than 167 percent of the transformer primary current rating [450-3(b)].

Question. What is the maximum size primary protection device permitted for a 250 VA transformer, if the primary voltage is 120 volt?

Answer.   3 ampere
$I = P/E$, $I = 250$ VA/120 volt = 2.08 ampere
2.08 ampere $\times$ 1.67 = 3.5 ampere
Next size down, 3 ampere [240-6(a)]

(2) Transformers 9 Ampere or More. Transformers whose primary current rating is 9 ampere or more must have primary overcurrent protection set no more than 125 percent of the transformer pri-

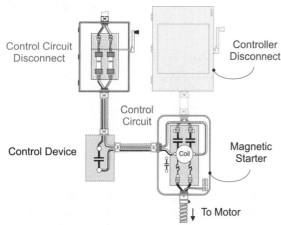

Control Circuit Disconnect

Controller Disconnect

Control Circuit

Control Device

Coil

Magnetic Starter

↓ To Motor

A separate disconnect for the motor control circuit is required when the control circuit is not tapped from the controller disconnect. The control circuit disconnect and the controller disconnect must be adjacent to each other.

Fig. 41-14 Disconnect Requirements for Motor Control Circuits

mary current rating [450-3(b)]. If the value of 125 percent of the primary current rating does not correspond with the standard rating of overcurrent protection devices [240-6(a)], the next higher standard rating can be used [450-3(b) Note 1].

Question. What is the maximum primary protection permitted for a 2.5 kVA transformer at 240 volt?

Answer.   15 ampere
I = P/E, I = 2,500 VA/240 volt = 10.42 ampere
10.42 ampere × 1.25 = 13 ampere
Next size up, 15 ampere

(3) Transformers Less than 50 VA. Transformers rated less than 50 VA and integral with the motor controller are considered protected by the branch circuit protection device.

(4) Transformers Less than 2 Ampere. Transformers rated less than 2 ampere can have the primary protection device set no more than 500 percent of the primary current rating.

Question. What is the maximum primary protection permitted for a 750 VA transformer if the primary voltage is 480 volt?

Answer.   6 ampere
I = P/E, I = 750 VA/480 volt = 1.56 ampere
1.56 ampere × 5 = 7.8 ampere
Next size down, 6 ampere [240-6(a)]

430-73 **Protection from Physical Damage**

Where damage to motor control conductors would introduce a hazard, the control conductors must be installed in a raceway or protected from physical damage.

Note. Class 1 motor control conductors must always be installed in wiring methods listed in Chapter 3 [725-25].

Accidental Grounding. Where one conductor of the control circuit is grounded (neutral), the control circuit must be installed so that the accidental grounding of a conductor will not start the motor nor bypass automatic safety shutdown devices.

430-74 **Disconnect for Control Circuit**

Motor control circuit conductors must have a disconnect that opens all conductors of the motor control circuit.

(a) Control Circuit Disconnect. The controller disconnect can also serve as the disconnecting means for control circuit conductors [430-102(a)].

If the control circuit conductors are not tapped from the controller disconnect, a separate disconnect is required for the control circuit conductors. The control circuit disconnect and the controller disconnect must be adjacent to each other, Fig. 41-14.

CAUTION: The *NEC* does not require the control circuit disconnect to be readily accessible, but Section 380-8(a) would require it to be located so that it may be operated from a readily accessible place and not higher than 6 feet 7 inches above the floor or working platform. Exception No. 2 to Section 380-8(a) permits the control circuit disconnect to be installed higher than 6 feet 7 inches if located adjacent to the motor control equipment.

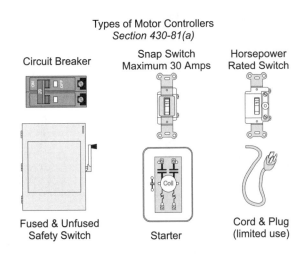

Fig. 41-15 Types of Motor Controllers

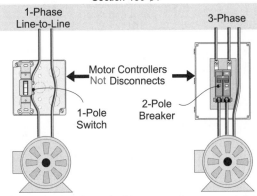

Fig. 41-16 Controllers Need Not Open All Conductors

PART G. MOTOR CONTROLLERS

430-81 General

(a) Controller Definition. A motor controller is a switch or device that is capable of interrupting the motor locked-rotor current [430-82] and it is used to start and stop a motor such as a horsepower rated switch, snap switch, or circuit breaker [430-83], Fig. 41-15.

Note. A pushbutton limit switch or other control circuit device used to operate an electromechanical relay is not a controller.

(c) Cord- and Plug-Connection Controller. Listed cord- and plug-connected portable motors rated 1/3 horsepower or less can use the attachment plug and receptacle as the controller [430-109(f)].

430-82 Controller Design

(a) Rated to Interrupt Locked-Rotor Currents. The motor controller must be capable of interrupting (opening) the motor at locked-rotor current.

Note. Table 430-151 lists motor locked-rotor current ratings.

430-83 Controller Rating

(a) General
(1) Horsepower Rating. Controllers other than circuit breakers and molded case switches shall have a horsepower rating not less than the motor. In addition the motor controller must have a current rating not less than the available short-circuit current [110-10].

(2) Circuit Breakers. Circuit breakers can serve as the motor controllers and they are not required to be horsepower rated.

Note. A switch or circuit breaker used as the controller also can serve as the disconnect for a motor if it opens all ungrounded conductors simultaneously [430-111].

430-84 Open All Conductors

The motor controller must open only as many conductors as necessary to stop the motor, for example, one conductor must be opened for single-phase motors and two conductors must open for three-phase motors, Fig. 41-16.

430-87 Controller for Each Motor

Each motor must have an individual motor controller.

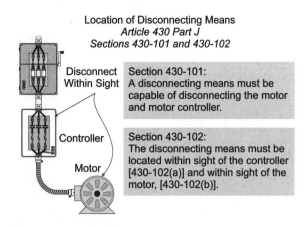

Location of Disconnecting Means
Article 430 Part J
Sections 430-101 and 430-102

Disconnect
Within Sight

Section 430-101:
A disconnecting means must be capable of disconnecting the motor and motor controller.

Controller

Section 430-102:
The disconnecting means must be located within sight of the controller [430-102(a)] and within sight of the motor, [430-102(b)].

Motor

Fig. 41-17 Controller and Motor Disconnecting Means Requirements

430-91 Motor Controller Enclosure Types

Motor controllers are installed in a variety of environments: rain, snow, sleet, corrosive agents, submersion, dirt, liquids, dust, oil, and so on. The selection of the enclosure for these environments must be according to the rules contained in Table 430-91 of the NEC.

PART J.   DISCONNECTING MEANS

430-101   General

Disconnecting means is required for the control circuit conductors [430-74], the motor controller [430-102(a)], and the motor itself [430-102(b)]. The disconnecting means must be capable of disconnecting the motor, the motor controller, and it must

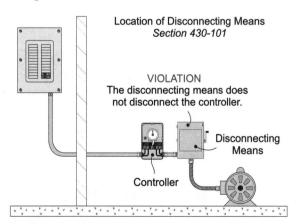

Location of Disconnecting Means
Section 430-101

VIOLATION
The disconnecting means does not disconnect the controller.

Disconnecting
Means

Controller

Fig. 41-18 Controller Disconnect Must Disconnect the Controller

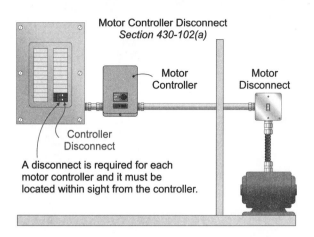

Motor Controller Disconnect
Section 430-102(a)

Motor
Controller

Motor
Disconnect

Controller
Disconnect

A disconnect is required for each motor controller and it must be located within sight from the controller.

Fig. 41-19 Motor Controller Disconnect Must Be within Sight of the Controller

open all ungrounded conductors of the circuit [430-74 and 430-103], Fig 41-17 and 41-18.

Note. With the proper layout, a single disconnect could be used for the control circuit conductors, the controller, and the motor.

430-102   Disconnect Requirement

(a) Controller Disconnect. A disconnect is required for each motor controller and it must be located within sight from the controller, Fig. 41-19.

The disconnect for the controller can serve as the disconnect for tapped control circuit conductors [430-74] and for the motor [430-102(b) Ex.].

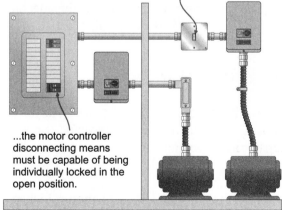

Motor Disconnecting
Section 430-102(b) and Exception

A motor disconnecting means must be within sight of the motor or...

...the motor controller disconnecting means must be capable of being individually locked in the open position.

Fig. 41-20 Motor Disconnect Must Be within Sight of the Motor, or Be Capable of Being Locked Open

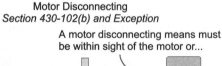

(b) Motor Disconnect. A disconnect is required for each motor and it must be located within sight of the motor, Fig. 41-20.

Exception. A motor disconnect is not required, if the controller disconnect [430-102(a)] is capable of being individually locked in the open position, Fig. 41-20.

Note. A circuit breaker or switch within a locked panelboard or locked equipment room does not satisfy the "individual" locking requirement of this exception.

FPN: Lockout/Tagout procedures are contained in NFPA 70E *Electrical Safety Requirements for Employee Workplaces and* OSHA Part 1910, Subpart S, Part II.

430-103 Open All Supply Conductors

The disconnect for the motor controller and motor must open all conductors and it must be designed so that no pole can be operated independently, Fig. 41-21.

430-104 Marking and Mounting

The disconnects required by Article 430 must indicate whether they are in the "on" or "off" posi-

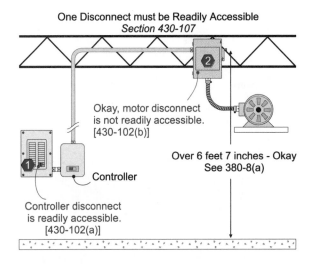

Fig. 41-22 Either the Controller or Motor Disconnect Must Be Readily Accessible

tion, and they must be legibly marked to identify their intended purpose [110-22 and 384-13]. When the disconnect is operated vertically, it must be installed so the "up" position is the "on" position [240-81, 380-7].

430-107 One Disconnect Must Be Readily Accessible

One of the disconnects required for the motor and motor controller required in Section 430-102 must be readily accessible, Fig. 41-22.

Note. Switches (disconnects) installed adjacent to motors shall be permitted to be located higher than 6 feet 7 inches if accessible by portable means.

430-109 Disconnect Rating

(a) General. The disconnects for the motor controller and motor must be:

(1) A listed horsepower rated motor-circuit switch.
(2) A listed molded case circuit breaker.
(3) A listed molded case switch.

(c) Snap Switch. General-use AC snap switches can serve as the disconnect for motors rated 2 horsepower or less [430-83(c)], Fig. 41-23.

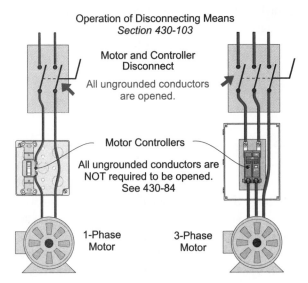

Fig. 41-21 Disconnecting Means Must Open All Conductors of the Circuit

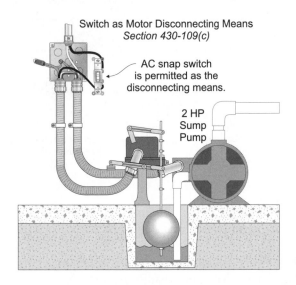

Switch as Motor Disconnecting Means
Section 430-109(c)

AC snap switch
is permitted as the
disconnecting means.

2 HP
Sump
Pump

Fig. 41-23 Snap Switch Can Be Used for the Motor Disconnecting Means

(f) Cord- and Plug-Connected Motors. Listed cord- and plug-connected motors rated 1/3 horsepower or less can use a horsepower rated attachment plug and a receptacle as the disconnect.

430-111 Combination Controller-Disconnect

A horsepower rated switch or circuit breaker complying with Section 430-83 can serve as a both the motor controller and the disconnecting means, if it opens all ungrounded conductors to the motor.

Table 430-148 Full-Load Current, Single-Phase Motors

Table 430-148 lists the full-load current for single-phase alternating-current motors. The values are used to determine motor conductor sizing, ampere ratings of disconnects, controller rating, and branch circuit and feeder protection [430-6(a)(1)].

Table 430-150 Full-Load Current, Three-Phase Motors

Table 430-150 lists the full-load current for 3-phase alternating-current motors. The values are used to determine motor conductor sizing, ampere ratings of disconnects, controller rating, and branch circuit and feeder protection [430-6(a)(1)].

Table 430-151 Locked Rotor Currents

Table 151A contains locked-rotor current for single-phase motors and Table 151B contains the locked-rotor current for 3-phase motors. These values are used in the selection of controllers and disconnecting means when the horsepower rating is not marked on the motor nameplate.

Table 430-152 Branch Circuit Short-Circuit and Ground-Fault Protection Size

Table 430-152 percentages are used to determine the rating or setting of motor branch circuit short-circuit and ground-fault protective devices according to the requirement contained in Section 430-52(c)(1).

SUMMARY

Part A. General ☐ The motor full-load current ratings listed in Tables 430-147 through 430-150, shall be used (instead of the motor nameplate rating) for sizing motor circuit conductors and short-circuit protection devices. ☐ The motor nameplate current rating is used to size the motor overload protection device. ☐ Unless identified otherwise, only copper conductors can be used with motor controllers. ☐ When installing motors, ventilation and maintenance clearances must be considered.

Part B. Conductor Size ☐ The branch circuit conductors to a single motor must have an ampacity not less than 125 percent of the motor full-load current sized according to the terminal temperature ratings listed in Section 110-14(c). ☐ Conductors that supply several motors must have an ampacity of not less than 125 percent of the highest rated motor full-load current of the group, plus the sum of the other motor full-load currents in the group.

Part C. Overload Protection ☐ Overload protection devices are intended to protect the motors, motor control equipment, and branch circuit conductors against excessive heating, but not against short-circuits or ground-faults. ☐ Fuses sized for overload protection and installed in each ungrounded conductor can serve as short-circuit and ground-fault protection. ☐ Overload protection is not required for motors listed for cord- and plug-connection and automatically started. ☐ If orderly shutdown is necessary (to reduce hazards to persons), the overload sensing devices can be connected to a supervised alarm instead of causing the motors to shut down.

Part D. Branch Circuit Protection ☐ The size of the motor branch circuit short-circuit and ground-fault protection device is dependent on the motor and protection device type. ☐ When the branch circuit protection device is not sufficient to carry the starting or running currents of the motor, the next larger size shall be permitted.

Part E. Feeder Circuit Protection ☐ Feeder conductor protection must not be greater than the largest branch circuit protection device for any motor of the group, plus the sum of full-load currents of the other motors in the group.

Part F. Motor Control Circuits ☐ Motor control circuit conductors require a disconnect that opens all power supply conductors of the motor control circuit.

Part G. Motor Controllers ☐ A controller is any switch or device that is capable of interrupting the motor locked-rotor current and is used to start and stop the motor. ☐ A pushbutton, limit switch, or any other control circuit device that operates an electromechanical relay is not considered to meet the requirements of a controller. ☐ The controller must be capable of interrupting (opening) the motor locked-rotor current. ☐ The controller must have a short-circuit current rating not less than the short-circuit current available. ☐ The controller must open only as many conductors as necessary to stop the motor.

Part J. Disconnecting Means ☐ A disconnect is required for each controller and must be located within sight from the controller location. ☐ If the disconnect for the controller is capable of being individually locked in the open position, then no disconnect must be installed within sight of the motor. ☐ The disconnect for the controller and motor must open all power supply conductors. ☐ The disconnect must indicate "on" or "off" position, be legibly marked to show its intended purpose, and when operated vertically, the "on" position must be the "up" position.   ☐ One disconnect within the motor circuit must be readily accessible. ☐ The disconnect must be a horsepower rated switch or circuit breaker; this switch or circuit breaker can serve as both the controller and disconnect.

REVIEW QUESTIONS

1. Explain the requirement for a motor control disconnect.

2. Give a brief summary of the requirements for a motor controller.

3. Give a brief summary of the requirements of a controller disconnect.

4. Explain the requirements of a motor disconnect.

5. Size the branch circuit conductors (60°C terminals), overload, dual-element protection short-circuit and ground-fault protection device for a 3 hp, 120 volt, single-phase motor. Nameplate rating of 31 ampere.

6. Size the branch circuit conductors (75°C terminals), overload, one-time fuse, and short-circuit and ground-fault protection device for a 5 hp, 208 volt, single-phase, S.F. 1.12. Nameplate of 28 ampere. Note: Branch circuit protection is not sufficient to start motor.

7. Size the branch circuit conductors (75°C terminals), overload, inverse time short-circuit and ground-fault protection device for a 10 hp wound-rotor, 208 volt, 3-phase motor. Temperature rise not over 40°C. Nameplate of 28 ampere. Note: Overload protection device is not sufficient to start the motor.

8. Size the branch circuit conductors (75°C terminals), overload, inverse time short-circuit and ground-fault protection device for a 20 hp synchronous type, 240 volt, 3-phase, motor. Nameplate rating of 48 ampere.

9. Size the feeder conductor (75°C terminals) and dual-element feeder protection device for two 20 hp, 208 volt, 3-phase motors.

10. Size the feeder conductor (75°C terminals) and feeder (inverse time breakers) protection device for the following 6 motors: Three 3 hp, 120 volt, single-phase motors; two 5 hp, 208 volt, single-phase motors; and one 10 hp wound-rotor, 208 volt, 3-phase motor.

Unit 42

Air-Conditioning and Refrigeration Equipment

OBJECTIVES

After studying this unit, the student should be able to understand:
- which Code Section applies to sizing the conductor and protection to central electric space heating equipment.
- which Code Section applies to sizing the circuit conductors and protection for a 5 hp motor used as a blower.
- the differences in the rules between motors and air-conditioning when installing a disconnecting means.
- how you size the circuit protection device and conductors to an individual air-conditioning unit.
- testing laboratories' requirements for short-circuit and ground-fault protection.
- the requirements for room air-conditioners.

PART A. GENERAL

440-1 Scope

This Article applies to electrically driven air-conditioning and refrigeration equipment that has a hermetic refrigerant motor compressor.

440-3 Other Articles

(a) Article 430. The rules in this Article are in addition to, or amend, the rules contained in Article 430 and other Articles.

(b) Equipment with No Hermetic Motor Compressors. Air-conditioning and refrigeration equipment that does not have hermetic refrigerant motor compressors such as furnaces with evaporator coils must comply with Article 422 for appliances, Article 424 for electric space heating, and in some cases Article 430 for motors.

(c) Household Refrigerant Motor-Compressor Appliances. Household refrigerators and freezers, drinking water coolers, and beverage dispensing machines are considered appliances and their installation must comply with the requirements contained in Article 422. Room air-conditioners

345

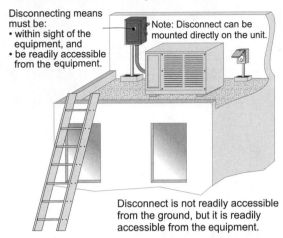

Fig. 42-1 Disconnecting Means Must Be Readily Accessible from the Equipment

must comply with Part G of this Article, as well as the requirements contained in Article 422.

PART B.   DISCONNECTING MEANS

440-13   Cord- and Plug-Connected Equipment

Listed cord- and plug-connected equipment such as room air-conditioners [440-63], household refrigerators and freezers, drinking water coolers, and beverage dispensers can use the attachment plug and receptacle as the disconnecting means.

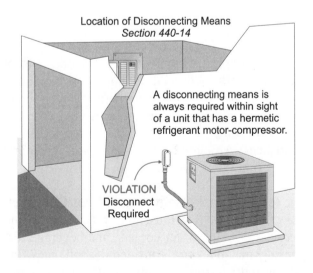

Fig. 42-2 Disconnect Must Be Within Sight of Air-Conditioning and Refrigeration Equipment

440-14   Location

The disconnecting means for air-conditioning and refrigeration equipment must be readily accessible and within sight from the equipment. The disconnect can be installed in, on, or at the equipment, Fig. 42-1 and 42-2.

Note. Some inspectors interpret readily accessible as not requiring a tool to remove a screw or bolt to gain access to the built-in circuit breaker disconnect.

Exception No. 2: Cord Connected Appliances. Cord- and plug-connected appliances [440-13] do not need the disconnect to be readily accessible or within sight of the appliance.

PART C.   CIRCUIT PROTECTION

The size and type of short-circuit and ground-fault protection device required for most air-conditioning and refrigeration equipment is marked on the equipment nameplate.

Note. If the equipment nameplate specifies "Maximum Fuse Size," then the equipment must be protected by a fuse. If the nameplate specifies "HACR Circuit Breaker," then the equipment must be protected by a heating air-conditioning and refrigeration rated circuit breaker [110-3(b)].

440-21   General

A protection device must protect the circuit conductors, control equipment, and motors against short-circuits and ground-faults, but not overload.

440-22   Short-Circuit and Ground Fault Protection Device Size

If the equipment does not have a nameplate specifying the size and type of protection device, then the protection device must be sized as follows:

(a) One Motor-Compressor. Motor compressor conductors must be protected against short-circuits and ground-faults by a protection device sized no greater than 175 percent of the motor compressor current. If the protection device setting is not capable of carrying the starting current of the motor compressor, the protection device can be sized up to 225 percent of the motor compressor current.

Note. The circuit conductors are sized at 125 percent of the motor compressor current rating [440-32].

NOTICE:
The branch circuit conductors are sized at 125 percent of the motor compressor current, and the protection device is sized from 175 to 225 percent. The conductor is not required to be protected at its ampacity.

Question. What size THHN conductor and fuse protection is required for a 24 ampere, 230 volt motor compressor, Fig. 42-3?

Answer.   No. 10, 40 ampere protection
Step 1. Conductor Size.
[110-14(c) and Table 310-16]
24 ampere × 1.25 = 30 ampere, No. 10

Step 2. Protection.
24 ampere × 1.75 = 42 ampere
The next size down, 40 ampere [240-6(a)]

If a 40 ampere protection device is not capable of carrying the starting current, then the protection device can be sized up to 225 percent.
24 ampere × 2.25 = 54 ampere or 50 ampere

Note. The required overcurrent protection for most refrigeration equipment is marked on the equipment nameplate.

PART D. CONDUCTOR SIZING

The minimum size circuit conductor for most air-conditioning and refrigeration equipment is marked on the equipment nameplate or installation instructions.

440-32 Conductor Size - One Motor Compressor

If equipment is not marked with minimum circuit ampacity, the branch circuit conductors to a single motor compressor must have an ampacity of not less than 125 percent of the motor compressor current, Fig. 42-3.

Note. Branch circuit conductors are protected against short-circuits and ground-faults, between 175 percent and 225 percent of compressor current [440-22(a)].

440-33 Conductor Size - Several Motor Compressors

Conductors that supply several motor compressors must have an ampacity of not less than 125 percent of the highest rated motor compressor current of the group, plus the sum of the other motor compressor currents of the group.

Note. These conductors must be protected against short-circuits and ground-faults, according to Section 440-22(b).

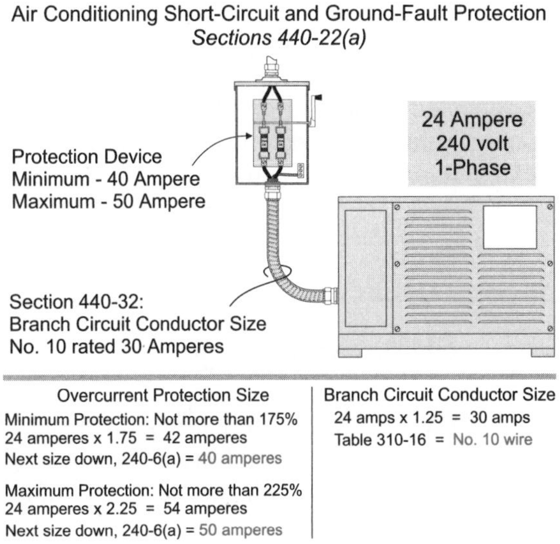

Air Conditioning Short-Circuit and Ground-Fault Protection
Sections 440-22(a)

Protection Device
Minimum - 40 Ampere
Maximum - 50 Ampere

24 Ampere
240 volt
1-Phase

Section 440-32:
Branch Circuit Conductor Size
No. 10 rated 30 Amperes

Overcurrent Protection Size	Branch Circuit Conductor Size
Minimum Protection: Not more than 175%	24 amps x 1.25 = 30 amps
24 amperes x 1.75 = 42 amperes	Table 310-16 = No. 10 wire
Next size down, 240-6(a) = 40 amperes	
Maximum Protection: Not more than 225%	
24 amperes x 2.25 = 54 amperes	
Next size down, 240-6(a) = 50 amperes	

Fig. 42-3 Short-Circuit and Ground-Fault Protection Sized between 175 to 225 Percent

440-34 Conductor Size - Motor Compressor(s) with Other Loads

Conductors that supply motor compressors and other loads must have an ampacity of not less than 125 percent of the highest rated motor compressor current of the group, plus the sum of the other motor compressor currents of the group, plus the calculated load according to Article 220.

PART G. ROOM AIR-CONDITIONERS

440-60 General

The rules in Part G apply to window or in-wall type units that incorporate a hermetic refrigerant motor compressor(s). Room air-conditioners rated not over 40 ampere, single-phase, 250 volt can be cord- and plug-connected.

440-62 Branch Circuit Requirements

(a) Conductor and Protection Size. Branch circuit conductors for a cord- and plug-connected room air-conditioner rated not over 40 ampere at 250 volt, must have an ampacity of not less than 125 percent of the motor compressor current [440-32]. The circuit short-circuit and ground-fault protection device must not be greater than the conductor or receptacle ampacity, whichever is less.

(b) Maximum Load on Circuit. If a cord- and plug-connected room air-conditioner is the only load on a circuit, the marked rating of the air-conditioner shall not exceed 80 percent of the rating of the circuit.

Room Air-Conditioners - Disconnecting Means
Section 440-63

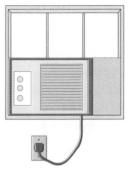

An attachment plug and receptacle can serve as the disconnecting means if:
• rated 250 volts or less,
• the A/C manual controls are readily accessible and within 6 feet of the floor,
• an approved manually operated switch is readily accessible and within sight of the A/C.

Fig. 42-4   Conductor Size, Short-Circuit and Ground-Fault Protection

440-63 Disconnecting Means

An attachment plug and receptacle can serve as the disconnecting means for a room air-conditioner if the manual controls are readily accessible and within 6 feet from the floor.

If the manual controls are located over 6 feet, a readily accessible switch must be installed within sight of the room air-conditioner, Fig. 42-4.

440-64 Supply Cord

The supply cord for room air-conditioning equipment shall not exceed 10 feet in length for 120 volt units, or 6 feet for equipment rated 208 through 240 volts.

SUMMARY

Part A. General □ Article 440 applies to electrically-driven hermetic equipment. □ Air-conditioning and refrigeration equipment that does not have a hermetic refrigerant motor compressor must comply with Article 422, 424, or 430.

Part B. Disconnecting Means □ Room air-conditioners, household refrigerators and freezers, drinking water coolers, and beverage dispensers can use the attachment plug and receptacle as the disconnecting means. □ The disconnect for other than cord- and plug-connected equipment must be readily accessible and within sight from the equipment.

Part C. Branch Circuit and Feeder Circuit Protection ☐ Air-conditioning equipment nameplates indicate the size and type of short-circuit and ground-fault protection required. ☐ If the nameplate specifies "Maximum Fuse Size," then the equipment must be protected by a fuse. ☐ If the nameplate specifies "HACR Circuit Breaker," then the equipment must be protected by a circuit breaker rated for heating air-conditioning and refrigeration equipment. ☐ The protection device rating must not exceed the manufacturer's values marked on the equipment. ☐ Feeder conductor protection must not be greater than the largest branch circuit short-circuit and ground-fault protection device for any motor compressor of the group, plus the sum of the other motor compressor currents.

Part D. Conductor Sizing ☐ Branch circuit conductors to a single motor compressor must have an ampacity not less than 125 percent of the motor compressor current. ☐ Conductors that supply several motors must have an ampacity of not less than 125 percent of the highest rated motor compressor current of the group, plus the sum of the other motor compressor currents of the group.

Part G. Room Air-Conditioners ☐ Room air-conditioners not rated over 40 ampere, single-phase, 250 volt, can be cord- and plug-connected. ☐ Branch circuit conductors to cord- and plug-connected room air-conditioners must have an ampacity of not less than 125 percent of the motor compressor current. ☐ The branch circuit short-circuit and ground-fault protection device must not be greater than the conductor or receptacle ampacity. ☐ Cord- and plug-connected room air-conditioners on an individual circuit must not exceed 80 percent of the circuit rating. An attachment plug and receptacle can serve as the disconnecting means. ☐ The supply cord for a room air-conditioner must not exceed 10 feet for a 120 volt unit or 6 feet for units rated 208 through 240 volts.

REVIEW QUESTIONS

1. Which *Code* Section applies to sizing the conductors and protection to central electric space heating equipment?

2. Which *Code* Section applies to sizing the conductors and protection to a 5 hp motor that is used as a blower for air-conditioning equipment?

3. What are the differences in the rules for motors and air-conditioning when installing a disconnecting means?

4. How do you size the circuit protection and conductor size to an individual air-conditioning unit?

5. What are the product listing requirements for short-circuit and ground-fault protection?

6. Give a summary of the requirements for room air-conditioners.

Unit 43

Article 450
Transformers

OBJECTIVES

After studying this unit, the student should be able to understand:
- size the primary conductors and protection device for a transformer.
- size the secondary conductors for a transformer.
- size the primary protection device for a transformer.

450-1 Scope

Article 450 covers the installation requirements of transformers and transformers vaults, Fig. 43-1.

Exception No. 4: Control Circuits. Class 2 and Class 3 control circuit transformers must comply with the requirements of Article 725.

Exception No. 7: Fire Alarm Circuits. Power-limited fire protective signaling circuit transformers must comply with the requirements of Article 760, Part C.

Note. Motor control circuit transformers must comply with the requirement of Sections 430-72(c) and 430-74(b).

450-3 Overcurrent Protection

Transformers must be protected against overcurrent in accordance with the percentages listed in Table 450-3 and applicable notes.

Note. Lighting and appliance panelboards on the secondary side of transformers must have overcurrent protection in accordance with Section 384-16(a) and (d).

FPN No. 2: Transformers that supply 120 or 277 volt nonlinear loads can cause excessive heating of the transformer's primary winding without opening the primary overcurrent protective device. This is caused by circulating odd triplen harmonic currents (3rd, 9th, 15th, 21st, etc.) that are reflective from the secondary onto the delta primary

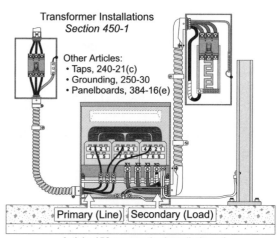

Transformer Installations
Section 450-1

Other Articles:
- Taps, 240-21(c)
- Grounding, 250-30
- Panelboards, 384-16(e)

Primary (Line) Secondary (Load)

Article 450 covers the installation of transformers and transformer vaults.

Fig. 43-1 Transformer Installation Requirements

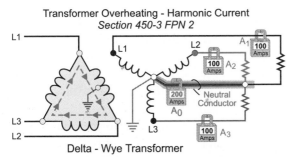

Fig. 43-2 Transformer Overheating—Due to Harmonic Currents

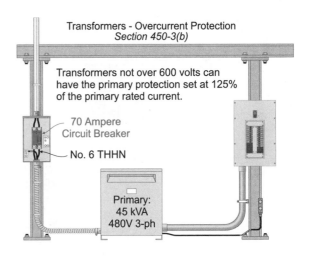

Fig. 43-3 Transformer Overcurrent Protection Not to Exceed 125 Percent of Primary Current Rating

winding. Circulating harmonic currents within the primary winding generates heat which is not detected by the primary overcurrent protection device, Fig. 43-2.

(b) Transformers 600 Volt or Less. Transformers must be protected against overcurrent, and the overcurrent protection must be provided according to the values and notes listed in Table 450-3(b).

WARNING:
The primary overcurrent protection device for 3- or 4-wire transformers does not protect the secondary conductor [240-3(f)], nor does it protect the panelboard on the secondary side [384-16(e)]. See Section 240-21 for the rules on secondary transformer conductors.

Exception: Motor Control Transformers. Primary protection for motor control transformers must be sized in accordance with Section 430-72(c)(4).

Primary Protection Only
Current Less than 2 Ampere. The primary protection device can be set at not more than 300 percent of the primary current rating.

Current between 2 Ampere through 8.99 Ampere. The primary overcurrent protection

device can be set at not more than 167 percent of the primary current rating. See Section 430-72(c).

9 Ampere or More (Note 1). Transformers with primary current ratings of 9 ampere and more must have the primary overcurrent protection device set no more than 125 percent of the transformer primary current rating.

When 125 percent of the primary current does not correspond with the standard ratings of overcurrent protection devices [240-6(a)], the next higher device rating can be used.

Question. What is the primary protection required for a 45 kVA, 480 volt 3-phase transformer, Fig. 43-3?

Answer. No. 6, 70 ampere
Step 1. Primary Current
I = VA/(E × 1.732)
I = 45,000/(480 volt × 1.732) = 54 ampere

Step 2. Primary Protection
54 ampere × 1.25 = 68 ampere

Step 3. Note 1, Table 450-3
Next-size-up, 70 ampere [240-6(a)]

Note. A poster containing overcurrent protection, conductor sizing, raceway sizing, and equipment grounding size for 15 ampere up to 320 ampere circuits is available at *www.mikeholt.com.*

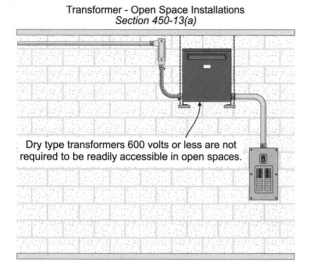

Fig. 43-4 Transformers Not Required to Be Readily
Accessible When Installed in Open Visible Spaces

450-9 Ventilation

The ventilation of transformers must be adequate so that when the transformer is fully loaded, the temperature will not rise over its maximum temperature rating. Transformer ventilating openings must not be blocked by walls or objects, and transformers must be installed according to the instructions marked on the transformer [110-3(b)].

> FPN: 120 and 277 volt nonlinear loads can increase the heating within a transformer beyond its temperature rating.

Note. Inductive harmonic heating of the transformer winding is in proportion to the square of the harmonic current frequency. For example, third harmonic inductive heating is 9 times the fundamental current heating. Because of high-frequency inductive heating, transformers supplying

nonlinear loads must be "K-Rated." For more details call 1-888 NEC-CODE.

450-13 Transformer Accessibility

All transformers and transformer vaults shall be readily accessible to **qualified personnel** for inspection and maintenance or meet the requirements of (a) or (b) below:

(a) Open installations. Dry-type transformers 600 volt, nominal, or less, located in the open on walls, columns, or structures, are not be required to be readily accessible, Fig. 43-4.

(b) Suspended Ceilings. Dry-type transformers rated not more than 50 kVA and not over 600 volt are not required to be readily accessible and they can be installed within a hollow space of s suspended ceiling not used for environmental air [300-22(c)], Fig. 43-5.

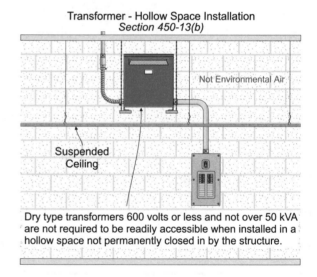

Fig. 43-5 Transformers Not over 50 kVA Can Be
Installed above Suspended Ceilings

SUMMARY

☐ Transformers having a primary current rating of 9 ampere and more must have a primary protection device set at not more than 125 percent of the primary current rating. ☐ Transformers rated 2 ampere through 8.99 ampere can have the primary protection device set at not more than 167 percent of the primary current rating. ☐ Transformers rated less than 2 ampere can have the primary protection device set at not more than 300 percent of the primary current rating except for motor control transformers. ☐ Transformers having a secondary protection device set at not more than 125 percent of the secondary current can have the primary protection device set up to 250 percent of the primary current rating. ☐ The primary protection device for 3- and 4-wire transformers cannot be used to protect the secondary conductors. ☐ Ventilation must be adequate so that when the transformer is loaded the temperature will not rise over the transformer temperature rating.

REVIEW QUESTIONS

1. Size the primary conductors (75°C terminals) and protection device for a 112.5 kVA transformer, primary 480 volt, and secondary 208Y/120 volt, 3-phase.

2. Size the secondary conductors (75°C terminals) for a 37.5 kVA transformer, primary 480 volt, and secondary 208Y/120 volt, 3-phase. Secondary conductors are not longer than 10 feet and terminate in a panelboard rated 125 ampere.

3. Size the primary protection device for a 75 kVA transformer, primary 480 volt and secondary 208Y/120 volt, 3-phase. The secondary conductors terminate in an overcurrent protection device sized not greater than 125 percent of the secondary current rating.

4. Size the primary conductors (75°C terminals) for a 75 kVA transformer, primary 480 volt and secondary 208Y/120 volt, 3-phase. The secondary conductors terminate in an overcurrent protection device sized not greater than 125 percent of the secondary current rating. The primary and secondary conductors are not longer than 25 feet.

GLOSSARY

The following definitions are terms that are contained in this book.

AIC: Abbreviation for Ampere Interrupting Capacity. All overcurrent protection devices must have an interrupting rating. Fuses are rated for 10,000 AIC [240-60(c)], and circuit breakers are rated for 5,000 AIC [240-86], unless marked otherwise, see interruption capacity.

Al: Abbreviation for aluminum, see Section 110-14.

Al-Cu: Abbreviation for aluminum and copper commonly marked on terminals, lugs, and other electrical connectors to indicate that the device is suitable for use with either aluminum or copper conductors.

Alternating current: The type of electrical current produced in a rotation generator (alternator), which changes in both magnitude and direction.

Ambient temperature: The temperature of the air, water, or surrounding earth. Conductor ampacity is corrected for changes in ambient temperature including temperatures below 86°F. The cooling effect can increase the current-carrying capacity of the conductor. These rules are covered in Section 310-15(b) of this book.

American Wire Gauge (AWG): The standard for measuring wire in America, see Section 110-6.

Ampere: Unit of electrical current.

ANSI (American National Standards Institute): An independent organization that coordinates the U.S. voluntary standard system and operates an approval process for standards such as the *National Electrical Code.*

Arc: A flow of current across an insulated medium.

Authority Having Jurisdiction (AHJ): For this book we will refer to this individual as the Inspector. This is the person who is responsible for interpreting the rules, approving equipment and materials, and granting special permission or alternate methods, see Sections 90-4, 90-6, 90-7, and 110-2.

Available short-circuit current: The amount of current that is available under a short-circuit condition. The current is limited by the impedance of the transformer, circuit conductors, and other components, see Section 110-9 and 110-10.

Backfill: Loose earth for filling and grading, see Section 300-5(f).

355

Back-fed: A term to describe a plug-in type breaker installed in a panelboard to provide overcurrent protection for the panelboard. Instead of the conductors terminating in the main lugs (MLO) of the panelboard, the energized conductors terminate in the breaker.

Balanced system: When all phase conductors carry approximately the same current. For delta systems, this applies to two-phase conductors, and for 3-phase wye systems, this applies to 3-phase conductors, see Section 310-15(b)(4)(a).

Bond wire: A slang term used to describe what the *NEC* calls the equipment grounding conductor, see Section 250-119.

Bonding bushing: A conduit bushing equipped with a conductor terminal for connecting a bonding jumper, see Section 250-94(4), 250-97, 250-98, 501-16(a), 502-16(a), and 503-16(a).

Bonding locknut: A threaded locknut used on the end of a conduit equipped with a screw. When the locknut is installed, the screw is tightened into the wall of the enclosure, see Section 250-94(4), 250-98, 501-16(a), 502-16(a), and 503-16(a).

Boss: An enclosure with threaded entry such as a box, enclosure, or conduit body, see Section 250-94(2) and 300-4(f).

Busbar: The conductive parts used as the main current supplying elements of panelboards or switchboards, see Section 384-3(e).

Bushing: A fitting for protecting conductor insulation from abrasion, see Section 300-4(f), 300-5(h), 300-16(b), 333-9, and 373-6(c).

BX: Nickname for armored cable, registered trade name of the General Electric Company [Article 333].

CATV: Acronym for Community Antenna Television.

Circular mil: The cross-sectional area of a conductor as listed in Chapter 9, Table 8 of the *NEC*.

Coil: A wire wound helix or spiral in series closed loops that concentrates the magnetic field produced by electric current flow.

Color Code: A system of circuit conductor identification by the use of solid colors such as phase tape or spray paint. Generally the *NEC* does not require conductors to be color coded, except the grounded (neutral) conductor must be white [200-6 and 310-12(a)] and the grounding conductor must be bare, green, or green with a yellow stripe [250-119].

Conductor impedance: All conductors have resistance, and when alternating current flows through the conductor, *eddy current* induced in the conductors opposes the flow of current (counter-electromotive force). This increased impedance is due to the opposition to the change of current and the conductors resistance.

Corrosion: The deterioration of a substance (usually metal) because of reaction with its environment, see Section 300-6.

Crawl space: Shallow space between the first tier of beams and the ground, see Sections 210-8(a)(4) and 210-63.

Cu: Abbreviation for copper, see Section 110-5.

Current: The rate of flow of electrons, measured in ampere.

Current rating: The amount of current flow a device is designed to withstand.

Cutout box: A surface mounting enclosure with a cover equipped with a swinging door, see Article 373.

Dedicated equipment space: The space above and below a panelboard or switchboard used exclusively for electrical wiring, see Section 110-26(f).

Delta-connected: Interconnection of three electrical equipment winding in series to form a closed loop (triangle).

Derating factor: A factor used to reduce conductor ampacity because of ambient temperature and conductor bundling, see Sections 310-10 and 310-15(b).

Different systems: Different voltage and frequencies such as 480Y/277 or 120Y/208 volt, and 50, 60, or 400 Hz, see Sections 200-6(d), 210-4(d), and 230-2. However, it could mean the difference between ac power, telephone, CATV, etc.

Drip loop: An intentional sag placed in conductors that connect to overhead conductors; the drip loop prevents rainwater from entering the weatherhead, see Section 230-54(f).

Eddy currents: Circulating currents induced in conducting materials by varying magnetic fields; eddy currents cause losses of energy by heating up the metal parts and opposing the current flow.

Elbow: A curved section of a raceway intended to change the direction of the run.

Electric-discharge lighting: A lamp in which light is produced by the passage of an electric current through a vapor or gas such as mercury, sodium, neon, argon, and so on, see Section 210-6 and Article 410, Parts P and G.

Electrical continuity: An unbroken low impedance path through which electrons flow. This applies to metal raceways and cables [300-10], conductors in general [300-13(a)], grounding conductors [250-2(d)], bonding conductors [250-119], and the grounding electrode conductor [250-64(a)].

Electrically connected: Connected so as to be capable of carrying current; not a connection through electromagnetic induction.

Electricity: The energy produced by the flow of "free" or valance electrons moving from one atom to another in a conductor.

Electrocution: Death caused by electrical current through the heart, usually over 50 milliampere.

Electrolysis: The action of a chemical change resulting when electrical current passes through an object.

Exothermic welding: Term to describe the welding of metal parts, commonly called "Cad Weld," see Sections 250-102(b) and 250-8 and 250-70.

Extension Ring: Used to extend the box when it is recessed back in a wall. Also provides more cubic inch capacity for the box if needed.

Fault-current: An abnormal, and often dangerously high, current flowing between conductors or conductors to a grounded surface, see Sections 110-9, 110-10, 250-2, 250-90, and 250-96(a).

Ferrous: A metal that contains iron.

Field: The effect produced in surrounding space by an electrically charged object, by electrons in motion, or by a magnet.

Fire-resistance rating: The time that material or construction will withstand fire exposure as determined by certain standards, see Section 331-3(2).

Fished: The drawing in of conductors in concealed spaces.

Flex: Common term used to refer to flexible metal conduit. *See* Article 350.

Four-way switch: Device used in conjunction with two 3-way switches to control utilization equipment from 3 or more locations, see Sections 200-7(c)(2) and 380-2(a).

Fuse: A fusible member that opens a circuit when the fusible element is heated directly and destroyed by overcurrent passing through, see Sections 240-6 and Article 240 Parts E and F.

Gray area: Slang term to describe a *Code* rule that is subject to interpretation.

Greenfield: Another name used to refer to flexible metal conduit, see Article 350.

Ground-fault: A undesirable connection (electrically conductive) between any of the conductors of the electrical system and any object that is grounded. During the period of a ground-fault dangerous voltages and currents exist and must be cleared, see Sections 110-7, 240-1, and 250-2.

Gutter: The space along the sides, top, and bottom of switches and panels to provide for conductors to terminate at the lugs or terminals of the enclosed equipment. Also used to refer to wireways and auxiliary gutters, see Articles 362 and 374.

Guy wire: A tension wire, or other mechanical member to add strength to the structure, see Section 230-28.

Harmonic current: An oscillation whose frequency is a multiple of the normal frequency. Third harmonic is equal to 180 Hertz; since normal frequency in the U.S. is 60 Hz, see Sections 210-4(c), 220-22, 300-13(b), and 300-20.

Hertz (Hz): International unit of frequency, equal to 1 cycle per second for alternating current.

Hickey: A funny term, but it means a fitting for hanging lighting fixtures from an outlet box [370-6(a)(1)]. It is also the trade name for a rigid conduit bender.

High hat: A recessed incandescent lighting fixture in the shape of a man's high hat, see Section 410-65.

High-leg (bastard leg): The conductor that has the highest voltage to ground of a 3-phase, 4-wire, delta-connected secondary. There are special rules for the location and identification of the high-leg conductor, see Sections 215-8, 230-56, and 384-3.

Horsepower: Motor rating that measures the time rate of work.

Hub: Fitting to attach threaded conduits to boxes, see Section 250-94(2).

Hysteresis: The internal friction of molecules caused by the time lag exhibited by a body in reacting to changes in magnetic forces.

Impedance (symbol: Z): A measure (in ohm) of the response of an electric circuit to an alternating current. The total opposition to current flow due to capacitance, inductance, and resistance. Low impedance is important for the grounding system so that it will carry fault-current to trip the overcurrent protection device. High impedance will reduce the fault-current flow, resulting in the slowing down or elimination of the opening of the circuit protection device, see Section 250-2.

Induction: The electromagnetic field caused by the pulsations of alternating current. This field can pass through any material. Under certain wiring conditions, induction can generate heat in metal enclosures [300-20], choke (reduce) the conductor's current-carrying capacity [250-64(d)], or cause electrical noise, or static, on sensitive communications and data processing circuits and equipment, see Sections 250-146(d), 250-96(b), 385-20, and 410-56(c).

Inductive Choke: The inductor effect of limiting ac current through a conductor that is installed within a metal enclosure that is not bonded to the conductor at both ends, see Sections 250-92(b), 280-85, 300-5(i) and 300-20.

Inrush current: The initial current when an electric coil or motor is energized. Often the initial current is equal to 6 times the normal current, see Sections 430-52, 430-63, and 430-72.

Insulation: A material having high resistance to the flow of electric current, often called dielectric.

Interruption capacity: The highest short-circuit current at rated voltage that the device can safely interrupt such as 10,000, 22,500, or 65,000 amperes, see Section 110-9 and AIC.

Intrinsically safe: Incapable of releasing sufficient electrical or thermal energy under normal or abnormal conditions to cause ignition of a specific hazardous atmospheric mixture in its most ignitable concentration, see Article 504.

Junction box: An enclosure for the connection or branching of one or more electrical circuits.

kcmil: Abbreviation for thousand circular mils, to describe a conductor that is larger than 4/0.

Knockout: A circular scored (cut) portion in the wall of a box or cabinet that can be removed easily to accommodate a conduit or cable termination.

Knockout (concentric): A knockout with two or more circular cuts that have a common axis (center point). This allows different standard size conduits or cable connections to be installed at the same point of a box or cabinet, depending on how many circles or rings are removed.

Knockout (eccentric): Same as concentric, except that the rings do not have a common axis or center point. Bonding jumpers are required around concentric and eccentric knockouts for service raceways [250-94(4)] and for over 250 volt circuits [250-97 Ex.].

Knife switch: A switch in which the circuit is closed by a moving blade engaging contact clips, see Section 380-6.

kVA: Kilovolt-ampere; the product of volt and ampere divided by 1,000.

kW: Kilowatts; the product of volt and ampere of resistive loads divided by 1,000. For example, 1kW is equal to 1,000 watts.

Lamp: A device that converts electrical energy to visible light.

Lampholder: Device that is intended to mechanically support and that connects the lamp to the circuit conductors. The grounded (neutral) conductor must be connected to the screw shell of the lampholder, 200-10(c) and 410-47.

Limit switch: A mechanically operated device that stops a motor from revolving or reverses it when certain limits have been reached.

Line side: The line side is the supply side of any equipment such as the power to the switch. When referring to services, the line side is referred to as the supply side, see Sections 230-7 and 250-142].

Load side: The load side is the conductors leaving a switch.

Locked-rotor: The circuit when the motor rotor is prevented from turning, see Table 430-151.

Lug: A terminal device for terminating a conductor, see Sections 110-14, 250-8 and 250-70.

Manual: Operation by personal intervention.

Mil: A unit used in measuring the diameter of a wire or thickness of insulation over a conductor. One one-thousandth of an inch. (.001")

Mechanical continuity: Raceways and cables must be mechanically continuous (secured to enclosures) [300-10, 300-12, and 370-27(b) and (c)].

Neutral (wire) Conductor: Grounded Conductor (not to be confused with the grounding conductor), this conductor permits the use of line-to-neutral loads (120 and 277 volt loads) and will serve as a current-carrying conductor to carry any unbalanced currents, and provide a low impedance path for the flow of fault-current to facilitate the operation of the overcurrent protection devices [250-2(d)]. During a ground-fault condition, dangerous voltage (potential) exists and it is very important to clear this fault in the shortest practical period of time. The grounded (neutral) conductor is used to assist in the clearing of fault-current, see Articles 200.

NFPA (National Fire Protection Association): An organization devoted to promoting the science and improving the methods of fire protection. Publishes the *National Electrical Code* and other safety related standards.

Nipple: Pipe less than 2 feet in length. It does not have to be straight, see Section 300-17, Section 310-15(b)(3) Exception No. 3, and Chapter 9, Table 1, Note 3.

Nonferrous: Without iron content such as copper, aluminum, brass, gold, silver, and so forth, see Section 250-52(c)(2).

Ohm: A unit of resistance; a resistance of 1 ohm sustains a current of 1 ampere when 1 volt is applied across the resistance, see Section 250-56.

Open: A condition of a circuit that prohibits current to flow.

Overload protection: A device that opens as a result of excessive current, but does not respond to short-circuit or ground-fault current.

Oxidation: To unite with oxygen resulting in rust or tarnish.

Parallel: Connections of 2 or more devices between the same 2 terminals of a circuit or a circuit that has more than 1 path for current flow, see Section 310-4.

Pendant: An overhead fitting suspended by a flexible cord that may provide electrical connection, see Sections 210-50(a) and 400-7.

Phase: One of the phase conductors (ungrounded, hot conductor) of an electrical system.

Pigtail: The splicing of 2 or more conductors with a lead wire for termination to a device or fixture, see Section 300-13(b).

Plaster Ring: All rings are called mud rings regardless of the finish wall material used. There are two styles of rings, one for mounting a switch or receptacle and one for mounting lighting fixtures.

Plenum rated cable: Cable approved by Underwriters Laboratories for insulation in plenums without the need for conduit because the insulation and jacket compounds used have low flame-spread and low smoke characteristics.

Polarity: Distinguishing 1 conductor or terminal from another. Identifying how devices are to be connected such as + or -, see Section 200-11. Assuring the mating of plugs and receptacles in the only correct position, see Section 200-11.

Primary: The power conductors that supply a transformer.

Pull box: A box with a blank cover used in a run of conduit to make it easier to pull in the conductors, see Section 370-18.

Pushbutton: A manually operable plunger or button for closing or opening a set of contacts. A pushbutton is not considered a disconnect or controller.

PVC: An acronym for Polyvinyl Chloride. A thermoplastic material composed of polymers of vinyl chloride from which nonmetallic raceways are made.

Reactance: The opposition to current flow in an ac circuit.

Receptacle (isolated ground): These receptacles are generally used for computers, medical, communication, and laboratory type electronic equipment. When this type of equipment is grounded through the conventional building ground, transient signals often cause their circuits to malfunction. The solution is the use of isolated ground receptacles that provide a pure grounding path separate from the conventional grounding circuit within the building [250-96(b), 250-146(d), 384-20 Ex., and 410-56(c)].

Receptacle (multioutlet): A receptacle with more than one outlet such as a duplex receptacle.

Reducing washers (Chinese money): A fitting used when the raceway fitting is too small for the enclosure opening. This fitting is not approved as an equipment grounding path, see Section 250-96(a).

CAUTION: The grounding terminal must be connected to the system grounding terminal of the service or separately derived system, see Sections 250-6(d).

Resistance: The opposition offered by a substance or body to the passage through it of an electric current; resistance converts electrical energy into heat.

Romex: General Cable Company trade name for nonmetallic-sheathed cable, Type NM, see Article 336.

Sag: The difference in elevation of a suspended conductor.

Secondary: The power conductors that leave a transformer.

Service factor: The amount of allowable overload indicated by a multiplier which, when applied to a horsepower rating, indicates the permissible loading, see Section 430-32(a)(1).

Shall: Mandatory requirement of the *Code*, see Section 110-1.

Short-Circuit: A connection (electrically conductive) between any of the conductors of the electrical system from line-to-line (one phase to another phase) or from line to grounded (neutral) conductor (one phase to the neutral). An often unintended low resistance path through which current flows around, rather than through, a component or circuit, see Section 110-7.

Short-circuit and ground-fault protection device: A protection device that does not protect against overload, see Sections 430-52 and 430-62.

Short-circuit current rating: The highest current (at a specified voltage) at the equipment line terminals that the equipment can safely carry until the overcurrent device opens the circuit, *see* Section 110-10.

Short-circuit current on the load side of electrical equipment increases to extremely high levels and can do extensive damage to electrical components in a very short period of time. Short-circuit current in excess of the equipment rating can damage the equipment because of thermal and magnetic forces. Every electrical component has a maximum short-circuit withstand rating which should not be exceeded, see Section 110-9.

All circuit components that are affected by shorts and grounds must be able to handle the conditions of a fault (heat and magnetic effects from high currents) from the time that the fault occurs to the time the overcurrent protection device (fuse or circuit breaker) opens. This current rating is generally marked on the equipment; if not, the equipment is rated for a maximum of 5,000 ampere, see Section 110-10.

Shunt-trip circuit-breaker: A breaker that is capable of being opened through the use of an electromechanical relay, which is controlled through a remote switch. The shunt trip pushbutton is not a disconnect, see Article 100 Disconnect and Section 230-70.

Single-phase: A single voltage and current in the supply.

Splice: The electrical and mechanical connection between two pieces of cable, see Section 110-14(b).

Split-bolt connector (bug or lug): A device that is designed for only 2 conductors. Bugs must be insulated after being made up [110-14(b)]. Many electricians think that they can be used for more than 2 conductors, but this is not the case. Be sure you check with the manufacturer on the number of conductors under a bug.

Starter: A controller designed to accelerate a motor to normal speed. A starter contains overload protection devices, whereas a lighting contactor does not.

Studs: Vertical skeleton members of a partition or wall to which sheet rock or wallboard is screwed or nailed.

Switch leg: Slang term to mean the conductor from the load side of the switch to the equipment. The *NEC* refers to this as a switch loop, see Section 200-7(c).

Terminal: Point where a conductor is intended to be connected. Receptacles and switches are designed for two types of different termination: side wiring (screw head locks wire to terminal) or quick wire connection (inserting wire into wire well automatically ensures good electrical connection), see Section 110-14.

Three-phase: Three electrically related voltages for which the phase difference is 120 degrees electrical separation, see Section 110-14 FPN.

Three-way switch: Device that changes the connection of one conductor, and which is normally used in pairs to control utilization equipment from two locations, see Sections 200-7(c)(2) and 380-2(a).

Torque: Twisting or turning force, see Section 430-9(c).

Transformer: A device that operates by electromagnetic induction consisting of winding(s) with or without a magnetic core, for the purpose of ping voltage or current up or down, see Article 450.

Travelers: A slang term to describe conductors between 3-way and 3-way or 4-way switches. These conductors are not the hot (ungrounded) or neutral (grounded), see Section 200-7(c)(2).

UL: An acronym for Underwriters Laboratories, a nonprofit independent organization which operates a safety testing and listing service for electrical and electronic equipment and materials.

Voltage drop: Voltage drop is in direct proportion to the conductors resistance and the magnitude (size) of the current, VD = I × R.

When a conductor resistance causes the voltage to drop below an acceptable point, the conductor size should be increased. See Sections 210-19 FPN No. 4, 215-2(d) FPN No. 3, and 310-15(a)(1) FPN No. 1.

Watt: A unit of electrical power. One watt is equivalent to the power represented by one ampere of current under a pressure of one volt in a DC circuit.

Weatherhead: The fitting at a conduit used to allow conductor's entry, but prevent weather entry, see Section 230-54.

Wirenut (wire connector): Electrician's term for a solderless wire connector, but "Wirenut" is a trademark of the Ideal Corporation.

Wye-connected: Interconnection of three electrical equipment windings in star (wye) fashion.

Yoke (strap): The metal or plastic strap on a switch and receptacle used to mechanically attach the device to the outlet box [370-16(a)(1), 380-10(b), and 410-56(f)].

Index

C

D

G